KU-565-266

Children's Games with Things

Children's Games with Things

MARBLES · FIVESTONES · THROWING AND CATCHING
GAMBLING · HOPSCOTCH · CHUCKING AND PITCHING
BALL–BOUNCING · SKIPPING · TOPS AND TIPCAT

BY

IONA AND PETER OPIE

WESTERN ISLES
LIBRARIES

39901797

793.0192

39143

Oxford   New York

OXFORD UNIVERSITY PRESS

1997

Oxford University Press, Great Clarendon Street, Oxford OX2 6DP

Oxford New York

Athens Auckland Bangkok Bogota Bombay Buenos Aires
Calcutta Cape Town Dar es Salaam Delhi Florence Hong Kong
Istanbul Karachi Kuala Lumpur Madras Madrid Melbourne
Mexico City Nairobi Paris Singapore Taipei Tokyo Toronto Warsaw
and associated companies in
Berlin Ibadan

Oxford is a trade mark of Oxford University Press

Published in the United States
by Oxford University Press Inc., New York

© Iona Opie 1997

All rights reserved. No part of this publication may be reproduced,
stored in a retrieval system, or transmitted, in any form or by any means,
without the prior permission in writing of Oxford University Press.
Within the UK, exceptions are allowed in respect of any fair dealing for the
purpose of research or private study, or criticism or review, as permitted
under the Copyright, Designs and Patents Act, 1988, or in the case of
reprographic reproduction in accordance with the terms of the licences
issued by the Copyright Licensing Agency. Enquiries concerning
reproduction outside these terms and in other countries should be
sent to the Rights Department, Oxford University Press,
at the address above

This book is sold subject to the condition that it shall not, by way
of trade or otherwise, be lent, re-sold, hired out or otherwise circulated
without the publisher's prior consent in any form of binding or cover
other than that in which it is published and without a similar condition
including this condition being imposed on the subsequent purchaser

British Library Cataloguing in Publication Data
Data available

Library of Congress Cataloging in Publication Data
Opie, Iona Archibald.
Children's games with things : marbles, fivestones, throwing and catching, gambling, hopscotch,
chucking and pitching, ball-bouncing, skipping, tops and tipcat / Iona
and Peter Opie.
p.   cm.
Includes bibliographical references (p.).
1. Games—Great Britain—History. I. Opie, Peter. II. Title.
GV1200.075 1997   793'.01'922—dc21   96–53210
ISBN 0–19–215963–1

1 3 5 7 9 10 8 6 4 2

Typeset by Best-set Typesetter Ltd., Hong Kong
Printed in Great Britain by
Biddles Ltd.
Guildford and King's Lynn

Preface

IN this book I am presenting the rest—and the last—of the material from our surveys of the 1950s and 1960s, thus completing the picture of the joys and entertainments of the mid-twentieth-century schoolchild. After the publication of *The Oxford Dictionary of Nursery Rhymes*, in 1951, we were left with some rhymes, sent by correspondents, which clearly did not belong to the nursery: they were too self-assertive and mocking, and belonged to children outside the home, in the robust world of the school yard (a world which was foreign to us, having been educated privately). Intrigued, we decided to find out what had survived of the traditions and pursuits of British schoolchildren—very little, we were assured. We wrote a letter to the *Sunday Times*, 6 November 1951, saying what we hoped to do, and had enough replies from interested schoolteachers to start a nationwide network—for those who replied introduced us to like-minded teacher friends in other parts of the country. Lists of contributing schools appear in *The Lore and Language of Schoolchildren* (1959), and *Children's Games in Street and Playground* (1969). No systematic collecting on a similar scale has been done since the end of those major surveys, although over the years a large number of people have kindly written to say what their children, grandchildren, neighbours' children, and children in the local playground were saying and doing: all this material will go—or has already gone—to the Bodleian Library, Oxford, to join the other Opie papers. I myself, of course, made notes of children's games whenever I had the opportunity, not least on my weekly visits to Liss Junior School playground.

Before starting this work, I had two options: to complete the games trilogy with a third volume which—using material already in the files—would span the three middle decades of the century, and thus round off a defined era of children's play; or to build up another national network and start a new survey. I knew that a survey would take at least four years (even supposing that present-day teachers, already overburdened, would be able to take part), and that more years would pass before I had finished selecting and entering the new contributions in the files. It seemed that I might exceed the traditional seven years' brewing needed for an Opie book. Life was running out, and—really just as a joke—I wanted to astonish Peter's obituarists, who said, 'In *Children's Games* we were promised a further volume, containing "the intricacies of ball-bouncing and other ball games, of skipping, marbles, fivestones, hopscotch, tipcat, and gambling . . . where they can be treated more fully, together with the singing games". What a pity that now we shall never see that volume.' I was encouraged by their interest. As always, the material proved too much to fit into a single further work; the singing and clapping games had to be dealt with separately before the 'games with things' could be dealt with.[1] *The Singing Game*, half-finished when Peter died, came out in 1985. And here is, probably, the final book by 'Iona and Peter Opie', for it is the last book that will contain a substantial amount of Peter's writing on children's lore.

I.O. and P.O.

West Liss in Hampshire
1989–96

---

[1] 'Games with things' were classified as games in which the objects used were the primary reason for the game. In many games an object is used (for instance the 'can' in 'Tin Can Tommy') but is subsidiary to the game itself. 'Conkers' and 'Soldiers' were borderline cases. Instead of giving them on their own as 'games with things' it was thought better to place them, as 'Duels by proxy', amongst games motivated by the same principle. They appear with other contests for two people in the 'Duelling Games' chapter of *Children's Games in Street and Playground*.

Acknowledgements

I AM very grateful to the many people who have written with contributions. Especially, I must thank Gareth W. Whittaker of Bradford, for an investigation he undertook in Denholme, Bradford; Jane Kearley of Hoylake, in the Wirral, who—having inherited an interest in children's lore from her father Frank Rutherford, author of *All the Way to Pennywell*—sent the tape of a long (and hilarious) radio interview she and her daughters gave on the subject, as well as other information; Adam T. McNaughtan of Glasgow, for sharing his knowledge and understanding of Scottish children's lore, as exemplified in the film he made in 1984 for BBC Scotland's *Collectors* series. Finally, my thanks to Graham Stephen, of Montrose, for his lasting friendship and encouragement. In the mid-1970s he enquired about current children's games through the letter columns of innumerable Scottish local newspapers, and obtained a mass of material which—after I had copied it into the Opie files—was lodged in the archives of the School of Scottish Studies, in Edinburgh.

Contents

List of Illustrations

Abbreviations

Abrahams	Roger D. Abrahams, *Jump-Rope Rhymes: A Dictionary* (1969).
Argyleshire	R. C. Maclagan, *Games and Diversions of Argyleshire* (1901).
Bluebells My Cockle Shells	Booklet produced by the pupils of Cumnock Academy, Ayrshire (1961).
Crofton MS	Revd Addison Crofton, 'Children's Rhymes', 2 MS vols; a fair copy, written out for his friend E. W. B. Nicholson, Keeper of Printed Books at the Bodleian Library, Oxford.
EDD	*English Dialect Dictionary.*
EETS	Early English Text Society.
EFDSS MSS	Manuscripts in the library of the English Folk Dance and Song Society.
Eifermann (1968)	Dr Rivka R. Eifermann, *School Children's Games* (Department of Psychology, Hebrew University of Jerusalem, 1968).
Eifermann (1971)	*Determinants of Children's Game Styles* (Jerusalem, 1971).
Golspie	E. W. B. Nicholson, *Golspie: Contributions to its Folklore* (1897).
Gomme	See *Traditional Games.*
Howard MSS	Games collected by Dr Dorothy Howard in the USA, 1931 onwards.
Husenbeth	F. C. Husenbeth, *History of Sedgley Park School, Staffordshire* (1856) (refers to *c*.1805).
Kellett MSS	Games collected by Rowland Kellett, chiefly in Leeds, 1920s onwards.
JAFL	*Journal of American Folklore.*
Jamieson	John Jamieson, *Dictionary of the Scottish Language* 2 vols (1808); Supplement, 2 vols (1825).
JLDS	*Journal of the Lakeland Dialect Society.*
London Street Games	Norman Douglas, *London Street Games* (1916).
Macmillan Collection	Collection of 243 children's letters written in 1922 to A. S. Macmillan, Juvenile Editor of *Pulman's Weekly News.*

Maclagan	R. C. Maclagan, *The Games and Diversions of Argyleshire* (1901).
Mactaggart	John Mactaggart, *The Scottish Gallovidian Encyclopedia* (1824).
N & Q	*Notes and Queries: a medium of communication for literary men, artists, antiquaries, etc.* (1849–).
Newell	W. W. Newell, *Games and Songs of American Children* (1883; augmented 1903).
OED	*Oxford English Dictionary.*
ODNR	*Oxford Dictionary of Nursery Rhymes.*
Opie, *Lore*	Iona and Peter Opie, *The Lore and Language of Schoolchildren* (1959).
Opie, *Games*	Iona and Peter Opie, *Children's Games in Street and Playground* (1969).
Opie, *Singing Game*	Iona and Peter Opie, *The Singing Game* (1985).
Pandrich	J. A. Pandrich, 'Child Lore . . . with Particular Reference to the North-East of England', thesis (1967).
Pitrè	Giuseppe Pitrè, *Giuochi fanciulleschi siciliani* (1883).
Rymour Club	*Miscellanea of the Rymour Club* (Edinburgh, 1906–28).
St Anne's Soho Monthly	Revd T. Allen Moxon, series of 'Games of Soho' (June–August 1907).
Strutt	Joseph Strutt, *Sports and Pastimes of the People of England* (1801; enlarged edition, 1903).
Sutton-Smith	Brian Sutton-Smith, *The Games of New Zealand Children* (1959).
Terrett MSS	Game rhymes collected by Miss Joyce Terrett in Swansea: Mayhill Junior School, 1937–9, and Glanmor Secondary School, 1952–64.
Those Dusty Bluebells	Booklet published by Cumnock Academy, Ayrshire (1965).
Traditional Games	Alice Bertha Gomme, *The Traditional Games of England, Scotland, and Ireland* 2 vols. (1894–8).
Turner	Ian Turner, *Cinderella Dressed in Yella* (Melbourne, 1969; 2nd edn. 1978).
Turner MSS	Rhymes collected by Ian Turner before publication of *Cinderella Dressed in Yella*.

Poster issued by the Chief Constable of Great Yarmouth on
1 February 1889, timed to suppress the top and marbles seasons.

I

Seasons, and Other Social Issues

SEASONS

No one writing about 'games with things' can avoid the subject of 'seasons', those periodic returns of such games as marbles, skipping, tops, fivestones, all of which are played with 'things'—for who has ever heard of a hide-and-seek or tig season? It makes sense for everyone to bring their marbles or skipping-rope to school and to play that game until the craze has passed; it would make less sense, and be far less exciting, for each child to bring equipment for half a dozen different games, in case someone else wanted to play.

It is no new observation that children's games are subject to waves of enthusiasm. In *Memoirs of an Oxford Scholar* (1756), the author writes: 'My Amusements were boyish, playing at Taw, whipping of Tops, and all the Train of Plays which succeed each other through the various seasons of the year.' The Revd Joseph Hunter, who was at school in Sheffield *c.*1790, noted in a copy of Brand's *Antiquities*, ii. 273, 'I dare say it is somewhere mentioned in this book [it is not] that Boys Sports have their cycle. Thus I remember when a School-boy that we used to say such a game is not in season, or such expression:- as when Marbles went out Tops came in &c.' William Curtis talked of 'rages', rather than 'seasons'. Writing home from Epping on 1 April 1817, he says, 'Uncle and Alfred went to London and the peg-top rage began again'; and later in life, when writing about his schooldays for his own children, he says, 'When a "rage" for hoops came up they were our constant companions . . . Another "rage" we had was pump making.' In *Blackwood's Magazine*, 21 August 1821, p. 34, the expression is 'periodical return':

The games among the children of Edinburgh have their periodical returns. At one time nothing is to be seen in the hands of the boys but cleckenbrods [wooden hand bats]; at another, dosing of taps [spinning of tops], and piries [peg tops] and pirie cords form the prevailing recreation; and at a third, every retired pavement, or unoccupied area, swarms with the rosy-faced little imps playing at bowls [marbles].

To a child, the arrival of a games season can seem supernatural. Robert Louis Stevenson, who was never far away from his own childhood, said that 'boys and their pastimes are swayed by periodic forces inscrutable to man; so that tops and marbles reappear in their due season, regular like the sun and moon'.[1] Thomas Burke linked games seasons even more firmly with the forces of nature:

All those street-games had their peculiar and unaccountable seasons . . . nobody knew just when those seasons opened and closed. There was no sporting calendar in which their dates were set . . . nor was there any unwritten but mutually-agreed code. It was something spontaneous which moved in the blood and set thousands of children, individually, in one particular week of the year, to produce marbles or whatever it might be . . . It was such an instinct as moves a hundred individual apple-trees at one time to blossom, or moves a million individual swallows in the same week to make a flight to the south.[2]

Crazes are easy to accept. More difficult to confirm is the strong popular conviction that street games had fixed calendar seasons and that it was wrong—almost illegal—to play a game out of its proper season.

Grown folk, looking back at the seemingly authoritative progression of games seasons through the year, are sure there was a definite order, and that each game appeared only once in the year; but, as remembered in books and newspaper articles and by correspondents, the order is never the same. (The belief in fixed annual seasons is deep-rooted; 'Why were children skipping here on 22 February, when I last saw them skipping in *July*?' asked a bewildered correspondent from Exmouth.) An adult normally thinks of the word 'season' in connection with spring, summer, autumn, or winter, or at any rate believes that it is an annual occurrence and fixed by dates, as suggested by the grouse-shooting season, fruit being 'in season', the 'seasonal customs' of folklore, and so forth. 'Season' is a word familiar to schoolchildren, but to them the meaning is nearer 'a certain period of time, not very long'. When they say 'marbles is not in season', they mean 'marbles is not being played much at the moment'. (It would be less confusing if 'periodic games' were the accepted term, rather than 'seasonal games'.) A 9-year-old Glastonbury boy discoursed helpfully on the subject of 'seasons' in June 1960:

We have marbles seasons, you know—we have seasons for everything. A season just starts when somebody brings something to school and the others start playing it. They've been playing something else and they're sick of it. No, it's not got anything to do with the weather or the time of year. There's not just one

---

[1] In 'The Lantern-Bearers', *Scribner's Magazine*, Feb. 1888, pp. 251–6.
[2] *Son of London* (1946), 24.

season for a game in the year, there can be several seasons. There's a marbles season going on now.

No other child who took part in our surveys said or wrote anything to contradict this general statement.

The sudden onrush of a 'season' can seem miraculous to a 7-year-old, but to 10- and 11-year-olds the process is more mundane. Thus girls at Driffield, Yorkshire, explained (May 1975), in a kind of litany:

Solo: It's the rounders season now. Say somebody starts playing rounders—
Chorus: *Everybody* else starts playing it.
Solo: Somebody brings elastic—
Chorus: And *everybody* plays elastic.
Solo: Somebody gets a skipping-rope—
Chorus: And *everybody* skips.
Solo: Somebody brings roller skates—
Chorus: And they *all* roller skate.

I.O.: You said the skipping season's over now. You mean that nobody's skipping at the moment?

1st solo: A few people might be, but not many. There won't be everybody. You'll have everybody playing rounders.

I.O.: What do you think's going to be next?

1st solo: Skipping.
2nd solo: Elastic.
1st solo: It won't be elastic, we've just had that.
Chorus: I think it'll be skipping next.
[*Voice offstage*: But sometimes you can just have nothing.]
1st solo: You think, 'Oh, what shall we do now'—
2nd solo: And you play on the bars.
1st solo: There's a Block season—that's a good season.
3rd solo: Or marbles.
2nd solo: There was a clacker season—
1st solo: But the main seasons seem to be roller skates, and elastic, and skipping, and rounders.

I.O.: What about two-balls, does that come in a season?

Chorus: No-o. A few people just play that, but it's not a main season.
2nd solo: I've just started playing that at home—two-balls.

The main deciding factors for the arrival of a new season thus seem to be: the length of time since a particular game was last played, and the personality of the girl who tries to start the new season, especially if she comes from 'outside'.[3] Add to these the state of the weather, and the

---

[3] For instance, an 8-year-old girl in Poole, Dorset, told us, 'I went to London and I played two-balls there. When I came back here there was no one playing two-balls 'cept me, and the next day everybody brunged some.'

appearance of the right toys—marbles, or skipping-ropes, or jacks—in the shops. Entrants for a *Farmer's Weekly* competition in 1952 acknowledged the influence of shops in deciding—or reinforcing—the games seasons: 'Shuttlecock for girls at Easter . . .Why Easter I don't know, but that was when they came into the shops' (near Carnforth, Lancashire, *c*.1900);[4] 'In our district the shops seemed to set the different seasons, and once the goods were displayed in the windows every child seemed to have one' (near Keighley, Yorkshire, *c*.1923); 'It seems to me . . . that the shops would decide the games for a particular season. Chalk was always available for hopscotch, but it was when marbles or fivestones or hoops or tops appeared in shop windows that we played those particular games' (Acton Town, *c*.1912; *London Lore*, i, pt. 4, p. 40). Or perhaps the children prompt the shops. A son of ours, who worked in W. H. Smith's wholesale branch in Worcester, said to us on 17 February 1969:

Of course you know the marbles season has begun. Children began asking for marbles about ten days ago, and the retailers have been coming in and asking if we have any. My consignment should arrive some time next week or I'll be up a gum-tree. We only get one consignment a year [from Cowan de Groot Ltd., Wakefield House, Chart Street, London N1, probably the best-known toy importers in the country]. There is another shipment in April, when we can get some more if we want them. Most of the marbles come from India now.

It is interesting that the marbles season had begun, for although the Worcester district had escaped the snow, the weather could hardly have been said to be fine and dry.

Even the doziest indoor adult knows when spring has come, and feels his vigour returning. But spring is not tied to the calendar; it can happen in February, or March, or April, whenever conditions are right for growth, and this is when children start playing games. Spring was presumably accountable for the quite closely synchronized openings of the major marbles seasons in 1961: Perth (where the weather was 'unusually warm') 26 February–4 March; Fulham, London, 9–14 March;

---

[4] Probably an example of an adult's faulty memory, for the beginning of the shuttlecock season was consistently said to be Shrove Tuesday: e.g. *Little Games for Little Players* (William Walker & Son, Otley, Yorkshire, *c*.1850), 4, 'Do you know when Pancake-Tuesday is? If you do, and you were to go out into the village streets on that day, you would see all the little boys and girls playing at shuttlecock'; *N & Q*, 5th ser., 12 (1879), 155: 'Years ago, in South Lincolnshire, Shrove Tuesday was the day for beginning the battle-door and shuttle-cock and top-whipping season. Some impatient spirits anticipated the festival, no doubt, but the nuisance was not full-blown or orthodox until the time devoted to *batter* was fully come.'

Wilmslow, 15 March; Bishop Auckland, 20 March; Liss, 20–7 March. It is the return of vitality, as much as the drying of the ground after winter, that causes the surge of children's games in the spring. John Clare celebrated 'The joys, the sports that come with Spring—The twirling top, the marble ring', in the May section of *The Shepherd's Calendar* (1827), 51. Perhaps that was poetic licence, for activity usually starts much earlier. Gilbert White included children in his natural history notes: '1782 28 December. Boys play at Marbles on the Plestor; 1785 23 Jan. Boys play on the Plestor at marbles, & peg-top. Thrushes sing in the coppices; 1788 Feb. 15 Taw and hopscotch come in fashion among the boys.' Spring takes precedence over dates in the calendar. It was most frequently said that February was the time for whipping tops (e.g. 'February, when the first Spring day was the signal for the opening of the top season', letter to *The Times*, 29 September 1952), and this may be because spring often does arrive in February. It may also be because Shrove Tuesday is usually in February, and the playing of games was once associated with Lent.

There is no evidence before the early sixteenth century of any special games season during the six weeks of Lent; and none of the later evidence suggests that those six weeks of licence had anything directly to do with the church, though it is tempting to think that games were encouraged as an alleviation to much stringent fasting and searching of conscience (rather as—so Herodotus claimed—the Lydians survived a famine by eating and playing games on alternate days). Naturally, people use holidays for play, as is explained in the 'prognostication' in *Poor Robin's Almanack* for 1701: 'The Spring Quarter according to the Astromical Account, begins the tenth Day of March. In this Quarter are very much practised the commendable exercises of Nine Pins, Pigeon-holes, Trap-ball, Barley-break, and Stool-ball, by reason Easter Holydays, Whitson Holydays, and May-day do fall in this Quarter.'

The two great holidays at either end of Lent, Shrove Tuesday and Good Friday (sometimes also Easter Monday) were occasions for communal games, some of which, such as the boisterous football games at Shrovetide, and the long-rope skipping on the Foreshore in Scarborough and in the Sussex village of Alciston on Good Friday, still exist. Seaside places, where fishermen had long ropes at hand, were natural venues for skipping (the skipping at Alciston migrated there from Newhaven, on the coast). The Good Friday long-rope skipping in London's dockland, and 'Long Line Day' in Brighton (near the fish market), seem to have been casualties of the Second World War, although the Brighton skipping was revived briefly in the early 1950s. However, Good Friday skipping was recorded in various inland places up until the war—most famously on Parker's Piece, in Cambridge—so perhaps it was once widespread.[5]

Tipcat, and its more sophisticated developments 'Knurr and Spell' and 'Trap Ball', were also widespread holiday games. On Shrove Tuesday, said Charlotte Burne (*Shropshire Folk-Lore* (1883), 319), *the* game at Newport was 'dog-stick', or 'trib and nur', otherwise known as 'knurr and spell', 'from which Shrove Tuesday itself was often called Dogstick Day'; and in 1624 a certain Father Milnes of Aislaby, near Whitby, Yorkshire, was whipped for so far forgetting himself as to 'play in the Churchyard . . . at a game called Trippett' on Easter Day in the time of afternoon service.[6] 'Trap' was one of the games played on Shrove Tuesday and other holidays in Finsbury Fields, London *c*.1642.[7] Bat and Trap was played on Parker's Piece, Cambridge, on Good Friday, and on Midsummer Common on Easter Monday.[8] A description of a match at Bury St Edmund's, Suffolk, dated 19 March 1825, was contributed to Hone's *Every-Day Book*, i, col. 430:

On Shrove Tuesday, Easter Monday, and the Whitsuntide festivals, twelve old women side off for a game at trap-and-ball, which is kept up with the greatest spirit and vigour until sunset. One old lady, named Gill, upwards of sixty years of age, has been celebrated as the 'mistress of the sport' for a number of years past.

The game was played at various places in Sussex, where its name is 'Bat and Ball'. The Prince Regent and his friends used to play trap ball on a piece of level ground, called the Level, in Brighton. He left the land in his will to the town of Brighton, to be used as a play place for children. The old public house called the Bat and Ball, which overlooked this playing-place, was pulled down in about 1969 and another was built in its place. A Good Friday match, played by teams from two breweries, and

---

[5] See esp. Ralph Merrifield, 'Good Friday Customs in Sussex', *Sussex Archaeological Collections*, 89 (1950), 87–8; he also had records of the skipping custom from Leamington, in Warwickshire, and Pontypool, Monmouthshire, to which can be added, from the Opie files, Haughmond Hill, near Shrewsbury; Millom, Cumberland; and Earl Shilton, near Leicester, where, *c*.1905, 'much of the village took part using enormously long ropes often across the street'. It must be remembered that it was common for young adults—factory girls, for instance—to skip at any time of year in the early years of the century.

[6] *North Riding Record Society*, 3: 2 (1885), 199.

[7] A. C. Generosus [probably Peter Hausted], *Satyre against Seperatists* (1642), 6, 'The Prentizes . . . if upon Shrove-tuesday or May-day Beat an old Baud, or fright poore whores . . . Their mounted high, contemne the humble play Of trap or footeball on an holyday In Finesbury-Fields.'

[8] See E. Porter, *Cambridgeshire Customs* (1969), 107, 108–9.

from Brighton and District Labour Club, was held on the Level for many years, and probably still is.[9] In Kent, Canterbury and District Bat and Trap League, founded in 1922, which boasted four divisions and 900 players in 1955, as well as a women's league of eight clubs, played throughout a long summer season, on a pitch 'usually in the back lawn or garden attached to a hostelry' (*Manchester Guardian*, 19 January 1955). Old-fashioned games and good ale go well together.

The Shrove Tuesday and Good Friday holidays may be seen as the opening and closing days of various games seasons that lasted through Lent. The emphasis was 'No shuttlecock before Shrove Tuesday', or 'No marbles after Good Friday'. This may be a survival from an ecclesiastical edict long since forgotten, and of which no evidence remains.[10] Lent was above all the time when tops were spun and whipped; the earliest evidence is from the sixteenth century, when whipping tops had already become a children's game. Alexander Barclay described the successive joys of childhood in his *Eclogues* (1521):

> Eche time and season hath his delite and ioyes,
> Loke in the stretes beholde the little boyes,
> Howe in fruite season for ioy they sing and hop,
> In lent is eche one full busy with his top.[11]

Boys can be seen whipping tops and playing with peg tops in Brueghel's picture of the 'Battle between Carnival and Lent', 1559; they are on the Lenten side of the picture. Richard Mulcaster, the first headmaster of Merchant Taylors' School, discourses on tops at some length in his book *Positions* (1581):

He that will deny the Top to be an exercise, indifferently capable of all distinctions in stirring, the verie boyes will beate him, and scourge him to, if they light on him about lent, when Tops be in time, as euerie exercise hath his season, both in daie and yeare, after the constitution of bodies, and quantities in measure . . .

---

[9] Merrifield, *Sussex Archaeological Collections*, 87–8; and *Daily Telegraph*, 23 Aug. 1979, p. 16 ('The continuing popularity of the tradition must be partly ascribed to the generosity of a firm of brewers').

[10] The only evidence linking the Church and Lent games is connected with the Roman Catholic ceremony of 'burying Alleluia' on Septuagesima Sunday, the third Sunday before Lent. The Alleluia chant was sung at all masses except those between Septuagesima and Easter. According to Hone's *Every-Day Book*, i (1826), col. 199, the story is that in one of the churches of Paris a choirboy used to whip a top, marked with *Alleluia*, written in golden letters, from one end of the choir to the other.

[11] *The Eclogues of Alexander Barclay*, ed. Beatrice White, EETS, original ser., 175 (1928), 184. I am indebted to Professor Nicholas Orme for this reference. He has pointed out that although Barclay was imitating the eclogues of Battista Mantovano (1448–1516), Mantovano simply remarked that every time of life had its joys, e.g. children enjoy games, and Barclay elaborated this to say that each season of the year had its enjoyments.

Ben Jonson alludes to the custom in *The Tale of a Tub* (1633), iii, pt. 4, 'I had whipp'd them all, like tops in Lent.' Thereafter, the practice was not noticed; although in *Folklore* 56 (1945), 270–1, a contributor was confident that there was still a Lenten games season up to the advent of the Second World War:

Before the war . . . on Shrove Tuesday, in the small shops of town and village that deal in such things, whip-tops and battledores and shuttlecocks made their appearance . . . no self-respecting child would dream of purchasing these toys at any other time of year, and by Easter the fashion had expended itself.

Correspondent after correspondent confirmed this: 'Shuttlecocks, like marbles and whip top, usually came out at Shrovetide' (Halifax); 'Whips and tops , and battledore and shuttlecock, were played on Shrove Tuesday' (Doncaster); 'Whip and top was the traditional plaything in the 1920s for boys at Shrovetide; the girls, more ladylike, played with battledore and shuttlecock' (Ripon). Also see, in Chapter 2, the calls legitimizing raids on other people's marbles, sometimes because the players were playing after Good Friday.

GAMES DISAPPEARING

It is characteristic of the human race that change is constantly deplored, and that 'the good old days' are believed to have been far better than the present day. In the realm of children's games the fixed idea is that children 'don't play games any more', or 'don't have the fun we used to have'. Pepys seemed to have been of this opinion; after a visit from his old school-fellow Jack Cole, he wrote (*Diary*, 25 July 1664): 'I made him stay with me till 11 at night, talking of old school stories, and very pleasing ones; and truly I find that we did spend our time and thoughts then otherwise then I think boys do now, and I think as well as my thoughts at the best are now.' A writer to *Notes & Queries*, 6 (1852), 242, sent a singing game to be 'rescued from oblivion', believing that 'National Schools are fast sweeping away all charms, fairies, folk lore and old village sports and pastimes'. J. S. Udal (*Folklore* 7 (1889), 203–4) blamed the railways for the decline of 'the old-world amusements of our grandparents' days'; he was often told by Dorset villagers, 'them gëames is a-dien out a'together'. A later contributor to *N & Q* (8 (1921), 355) thought it was 'the rival attractions of the cinema and the frequent passage of motor vehicles' that was causing the gradual disappearance of children's games. A writer in the *Staffordshire Mercury* (40 (1924), 169) said, 'Nowadays children have forgotten how to play organised games of their own devising—degenerate childhood now seeks amusement in the artificialities of the Cinema, the Radio Broadcasting, or the Gramophone.' Richard Church, in *Over the Bridge*, 123, wrote that 'such tribal activities [as French Knitting] are not seen in the modern, mid-twentieth century suburbs . . . to me it is a paradox that the democratic standardisation of life, especially urban life, should result in the disappearance of these folk games'; looking back to Battersea, *c*.1901, he regrets the passing of hoops, conkers, and cherry-oggs. The *Halifax Evening Courier* (2 September 1966) said, 'These street games belonged to a more energetic age. The coming of radio, television, and later, pop music sounded the death knell of the old street games.'

In the early 1990s teachers began to be worried that 'children appeared to have lost the art of playing games', claiming that they simply sat around bored, or played solitary games on their computers.[12] In some schools, dinner ladies were co-opted to teach the children games (but children have always learned occasional games from dinner ladies and other friendly grown-ups). The truth is that, unless they are seriously undernourished or in a state of fear, children will always play when they are on their own, unsupervised, in the freedom of an open space. Teachers on playground duty have their minds on things other than identifying games, and in any event cannot always recognize which game is which, amidst the criss-crossing mêlée of the playground. A chasing game superimposed on a game of 'May I?', and both games intersected by a diffuse game of 'War', can look like 'children running about aimlessly'. A girl roughly tugging at another girl is probably not molesting her but taking her to the witches' den in a game of 'Witches and Fairies'. When I first went into the playground of my local primary school, in 1960, the deputy-headmaster told me definitely that children did not play games any more. Later I asked him if he knew the local name for a game that was in progress on the far side of the playground. 'Oh, I shouldn't think that's *anything*,' he said. It turned out to be that old favourite

---

[12] See e.g. *Sunday Telegraph*, 31 Jan. 1993, on the successful introduction of hoops, tops, diabolo, and hopscotch to Dalbeattie Primary School, Kirkcudbrightshire, which had help from the Association for Fair Play for Children in Scotland; *Daily Mail*, 26 July 1993, report on Grange Primary School, Meir, Stoke-on-Trent. That year the scare spread to other schools in Stoke, to Liverpool, Birmingham, Poole, and probably many other places, and has still, in 1996, not gone away; for example, the headmistress of a Glasgow primary school became worried about the lack of play in the school playground, and asked a team of social workers to teach the children traditional games, some of them with softened rules and words so as to be less aggressive and intimidating (*Independent on Sunday*, 27 Oct. 1996). It may have been started by the government-supported report, *Children's Exercise, Health and Fitness*, publ. 6 Sept. 1988. Cassandra Jardine, in a down-to-earth article in the *Daily Telegraph* supplement, 30 July 1993, found that 'parents and primary teachers around the country disagree that play has died out. It just sometimes takes a slightly different form.'

Pieter Brueghel the Elder's painting 'Kinderspiele' ('Children's Games'), 1560. The marbles games of 'Castles' and 'Long taw' can be seen, and boys whipping tops under the portico.

'The Playground', engraved by F. Joubert after Thomas Webster, RA (Glasgow Art Union), 1858. A large game of peg top is in progress to the left, with marbles and football further back.

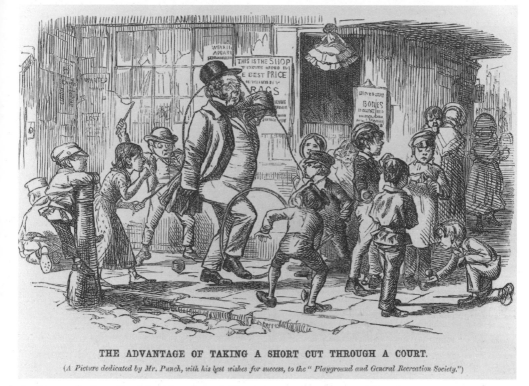

THE ADVANTAGE OF TAKING A SHORT CUT THROUGH A COURT.

(*A Picture dedicated by Mr. Punch, with his best wishes for success, to the "Playground and General Recreation Society."*)

Punch cartoon, 4 June 1859. The old gentleman 'taking a short cut through a court' is much impeded by a skipping rope, a hoop, tipcat, and a crowd of boys playing with tops.

Girls whipping tops in a Workington street, 1963.

'Grandmother's Steps', under the name of 'Peep behind the Curtain'. Now, in the 1990s, the case seems to be much the same. Fellow collectors of children's games, if they are allowed into primary-school playgrounds (not as easy of access as in the middle of the century), all say that the first thing they are told by a teacher is, 'Children don't play games any more.'

Change is an inevitable part of life, and it is true that, for a variety of reasons, some games have indeed died out. It is a truism that life has speeded up, that the huge increase in motor traffic has made the roads impossible to play on and the pavements dangerous. Today's smaller families and spaced-out houses, each with its own garden, mean a dispersed child-population. The large families and crowded housing of the days before the First World War provided a vibrant games environment (remembered by a correspondent (November 1952) as 'the joyous noisy crowd of fifty years ago, 61 children playing in the same back street'). This atmosphere still exists in certain parts of London where immigrant families have settled.

Most children no longer walk to school. The walk to school was an ideal time for playing games, and it seems, surprisingly, that the children were seldom late. In the eighteenth century they were frequently admonished not to linger over a game of marbles on their way to school. 'The Boy that is good', said *A Primer or Reading Book for Children* (*c.*1780), 'to the School straight himself betakes, And plays not, nor loiters; but Haste thither makes . . . Then hangs up his Hat, and repairs to his Place, With visible Tokens of Joy in his Face.'[13] Marbles (especially 'Follows'), whipping tops, hoops, or simply kicking stones, fir cones, or frozen horse-manure, were interesting ways of covering a mile or so. Flora Thompson (*Lark Rise*, ch. 11) described the hamlet children, in the 1880s, taking an hour and a half to walk the mile and a half to school, 'partly because they liked plenty of time to play on the road and partly because their mothers wanted them out of the way before house-cleaning began'. The delights of dawdling to school have not entirely disappeared. When Ian Duncan made his classic film *Playing Out* for BBC 2 (shown 1 November 1994), shooting it in his own home town of Keighley, he told me not to lose faith: 'They still do *exactly* the same things as I used to, walking the mile to school—feeding the horse with bits of apple, hanging head-down over the railway bridge, jumping off walls.' The whole film is a testament to the uproarious pleasures of childhood, complete with secret den and sausages cooked over a camp-fire.

In some cases it has been realized that games which once belonged to

---

[13] British Library C.40.a.78, pp. 58–9. Lacks title-page.

young adults are not suitable for children. The older ring games, played centuries ago by 15-year-old youths and maidens courting in the spring, were inherited by younger and younger people until they arrived in the kindergarten; but it never felt quite comfortable to 'chose the one you love the best', and none of the adaptations worked satisfactorily. The only courtship games to have survived are those that have an element of comedy (and they are becoming rare), and those that have successfully been turned into skipping songs.

Adults can be savagely critical of the supposed sophistication or inertia of contemporary schoolchildren; and equally self-righteous about their own childhoods. The much-reiterated phrase is, 'We used to make our *own* amusements.' At the same time they all but prevent their children from making their own amusements by supplying them with generous pocket-money, and giving them expensive toys. Often it was lack of money that caused children to play with home-made toys that cost nothing. Human nature being what it is, a child would rather play with glamorous glass marbles than with cherry-stones picked up from the gutter.

The changing fashions in children's games are also to some extent affected by the games played by their seniors. Children must have heroes to copy. The present-day heroes are footballers. Even the smallest boys worship famous footballers, watching them on television, knowing every detail of their careers, and having opinions about their prowess. No wonder they spend much of their leisure time trying to emulate these heroes. (In days gone by boys must have felt much the same about local champions of 'Knurr and Spell' and 'Trap Ball'.) When W. G. Grace was around, cricketers were the heroes, and in the cities every house-end had its wicket, drawn with chalk or, for more permanence, with tar scraped off the road. Now, we are told, only Pakistani immigrant boys play street cricket with the same enthusiasm—cricket being their national game. Role-models are of prime importance. All the traditional games once had their adult or nearly-adult exponents. Men played marbles; teenagers bowled hoops and whipped tops. But increasingly, through the nineteenth century, the public schools needed to organize 'manly', competitive sports, and to regulate the rules so that matches could be played between different schools, and young men from different schools could all play by the same rules when they reached university. Football and cricket were the chief of these prestigious games; in contrast, the old informal games were despised as outmoded, childish, or uncouth. From this point of view, the revival of some street games as world sports has been an excellent thing, although the romantically minded cannot help regretting a loss of informality and spontaneity. Double-rope skipping, with two long ropes turned in opposite directions, benefited from the advent of the first 'Double Dutch Skip Rope Championship' in New York, in March 1974, and the subsequent forming of teams in other American cities and other countries. Competition has raised the standard of double-skipping higher than it ever was before; and the age of the oldest competing skippers has risen to about 16. These much-publicized stars have been copied by the children on the sidewalks, and now if one asks them, 'Can you do Double Dutch?', the answer is likely to be, 'Of course'. The adult participants in the Ashton World Conker Championship, in Staffordshire, and the World Marbles Championship at Tinsley Green, West Sussex, may not have much influence on children's conker and marble play, but it is important that some people should be demonstrating the skill possible in these games.[14]

BANS

Circumstances rule all. It is a sad thing that the games whose passing is mourned by adults were largely suppressed, through necessity, by the adults themselves. The reasons are easy to see. The movement of population into the cities during the industrial revolution, the consequent uncontrolled urban spread, and the teeming child-population, meant that children played in the streets within call of home, 'to the serious inconvenience of the passengers'.[15] The enclosure of commons caused another shrinkage of play-space. A letter to the *Worcestershire Chronicle*, 1847, complained of this, saying that 'in Bromsgrove here, the nailor boys, from force of circumstances, have taken possession of the turnpike road to play games'. It was a long while before the authorities had the leisure to understand the need for playing-places, and the money to provide them; the public parks were too far from home for young children, and the policy was 'Please keep off the grass'.[16] The solution was to ban the more obstructive games from the streets.

---

[14] Other conker competitions have been held from time to time, the earlier ones for children, e.g. the Nottingham conker tournament (*Yorkshire Post*, 26 Oct. 1959), and those at Walton-on-Trent, Derbyshire (*Yorkshire Post*, 10 Oct. 1961) and Newport, Isle of Wight (Oct. 1995; on local TV news); more recently, 'contestants from rival firms' have battled in the annual 'City of London conker competition', in Finsbury Square, 'fortified by liquid refreshments' (*Independent*, 23 Oct. 1991). However, the Ashton championship has held the limelight, and grows bigger every year. Both that and the Tinsley Green marbles contest have graduated from being 'British' to being 'World' championships. Britain is the only country where the game of conkers is indigenous, so foreign contestants learn the game in order to compete.

[15] J. Brand, *Popular Antiquities*, ed. Sir H. Ellis (1849), ii. 433.

[16] W. H. Cremer, Jun., proprietor of Cremer's toy manufactory and toyshop in Regent Street, predecessors of Hamley's, wrote in *The Toys of the Little Folks* (1873), 44 n., 'I rejoice to see that

Bans were nothing new. Sometimes the best early evidence of the existence of a game is an ecclesiastical ban. A churchyard was a favourite place to play; the central and convenient flat space of the churchyard itself, the flagged cloisters, the walls of the church and angles of the buttresses, the flat tops of the tombs, were all excellent for 'games with things'. But bans never seemed to have any lasting effect. In 1287, games and secular business were forbidden in churchyards by the Synod of Exeter.[17] In 1385 the Bishop of London was forced to declaim against the ball-play about St Paul's; and in 1447 the Bishop of Exeter complained to the Mayor of 'yong peple' playing in the cloister, even during divine service, such games as 'the toppe, queke, penny prykke, and moste atte tenys, by the which the walles of the saide Cloistre have be defowled and the glas wyndowes all to brost'.[18]

Street play was undoubtedly an inconvenience to 'passengers'. Tipcat was the chief culprit. 'This has been sometimes objected to by a bloated constabulary and venal press, on account of a few individuals having caught the cat in their eye', said a 'town boy', in the *Boy's Monthly Magazine* (January 1864), 12–13, and added, 'Why don't they leave their eyes at home?' Even so sane a person as Frances Low, who considered tipcat 'a highly interesting game', was glad that it was being rigorously regulated by the police and thought it ought to be entirely abolished from the small back-streets of London (*Strand Magazine*, (1891), 513). A poster issued by the Chief Constable of Great Yarmouth on 1 February 1889, gave notice that 'All BOYS And others obstructing the public thoroughfares of the Town, by SPINNING TOPS or PLAYING AT MARBLES or other games upon pavements or footpaths, will be summoned before the Magistrates and fined.' Norman Douglas said: 'marbles are not played as they used to be. The police are getting more interfering every day; they tell the boys to move on and not block up the pavement . . . and if you don't clear off at once, they kick your marbles into the gutter where they get lost down a drain' (*London Street Games* (1916), 112). Alongside

actual play is now, thanks to a very sensible but much abused First Commissioner of Works, allowed in most of our parks. When I walk down Dudley-street, St. Giles, and count *one hundred and seventy-two* poor little children, chiefly from the cellars below, amusing themselves in the gutters as best they can, I cannot but lament that there is not an authorized playground in every parish. Why should not our own parish church-yards be utilized for this purpose?'

[17] D. Wilkins, *Concilia M. Britanniae et Hib*, ii. 140. In the 1960s and 1970s some London churchyards *were* converted into play-places. At St Luke's, Chelsea, for instance, the tombstones were set on end around the perimeter, and the churchyard asphalted over as a playground divided into two parts, one for younger children, with swings and roundabouts and a shelter for mothers, and the other for anyone who wanted to play football. Similarly, the churchyard of St Martin-in-the-Fields was turned into a netball court.

[18] *N & Q*, 7th ser., 3, p. 485; *Letters & Papers of John Shillingford, Mayor of Exeter (1447–1450)* (Camden Society, 1871), 10. 'Queke' is a game played on a chequer board.

complaints from the public, and the resultant police action, were charitable folk trying to organize alternative places to play. The cartoon
(*Punch*, 4 June 1859, p. 233) of an old gentleman 'taking a short cut
through a court', and getting entangled in a skipping-rope, hoop, and a
crowd of boys whipping tops, was 'dedicated by Mr Punch, with his best
wishes for success, to the "Playground and General Recreation
Society"'.

Children are law-abiding creatures. If they see a 'Trepassers will be
prosecuted' notice, the majority will not put one foot within the forbidden area; and memories of prohibitions linger on. The following conversation took place between a building worker and his 12-year-old son, on
a train journey from Liss to London in May 1973:

Father: They don't play whip top like they used to.
Son [*quick as a flash*]: They're illegal.
Father: Yes, that's true. Same with hoops. The police stopped them. They're
   not allowed on the road.
Son: Not allowed in the playground, either.

A certain self-righteousness emanates from children (especially girls)
when they talk about forbidden games. A headmaster says he does not
want any more writing on the school walls; the girls interpret this, with
a martyred air, as, 'We're not allowed to chalk our hopscotches on the
playground any more', and one has the impression they would not bring
chalk to school even if they were ordered to. Misunderstandings apart,
chalk has not been allowed in most primary schools since the 1960s, and
the girls played hopscotch only at home until the authorities had the idea
of painting permanent hopscotch diagrams and other patterns on playgrounds. Balls have also caused much dissension; schools place tempor-
ary or permanent bans on them, according to the number of windows
broken—the old schools, with high windows, were less vulnerable than
the modern 'glasshouses'. Conflict of interest is at the root of the problem. On council estates it is usually complaints from older tenants that
lead to the banning of ball games. (The record for the severest restriction
must surely be held by the by-law passed by Leicester City Council, in
October 1970, banning ball games on the open spaces of its council
estates by children of 8 years old and over, with a £10 fine for non-
observance.) The noise of children playing is no longer tolerated. Perhaps, after all, it is unwise to house the old and the young within earshot
of each other.

2

Marbles

The boy to play the man with a better grace, resolves to try if he cannot add profit to his pleasure. The most manly imitation for this purpose is the play at marbles, which is billiards in miniature.

(*The Instructor and Guide for Little Masters* (1772))

Marbles is a complicated game.

(Boy, 8, Birmingham, 1952)

SEVERAL games of the ancient world could be said to have been played like marbles. 'Omilla' ('The Circle'), described in the second century AD by Pollux in *Onomasticon* (ix. 102), was a game like 'Ringy': the aim was to lodge one's own knucklebones in the circle and dislodge those of one's adversary. In Ovid's poem *Nux*, a walnut-tree complains of ill-usage by boys who strip its branches of nuts for their games. Two of the games follow the general marbles' principle of rolling one object to hit another. In one of these, the players try to hit each other's nuts in succession. In the other, a nut is rolled down a sloping plank with the aim of hitting a nut on the ground; this game was still played in Sicily in the nineteenth century (see Pitrè, 'A la Sciddicalura'), and was being played in Jerusalem in 1965 (see Eifermann (1971), 48–9).[1] Nuts, however, have the disadvantage of not rolling very well.

In the British Museum there are many small spherical objects of unknown use: stone spheres, about half an inch in diameter, found in early Egyptian tombs; clay balls from Petsofia, in Crete, from the Minoan period, 2000–1700 BC, one red clay, one red-brown; 'trays full' of glass balls (some in ravishing mixed colours, blue, yellow, red, and green), clay balls, marble balls, from the Roman Imperial period, from places such as Capua, Syria, Ephesus, Oxyrhynchus, and Cnidus in Asia Minor. They look like marbles, and the curators say, 'We called them marbles because we couldn't see what else they could be. They didn't seem right for perfume-bottle stoppers.'

[1] A similar game, 'Rolduitje' ('Roll Farthing') must have been very popular in Holland in the early 19th c., for it was much depicted on tiles (see the catalogue of the *Kinderspelen op tegels* exhibition, Tegelmuseum It Noflik Sté, 1979). This must be related to the 'Roll-a-penny' game at fairs and fêtes.

With marbles thus strewn about the ancient world it seems almost inconceivable that no early references should yet have been found. Perhaps the game was not very popular, or perhaps it was exclusively a children's game, and beneath the notice of adults. (What word would have been used, for these small stony games-balls? In Italy the word for marbles is simply *palline*, little balls.) Two apparent references to marbles turn out to be mistranslations. One is the story Steele tells in the *Tatler*, 112 (1709), about the emperor Augustus, who 'is said to have passed many of his Hours with little *Moorish* Boys at a Game of Marbles, not unlike our modern Taw'. This is from Suetonius, ch. 83, and the word used was *talis*, i.e. knucklebones. The other is St John Chrysostom's *Discourses against Judaizing Christians, c.* AD 390 (trans. Paul W. Harkins (1979), 31), where a demon, stealing men's souls, is compared to kidnappers enticing little boys 'by offering them sweets, and cakes, and marbles'; here again, 'marbles' should have been 'knucklebones'.

Or perhaps the marbles of the ancient world were used on 'boards' incised on stone slabs. The authors of *Timgad: Une cité africaine sous l'empire romain* (1905),[2] comment that certain diagrams on the paving stones of the forum, a great number of which are also to be found in the forum at Rome, are likely to have been used for a game something like bagatelle. Within these oblong diagrams, which are about 50 cm long, are shallow holes—too shallow to accommodate ordinary balls—varying in number and arrangement from diagram to diagram. It seems possible that players tried to roll a marble from one end of the diagram to the other, avoiding the holes (in one example cited, a separate hole at the far end seems to be the goal), or that they tried to get their marbles into the holes, scoring in some way, or even that they aimed at successive holes in the manner of the modern marbles game of 'Three Holes'.

Evidence of marbles-play is slow to appear. Hugo von Trimberg, the German moralist, who wrote his *Renner* from *c.*1290 onwards, equates 'tribekugeln' and other children's games with the dissolute gaming of adults (1908–9, line 14905). In a sixteenth-century Flemish calendar (BM Add. MS 24098) 'September' has a border of children's games, in the centre of which four boys are playing at ring marbles. In Brueghel's picture of children's games, 1560, boys are shown throwing a marble at others arranged in a line (this is 'Long Taw', listed by Randle Holme in 1688).[3] The engraved illustration to Jacob Cats's poem 'Kinder-spel', in

---

[2] E. Boeswillwald, R. Cagnat, and Alb. Ballu, pp. 29–31.
[3] 'Long Taw' was still played until the late 19th c. Routledge's *Every Boy's Book* (1856), 60, says it is 'seldom played by London boys, but is very common in the different English counties': 'One boy places his marble on the ground at A, the other at B; then both retire to the spot C [A, B, and

the second edition of *Emblemata* (printed in the same year as the first, 1618), shows the same game in progress, with the player knuckling his marble in true championship style. In the Low Countries, children's games were a common subject for both tiles and popular prints: a mid-seventeenth-century set of twenty-four little games-scenes (the equivalent of 'scraps') has boys playing ring marbles with four target marbles spaced round inside the ring. In Comenius' *Orbis Pictus*, translated by Charles Hoole (1659), ch. 136, 'Boyes-Sports', there is a boy kneeling, apparently playing marbles; the game is listed as 'Bowling-stones'. Two small boys are playing marbles outside an inn, in Adriaen van Ostade's picture 'Children and Dog', dated 1673, but with no discernible layout. A short exchange in a school-book of colloquial sentences in Latin and English (*Familiaries Colloquendi Formulae* (1678)) dismisses the subject in a dozen words: 'Have you any Marbles? Ay, but I will not play with you.'

A pleasant spotlight is thrown on marbles-play in Daniel Defoe's *Life of Mr Duncan Campbell*, the deaf and dumb soothsayer, (1720) 21:

Our young prophet had taught most of his little Companions to converse with him by Finger . . . Marbles (which he used to call Childrens playing at Bowls) yielded him mighty Diversion; and he was so dexterous an Artist at shooting the little Alablaster Globe from between the end of his fore-finger and the knuckle of his Thumb, that he seldom missed hitting *Plum* (as the Boys call it) the Marble he aimed at, tho' at the distance of two or three yards.

In the mid-eighteenth century Coles Child, haberdasher, of the Blew Boar on London Bridge, was selling many other things besides needles, buckles, and buttons. His trade card advertises 'all sorts of English and Dutch Toys, Pill Boxes, Nest Boxes, Babies [dolls], Marbles, Alleys'.

CASTLES

This game makes time seem nothing. 'Castles' or 'Pyramids' was played with nuts in Rome *c*. AD 10 and thereafter in Italy certainly until *c*.1925; it was played with nuts in the Jewish community in Salford in the 1920s and was still played there, though with marbles, when we were collecting

C are in a straight line]. The first boy now shoots at B from a line marked at C. If he strikes it, he takes it and shoots at A; if he strikes A, he then wins the game. If, however, he misses B, the second boy then shoots at B; if he strikes it, he can then either shoot at the first boy's taw at the place at which it lies, or he can shoot at A. If he hits his opponent's taw, he is said to kill him, and wins the game, or if he shoots at A and hits it. The boy who hits the last shot has the privilege of shooting at the taw of the other, provided it has not already been killed. If he hits it, the taw is taken, or the owner must pay one, and the game ends; and if he misses it, the game is then at an end also.'

(276)

CXXXVI.

Ludi Pueriles.

Boyes-Sports.

(277)

Boyes	Pueri
use to play either with	ludere solent,
Bowling-stones ; 1.	vel *globis fictilibus* ; 1.
or throwing	vel jactantes
a Bowl 2.	*Globum* 2.
at Nine-pins ; 3.	ad *Conas* ; 3.
or striking a Ball	vel *Sparulam.*
thorow a Ring ; 5.	*Clavâ* 4. mittentes
with a Bandy; 4.	per *Annulum* ; 5.
or scourging a Top 6,	vel *Turbinem* 6.
with a Whip; 7.	*Flagello* 7.
or shooting with	versantes ;
a Trunck, 8.	vel *Sclopo*, 8.
and a Bow; 9.	& *Arcu* 9.
or going upon	jaculantes ;
Stilts; 10.	vel *Grallis* 10.
or tossing	incedentes ;
and swinging	vel super *Petaurum* 11.
themselves upon	se agitantes
a Merry-totter. 11.	& oscillantes.

John Amos Comenius, *Orbis Pictus*, 1659. Among the 'Boyes-Sports' are 'Bowling-stones' (marbles) and 'scourging a Top with a Whip'.

Boys playing marbles outside a tavern: Adriaen van Ostade, Haarlem, Netherlands, 1673.

'La Rangette' ('Castles') in Jacques Stella, *Les Jeux et plaisirs de L'enfance*, 1657.

Boys playing a game like 'Holie' on a terrace, from an engraving by the French artist Abraham Bosse, 1636.

games in the 1970s.[4] The game is vividly depicted on the side of a Roman
boy's stone coffin, where nuts are arranged in 'castles' of four, three in a
triangle and one a-top, and boys, in short tunics to the knee, aim to knock
them down and win the nuts.[5] 'Ludus castellorum' is one of the seven nut
games described by Ovid in *Nux*.

Random references through the centuries show that this game must
have been both international and popular, offering exciting opportunities
for subterfuge.[6] Jesus and his companions can be seen playing it in a
fourteenth-century panel now in the Kunsthalle, Hamburg (Adey
Horton, *The Child Jesus* (1976), pl. 69). Gargantua played 'au chastelet'
(Rabelais (1534), i. ch. 22). 'Castles' appears in Brueghel's painting
'Children's Games', 1560. It is described in Jacob Cats's poem 'Kinder-
spel', in *Houwelyck* (1625). It is shown, under the name of 'La Rangette',
in Jacques Stella's *Les Jeux et plaisirs de l'enfance* (1657). In Erasmus'
Colloquies, (1526; trans. N. Bailey (1725), 428), Apicius advises on how
the serving should be done at a great banquet: 'To every four Guests set
four Dishes, so that the fourth may be the middlemost, as Boys upon
three Nuts set a fourth.'

The English name for the game was 'Cobnut'. It is mentioned in Sir
Thomas More's *Second Parte of the Confutation of Tyndals Answere*
(1533), 103: a boy on his way to school, meeting some lads at play, 'falleth
to wurke wyth them at some suche prety playes of lykelyhed as chyldren
be wont to play, as . . . cobnutte'; and is given by Cotgrave, in his
Dictionarie of the French and English tongues (1611): '*Chastelet*, the childish game *cob nut*, or (rather) the throwing of a ball at a heape of nuts,
which done, the thrower takes as many as he hath hit or scattered.' It also
appears under *cob-nut* in Grose's *Provincial Dictionary* (1790).

Cherry-stones were used in Scotland. Sir Walter Scott—who was
better at the 'yards' (or playground) than in the class—said (*Minstrelsy of*

---

[4] The Jews have preserved many ancient games. 'Pupke' is played especially at Passover. Leila
Berg, as a child in Salford in the 1920s, knew the game as 'cupky': 'you balanced four nuts—three
points together and a nut on top—and the other children rolled their nuts to knock yours down'
(*Guardian*, 24 April 1965, p. 8).

[5] The sarcophagus is in the Vatican Museum. The game can also be seen on a similar sarcophagus
reproduced by Daremberg and Saglio, *Dictionnaire des antiquités*, 4, pt. 1 (1904–7), 115.

[6] In Capri, *c.*1923, for instance, where the game was played with Christmas hazel-nuts, the
players stood about 25 feet away from the row of 'castelli', each owned by a different boy, and a
difficult shot was made more difficult when the owner of a 'castello' had weighted his top nut (the
'cappone') with lead shot. A hole was made in the nut, filled with shot, covered with wax, and then
inked over to disguise the handiwork. If anyone suspected foul play he could challenge the owner of
the 'castello', who might respond, 'All right, you can look if you like, but if it is a good nut with no
shot in it I can take all the "castelli".' The nuts used by the Efik of Calabar Province, Nigeria, are
large round ones called *nyori* (which is also the name of the game) (*Folklore* (1958), 32). In
Yugoslavia the game is played with horse-chestnuts (Susan Adams, *Games Children Play Around the
World* (1979), 29).

the Scottish Border, ed. T. Henderson, 114): 'my companions at the High School of Edinburgh [1778–81] will remember what was meant by *herrying a sowie*'. The *sowie* was the small heap of cherry-stones. Jamieson (1808), defines *Paip* as 'A cherry-stone picked clean, and used in a game of children. Three of these are placed together, and another above them. These are called a *castle*. The player takes aim with a cherry-stone, and when he overturns this castle, he claims the spoil.' James Ritchie, when writing *The Golden City* (Edinburgh), in 1965, was told that 'long ago' children counted their paips in castles or 'caddles' of four, and tried to throw four at once into a hole. All that had survived of the original game was the knowledge that four was a 'castle' (p. 73).

In the games books of the nineteenth century, from the second edition of *The Boy's Own Book* (1829), onwards, the game usually has the more pedantic name of 'The Pyramid'. It is recommended as 'a good indoor game'. The instructions are that a small circle should be drawn on the ground and a pyramid of four marbles erected at the centre. If a shooter strikes the pyramid with his taw, as many marbles as are driven out of the circle belong to him. The illustration shows the owner of the pyramid sitting with his legs wide apart behind this set-up; his task is to keep the pyramid replenished; and, since it seems that his assets are being steadily whittled away, it is as well that a change of pyramid-keeper is suggested 'at stated intervals'. In *The Book of Games* (c.1837), where the game is called 'Castle', the essential detail is added that the castle-keeper takes all the marbles that have missed. In *Cassell's Book of Sports* (1888), the proprietor is allowed to charge a marble a shot, and this reinforces the game's great similarity to a side-show at a fair.

Boys in Salford, in 1970, were still knocking down little piles of marbles, though they had no special name for the game.

We 'ave it three on the bottom, two on the three, and then one on the top. We get some Plasticine—not a lot—and stick it on the marbles, 'cos the marbles don't stay, they keep coming down. We 'ave like six piles, there's three of mine and three of the other person's. There's two people playing, and you take it in turns. If you knock the pile over, you win all the marbles. We set it up where there's smooth ground, at the side of the road. Sometimes we set it up in the middle of the road, and the cars go right over it and you just get out of the way and then get back again.

In Sweden, marble-shies are set up in the same way. The owner makes a pile of three or four or however many marbles he is willing to risk, and sits behind it with his legs apart. The higher the stake, the further away the thrower must stand or kneel (Mullsjö, c.1965, and Mölndal, 1995).

MARBLE BOARD

This is an old European game of great simplicity. It is played with a long
piece of wood forming a bridge which has a number of arches cut in it.
The arches have a number over each, and small balls, or marbles, are
rolled along the ground towards them, scoring according to the archway
the ball goes through. Traditionally the French board had thirteen holes,
the Italian twelve, and the English usually eleven, and later nine.

The French name was 'Trou-madame', although Rabelais
(*Gargantua*, i (1534), ch. 22), a little earlier than the earliest reference to
'trou-madame', used the term 'croc-madame' (which Urquhart translates as 'grapple my Lady').[7] 'Trou-madame' was perhaps considered
more stylish than the English name of 'Trunks' ('Why say you not that
Munday will bee drunke, Keepes all unruly wakes, and playes at
trunkes?', *Christmas Prince*, the revels held at St John's College, Oxford
(1607–8)), and became 'Troll-madame' under the influence of 'troll', to
roll. Other names were 'Pigeon-holes', and 'Nine holes'. The boards
could be large, or small enough to stand on tables. The setting might be
a fashionable watering-place, as described in Dr John Jones's *Benefit of
the Auncient Bathes of Buckstones [Buxton]* (1572), 12, where 'the Ladyes,
Gentle Women, Wyves, and Maydes, maye in one of the Galleries walke:
and if the weather bee not aggreeable too theire expectacion, they may
haue in the ende of a Benche, eleuen holes made, intoo the which to
trowle pummetes, or Bowles of leade . . . or also, of Copper, Tynne,
Woode . . . the pastyme Troule in Madame is termed.' Or the game
could be a side-show at more disreputable gatherings. Autolycus lies
squirming on the ground (in Shakespeare's *Winter's Tale*, IV. iii) pretending to the Clown that he has been beaten up by a rogue called Autolycus;
this rogue, he says, 'I have knowne to goe about with Troll-my-dames'.
The Clown knows the fellow well, he 'haunts Wakes, Faires, and Beare-
baitings'.[8]

'Trou-madame' is certainly much older than the date when it was first
referred to, and was doubtless played by children as well as adults—the
generations sharing their pleasures without distinction. It was not until
books were produced specifically for children, in the eighteenth century,
that children's activities were better defined. Richard Johnson's *Juvenile*

---

[7] It seems likely that the Scottish court poet William Dunbar was referring to 'Trou-madame'
when, in *c.*1500, he wrote scathingly: 'So mony lordis, so mony naturall fulis, That bettir accordis
to play thame at the trulis.'

[8] Florio, in his *Dictionarie of the Italian and English Tongues* (1611), defines *Trucco*, as 'a Billiard-
boord. Also the play at Billiards. Also a game vsed at May-games in England in the high-waies, with
casting little bowles at a boord with thirteen holes in it, and numbers ouer them.'

Seventeenth–century gentlefolk playing 'Trou-madame', the forerunner of the marbles board.
Engraving by the German artist Mathäus Merian, *c.*1630.

Marbles board, using the ubiquitous plaything of the London streets—cherry stones. *Comic Cuts*, 1912.

Sports and Pastimes (1776), 99, puts 'marble board' firmly in the possession of schoolboys:

The marble-board affords tolerable good diversions in the house in wet or damp
weather, when we are not permitted to pursue our sports in the open air. The
marble-board is very easily made. Get a piece of smooth deal, of what length you
like, about two inches high when placed on the ground, and about an inch thick.
At the bottom of the board cut a number of little round holes, at the distance of
about two inches from each other, but observe that these holes must be cut very
little larger than the marble.

The holes were not numbered; a successful player simply won a marble
from each of the other players.

   Strutt (1801) included 'Nine Holes' among 'The popular Pastimes
among the Men imitated by the Children', and said, concisely, that it
'consists in bowling of marbles at a wooden bridge with nine arches' (p.
287). The game was also known as 'Bridgeboard' (Gomme, i (1894), 45;
London Street Games (1916), 111); 'Donkey', G. F. Northall, *Warwickshire Word-Book* (1896); 'Marble Alley', Swansea, 1925; 'Marble Board',
London, 1908 ('There was keen competition. Some boys gave better
odds on their boards'); 'Marble Knack', Scarborough, 1900–20 ('My
uncle made me one—you bored the holes with a red hot poker') 'Shanty',
Swansea, *c*.1930 ('One hole might be marked 12, but no ordinary marble
would go through it'). It was also played with the less expensive 'cherry
oggs', 'cherry wobs', or 'cherry bobs', and then often not with wooden
arches but, as Richard Church remembered from Battersea, *c*.1901, in
Over the Bridge, p. 121, with 'castle façades of cardboard', which had
numbered doors for the 'cherry oggs' to be rolled into. An alternative was
to use cardboard shoe-boxes, upturned, with numbered holes cut in the
bottom, into which the players pitched their cherry-stones. The children
in the large Jewish communities in London and Salford played thus, at
Passover, with Barcelona nuts if they could afford them, and cherry-
stones if they could not. This game survived at least into the 1970s. A boy
from Wilbraham Primary School, Manchester, introduced it to St Clem-
ent's School, Salford, when his family moved to Salford in 1970: 'It's
called Griddie. There's this box, and it's got, like, six, four, three, and
then two, and on the top it's got six alleys in one compartment, four in
the uvver, three in the uvver, and then two, and if you get an alley in one
of 'em you win six alleys and so on.'
   Marbles boards, made of wood, hardboard, or cardboard, were well
known in all the primary schools in Guernsey. They were called 'marble
boards', or 'arch boards', when the game was known as 'Arch Marbles'.
'At playtime the wall is lined with marble boards,' wrote a teacher in

December 1962. 'Some are inscribed "Try Your Luck", and frequently very elaborately made. Often a boy or girl will win fifty marbles in fifteen minutes and the marble-board owner finds himself "skint".'

On the mainland there was apparently no wild enthusiasm, though occasionally a child, asked about the different ways to play marbles, would answer, 'There's a gambling game, with a cardboard box with holes in' (Southam, Warwickshire, 1957); or, 'You need a marble wall which is made of wood with arches with numbers over them. You bowl a marble up to the arch, then the other person bowls his. If I got six and he got one he would have to give me five marbles' (Ponder's End, Enfield, 1954); or, 'Take a cardboard box and cut squares out and put numbers above . . .' (Sneyd Green, Burslem, 1966); or, 'Make a bridge and roll them in' (Harborne, Birmingham, 1970).

RINGY

'Ringy' is the most prestigious of the marbles games. 'Ring-taw for ever,' cries Tom, in *Philosophy in Sport* (1827), ii. 6, 'it is the only game of marbles worthy of being played.' (*Taw*, a versatile word of obscure origin, means, according to context, a shooting marble, the action of shooting, or the line from which a marble is shot.)

Randle Holme, author of *The Academie of Armory* (1688), who was blessedly interested in much else besides heraldry, begged his readers' leave to add, after 'the severall recreations and sports', 'a small catalogue more which are used by our countrey Boys and Girls, and some of them by people of riper years'. He lists 'Long Taw, circle Taw, Bank about Rubbers . . . played with round Balls, or Marble Bullets called Marvels'.[9]

'Those of riper years' may well have been playing 'circle Taw' through the eighteenth century, as we know they did in subsequent centuries, but the game was seen as a boys' game, and featured as such in publications like Newbery's *Little Pretty Pocket-Book*, (1744: 'Knuckle down to your *Taw*, Aim well, shoot away; Keep out of the *Ring*, And you'll soon learn to play'); and John Marchant's *Puerilia* (1751), where 'Master playing at marbles with his Brother' instructs 'before we begin, Place 'em round in a Ring; I'll fright 'em, And smite 'em While my Thumb has a Spring.' The illustrations in the *Little Pretty Pocket-Book*, and in *Youthful Sports* (1801), pl. 18, 'Playing at Taw', show that four marbles—more, if there were more than the usual four players—were placed just inside the

---

[9] 'Marvels', a word not confined to any one locality, is here recorded earlier than 'marble', whose first appearance in *OED* is from J. Houghton's *Collection [of letters] for the Improvement of Husbandry and Trade* (1694–5), no. 189, 'The next are marbles for boys to play with'.

The little b Play.

TAW.

KNUCKLE down to your *Taw*,
   Aim well, ſhoot away;
Keep out of the *Ring*,
   And you'll ſoon learn to play.

MORAL.

Time rolls like a *Marble*,
   And awes ev'ry State;
Then huſband each Moment,
   Before 'tis too late.

HOOP

The little f Play.

KNOCK OUT *and* SPAN.

STRIKE out your *Taw* ſtrong;
   For the very next Man
Will bear off the Prize,
   If you come to a *Span*.

MORAL.

This *Span*, my dear Boy,
   Shou'd your Monitor be;
'Tis the Length of a Life,
   As we oftentimes ſee.

HOP,

Left: 'The little b Play—TAW', in John Newbery's *A Little Pretty Pocket-Book*', 1744 (1767).
Right: 'The little f Play—KNOCK OUT and SPAN' from the same book.

perimeter of the circle (at the four points of the compass, so to speak), with sometimes another marble at the centre, or, in the case of the *Little Pretty Pocket-Book*, an additional cluster of marbles. The main rules of the game have subsisted to the present day: a player must endeavour to knock other people's marbles out of the ring without leaving his own inside, and players have to knuckle down and shoot with the thumb, in the classic fashion described below.

'Ring-taw' was the name of the game until the end of the nineteenth century, when it was beginning to change to 'Ringy' or 'The Ring' (see Maclagan (1901), 153–4). The game was known in two varieties, 'Big Ring' and 'Little Ring', from at least the period 1803–10, when these appear amongst the varieties of marbles games played at Sedgley Park School (F. C. Husenbeth, *History of Sedgley Park School, Staffordshire* (1856), 105). The two kinds were first described in *The Boy's Own Paper* (1887), 175. 'Little Ring' is the same small circle with target marbles placed round it at intervals and with a shooting-line a certain distance away as was depicted in the eighteenth century. 'Big Ring' consists of a much larger circle, with a small circle in the middle enclosing a cluster of target marbles; the players shoot from the perimeter of the large circle. The difference between the two is not as great as it seems, since the shooting line for 'Little Ring' can be seen simply as a segment of the large outer ring of 'Big Ring'.

The best description of the larger game is probably Alfred Elliott's in *Out-of-Doors* (1872), 62–3—he still calls it 'Ring Taw':

Draw a circle about 18 feet in circumference, and within it another 6 inches in diameter. The outer circle is called 'the offing'. Into the smaller one each player puts a marble, called 'the shot'. From the offing the players in turn shoot at the ring, and whoso knocks out a marble wins it, and is entitled to shoot again before his companions take their turns. When all have shot their marbles, they fire from the points where the marbles rested at the last discharge, and *not* from the offing. If that player's taw remains in the inner ring when shot, he is out, and must deposit a marble, and all the marbles won by previous discharges.[10] It is the law, moreover—and as unalterable as the laws of the Medes and Persians—that if one player's taw is struck by another's, the taw so struck is looked upon as 'dead', and its unlucky owner must give to the striker all the taws he has acquired during the game.

The first player breaks up the cluster at the centre of the ring, as in snooker.

---

[10] In a variation called 'Increase-pound', which featured in all the boys' books of games from *The Boy's Own Book* (1828) onwards, the owner of a marble left in the ring pays one to the 'pound' rather than being 'killed'.

Since this is the most skilful (and now 'championship') marbles game, it naturally employs the most skilful method of shooting. The shooting marble is held in the curled forefinger, the tip of the forefinger holding it lightly onto the top joint of the thumb. The thumb-nail is levered under the forefinger, and when the thumb is released, the marble shoots out as though propelled from a spring. (Players can produce a top spin by turning their wrist over to the right while shooting.) As the shot is delivered, the middle joint of the forefinger must touch the ground, that is, the player must 'knuckle down'. In Scotland and the northern counties shooting in this way is known as 'plunking' or 'plonking'; hence the shooting marble is a 'plunker', 'plunkie', 'plonker', 'plenker', 'penker', or 'penkie'. These words derive from *plunk*, *plonk*, *penk*, to hit, strike. A large, heavy marble is used, or, even better, a large ball-bearing. The action is known as 'shooting', 'firing', or—very widely—'flicking'. In *c*.1905, in Bredbury, Cheshire, and in north-west Derbyshire, the knuckle-down method was called 'flirting'.

In a serious game of knuckle-down ring marbles the worst crime is to thrust one's hand forward while shooting; this is considered to be deliberate cheating, and the rules of the Men's British Marbles Championship end with the words 'Definitely NO FUDGING'. On the schoolchild level, the term 'fudge' has survived in a few places, but—probably because the finer points of marbles technique are no longer understood—the meaning has sometimes become muddled. Boys in Gloucester, 1971, said that in a game of 'Ringy' it was 'strictly prohibited to move your hand forward when flicking'; all of them called this action 'shucking', except one, who called it 'fudging'. In Knighton, 1952, 'fudging' meant 'you have leaned over too far to shoot at the marble'. In Salford, 1970, leaning over too far (in a game of 'Ringy') had no special name, but evoked the comment, 'You're a right cheat, you are.' In Knighton and Liss the term had acquired a permissive meaning: a call of 'fudge' meant that a player could 'take a step or so from where the marble stops', or that he could 'take one step forward'.

Other English words for the action, with the same connotation of cheating, have been 'fobbing' (Burslem, Staffordshire, *c*.1910); 'fubbing' (see *EDD*) and 'fulking' or 'fullocking' (see *EDD* and *OED*, from the eighteenth century). 'Funch', to push or thrust, was also used for the marbles action (H. Harman, *Buckinghamshire Dialect* (1929), 149). 'Cave' was apparently the Yorkshire term (*EDD*). In Edinburgh, 'A "pussie" or a "cattie" is a shot that's pushed and no right "plonked" [knuckled]' (J. T. R. Ritchie, *Golden City* (1965), 62).

We had accounts of 'Ringy' from forty-five places in Britain (though

not from London) during the period 1950–80, and many of the players
were aged 13 and 14. The name of the game was usually 'Ringy', with
minor variations such as 'Ring Marbles' (Swansea, 1952); 'Ringsies'
(Guernsey, 1961); 'Ringers' (Wolstanton and Sedgley, Staffordshire,
1960 and 1970).[11] In Penrith, 1957, they still had the choice of 'Big Ringy'
or 'Little Ringy', as also in Cumnock, 'Big Ringy' or 'Wee Ringy', 1961.
Most accounts simply state that the target marbles are 'put in' the ring
('A ring is drawn with chalk or a finger in the dust. Each boy puts a
shottie in the ring'; Tunstall, Staffordshire); some specify 'the centre'
('The players put an equal amount of marbles in the centre of the circle',
Spennymoor). However, some show that the eighteenth-century place-
ment of the marbles was still in force in the twentieth century. 'The game
of marbles is comprised of a circle of marbles and a large ball-bearing',
said a lad from Nelson, Lancashire, in 1970. In Sneyd Green, Burslem,
1966, an 8-year-old boy, having specified 'a circle of marbles', drew a
diagram showing fifteen marbles arranged evenly just inside the perimeter of a circle, with one marble at the centre, a pattern also sent from
Cumnock, Ayrshire, in 1961. Instructions on 'How to play Ringy or
Flicks', from Gloucester, 1971, ran:

Draw a circle with chalk, and players each put a set of marbles into the ring (sets
can be 1, 2, 3, or more). Space them out inside the ring. Then from a line roll up
your own marbles [this is an underarm bowling movement]. The person with his
marble nearest to the ring has the first flick. If any person bowling up lands his
marble in the ring he sticks and is out of the game. The boys then take it in turns
to flick marbles from the ring. All they knock out they keep.

Maclagan, in 1901, gave the variation of 'Square Ringy', which is the
same game played in a square, each player placing a marble 'on one of
the lines'. 'Squarey', or 'Squaresey', was still played in a few places in the
1950s—for example, Alton, Bishop Auckland, and Parson Cross, near
Sheffield—the marbles being placed in the four corners of the squares,
with one in the centre.

Marbles was once a man's sport, usually played for money. William
Skene, remembering Aberdeen c.1850, says, 'During the winter months
the Barrack Hill used to be so crowded with players at the "bools"—

---

[11] *EDD* has '*Rink*, a circle, ring, esp. the ring in marble-playing'; and Arnold Bennett,
Clayhanger, ch. 1, referring back to Burslem, Staffordshire, 1872, describes 'the open gates of a
manufactory' disclosing 'six men playing the noble game of rinkers . . . these six men were . . . the
three partners owning the works, and three of their employees. They were celebrated marble-
players, and the boys stayed to watch them as, bending with one knee almost touching the earth, they
shot the rinkers from their stubby thumbs with a canon-like force and precision that no boy could
ever hope to equal.' 'The majestic "rinker", black with white spots' was, he says, 'the king of marbles
in an era when whole populations practised the game.'

sailors and boys—that numbers had to shift elsewhere' (*East Neuk Chronicles* (1921), 18). At the same period, the churchwardens of Warnham, near Horsham, Sussex, gave notice that 'the assemblage of idle and disorderly persons who molest and otherwise annoy females who pass along the highways of this Parish, and who also play marbles and other unlawful games thereon', on Sundays, 'will be punished with the utmost severity of the law' (*Daily Mail*, 22 December 1970). About 1910, the fishermen of St Ives played marbles in their idle moments, when bad weather prevented them going to sea (*Western Morning News*, 14 January 1975). In the late 1920s miners played marbles on Sundays, 'morning, noon, and night', for 3d or 6d a time. Liverpool men were especially keen players, at least until the 1960s; Frank Shaw assured us of this, and says in *My Liverpool*, p. 229, 'Grown men play the ollies game Segs, Lassies and Up [Three Holes], the competition being keen, and successful marbles being cherished by a player as he might have his own billiard cue or set of darts or bowls.'

Now that men have apparently given up the sport, the only expert play to act as a model for the younger generation is at marbles championships. Marbles championships may have been born of the Depression; the earliest I know of was in Philadelphia in 1926 (*Guardian*, 19 May 1926). The World Marbles Championship at Tinsley Green, Sussex, on Good Friday, has been in existence since 1930; the Good Friday marbles match on Battle Abbey Green, Sussex, between Battle Marbles Club and the men of Netherfield, was revived by Mr 'Pat' Waite and Mr Frank Anderson in 1936 (the custom was said, in the *Sussex Express*, 7 April 1966, to be more than a century old). For the record, the rules for the Tinsley Green championship are as follows: Ring to be 6 ft. diameter, made of concrete raised approximately 2 in. from the ground and liberally sprinkled with sand. Forty-nine glass marbles are clustered in the centre of the ring. The first team to knock twenty-five marbles out of the ring is the winner. Each team consists of six players. Referee draws a line in the sand. Captains hold their tolley (shooting marble) at nose-level, and drop it as near the line as they can—nearest earns first shot for his team. First player knuckles down at edge of ring and shoots his tolley to knock one or more of the marbles right out of the ring.[12] If he succeeds and his tolley remains in the ring he shoots again. If he fails, but his tolley remains in the ring, it stays there until his turn comes round again, when he shoots from wherever it happens to be. If in the meantime his tolley

---

[12] 'Tolley' is a received name for a player's individual shooting marble in parts of Sussex, Hampshire, Oxfordshire, and perhaps elsewhere. The word is probably a corruption of *alley*, cf. *olly*, a marble, Liverpool, 1910–70.

has been knocked out of the ring by his own or the opposing side, he is 'killed' and is out of the game.[13]

THREE HOLES

'Three Holes' is second only to 'Ringy' as the classic adult game of marbles. In 'Three Holes', as in 'Ringy', the experts shoot 'knuckle down'; and it, too, was the pastime of working men who had intervals of hanging about in their work—such as the car-men of Belfast at out-of-the-way car stands, c.1854–8. There, at that time and later, as also in Dublin, the game was known as 'Hole and Taw'. Three holes were made in the ground, 'three times up and down concluded the game', and 'sometimes the winners (those who completed the course) became "rovers" (like croquet) with killing powers' (N & Q, 9th ser., vol. 3).

Jamieson, giving the earliest description of 'Three Holes', in his *Dictionary of the Scottish Language*, Suppl. (1825), under 'Capie-hole', does not imply that boys are the only players; in his day the only impediment to playing marbles was the stiffness of joints. 'In Angus,' he says, 'three holes are made at equal distances. He, who can strike his bowl into each of these holes, thrice in succession, wins the game. There it is called *capie-hole*, or by abbreviation *capie*.'

J. E. Emslie provided a more detailed account, for *Traditional Games*, ii (1898), 256–7, of 'Three Holes' as he had played it in London as a boy:

Three holes were made in the ground by the players driving the heels of their boots into the earth, and then pirouetting. The game was played with the large marbles (about the size of racket balls) known as 'bouncers,' sometimes as 'bucks'. The first boy stood at 'taw', and bowled his marble along the ground into 1. (It was bad form to make the holes too large; they were then 'wash-hand basins,' and made the game too easy). Taking the marble in his hand, and placing his foot against 1, he bowled the marble into 2. He was now 'going up for his firsts.' Starting at 2, he bowled the marble into 3, and had now 'taken off his firsts,' and was 'coming down for his seconds.' He then bowled the marble back again into 2, and afterwards into 1. He then 'went up for his thirds,' bowling the marble into 2, and afterwards into 3, and had then won the game. When he won in this fashion, he was said to have 'taken off the game.' But he didn't often do this. In going up for his firsts, perhaps his marble, instead of going into 2, stopped at A; then the second boy started from taw, and having sent his marble into 1, bowled at A; if he hit the marble, he started for 2, from where his marble stopped; if he missed, or didn't gain the hole he was making for, or knocked his

---

[13] Good Friday is the day for all Lenten marbles championships except one (as far as I know), which was an informal affair at Halesowen, Worcestershire, on Shrove Tuesday. Men who remembered having the afternoon off school and going down to the Tunnel—really a bridge—for a marbles match, c.1905–10, were trying to revive the custom in 1971 (*Birmingham Post*, 22 Feb. 1971).

antagonist's marble into a hole, the first boy played again, hitting the other marble, if it brought him nearer to the hole he was making for, or else going on. In such a case as I have supposed, it would be the player's aim to knock A on to . . . some place between 2 and 3, so as to enter 2, and then strike again so as to near 3, enter 3, and strike on his way down for his seconds, and near 2 again.

According to an article in the *Evening News*, 21 December 1931, p. 11, the game was still very much alive in London:

Marbles in their due season, claim the skill of the London boy. Some people think marbles are dying out, but you ask the school-teachers! Some of the teachers have a rule that a boy mustn't bring more than eight marbles to school with him . . . And if you are still in doubt, just go down to the little streets. There you'll see young Arthur Ferguson standing over a cup-shaped depression in the ground and crying his challenge, 'Andy Or!'. There you'll hear Herbert Harvey, crouched beside the third of three holes, shouting 'Hit blood man dead!' before his opponent, young Nobby Clarke, can yell the counter-call, 'Hit blood man *not* dead!'.

As can be seen, the game needed skill. The earliest description is in *The Boy's Own Book* (2nd edn., 1828), 10, under 'Holes'.[14] The proced-ure is the same as in Emslie's account, and apparently the second player, after successfully entering the first hole, can hit his opponent's stranded marble in the same way. *The Boy's Own Paper*, 1887, p. 207, emphasizing that 'every time he hits the taw [marble] or gets into the hole he has another shot', points out that the strategy is to 'fall in with his adver-sary as soon as possible and take him along with him to get easily into each hole. If he cannot take him along with him he knocks him as far away as possible.' The game is regularly featured in boys' books of games throughout the nineteenth century. A. E. Baker gives the game as 'Cob', in her *Northamptonshire Glossary* (1854); and Maclagan (1901), 154–5, as 'Mushie', or 'Knucklie' when it involves the knuckle-bashing punishment for the loser (cf. 'Moshie' in Glasgow in the 1920s: Clifford Hanley, *Dancing in the Streets*, 53, 'Quite grown-up boys, fifteen or sixteen years old or more, joined the moshie schools in the coup [rubbish tip] behind the Honeymoon Building, where you could always find three decent holes set in the correct triangle by some previous moshie school').

A correspondent who played 'Three Holes' in Liverpool *c*.1905 said: 'it was a very important game. Men played it, and there were regular pitches set out.' In Merthyr Tydfil, Glamorganshire, *c*.1925, when a player had successfully played into the three holes and back

---

[14] Although the Revd Joseph Hunter, in a MS note to Brand's *Antiquities* (1813), ii. 295, said he remembered the game of 'Three Holes' from his boyhood in Sheffield, *c*.1790–5.

again, his marble became a free-ranging 'killer', and the game was called 'Killer'.

The game survived in Liverpool until at least 1954, when we had the following descriptions from 13-year-olds at Balliol Modern Secondary School:

A most common game of marbles is Knock Away Six. Any number of boys can play this game. The game starts off by each boy trying to ring a hole. When a boy rings [gets into] a hole he then shoots at another boy's marble and tries to knock it as far as he can. If he knocks the marble more than six times the length of the boy's feet and then rings the hole, the boy has to give him a marble, and he is out of the game.

In Killer a number of holes are dug about 8 feet apart. When a boy has ringed all the holes he can shoot at any marble and ring any hole he wants. When he is 'killer' any marble he hits is out of the game and the owner of the marble has to give the killer which hit his marble, a marble.

It would probably be true to say that 'Killer' was 'a most common game of marbles' in Scotland, northern England, and the Midlands in the 1950s and 1960s. In Bishop Auckland up to seven holes were used, and the expression 'to ring the hole' also obtained. In Stoke-on-Trent ('Killer is my favourite outdoor game') there was some interesting terminology:

Two lines are drawn, one at each end of the pitch. If you go past them when flirting you go to lamb, and behind there they can't hit you with their marble. The game begins and you flirt for the hole which is known as the killer. If you get in from lamb you are killer and you only have to hit your opponent once. He has got to get in the killer three times, and he has got to hit me three times. The three hits are known as tap, lob and dead. After I have hit his shottie once he is dead because I was killer from lamb. But if he hit me first he would be the winner. When the game has finished loser gives winner a shottie.

In some places the last hole is said to be the 'pongy hole' (Bishop Auckland) and when a player gets in this hole he is 'pongy' and can hit any other person in the game and demand a marble from him; or it is called the 'poye' (Spennymoor) and 'once he shoots into this hole he is poison to anyone that hits him, and he may knock his opponents off and they are out of the game and have to forfeit a muggle [marble]. Once someone goes past the by-line he is "bye" and cannot be knocked off until he shoots on to the pitch again.'

Orthodox versions of the game came from Ballingry, Fife ('Pout'); Kinlochleven, Argyllshire ('Holey'), Cumnock, Ayrshire ('Muggy'), Whickham, near Newcastle upon Tyne ('Killer'), and Trevethin, near Pontypool ('Killer').

HOLIE

There are two main ways of playing this game, plus one lesser-known way and another which is peculiar to Guernsey. Whichever method is used, it is important to decide beforehand how many marbles each player will be putting into the game: 'You can roll different numbers of marbles by calling "Onesie", "Twosie", "Threesie"—it could be as much as five' (Liss, 1980).

This first way is nothing more than a decayed form of 'Three Holes' or 'Killer' (see above):

Our playground is full of holes and we use these for kells [holes aimed at]. You take how many marbles you are going to play with. To start with you start from a starting line. Either one of you shoots at the kell, and then the other does. After you have been in kell you can shoot at the other person's marble. If you hit it you keep it. Then you can go on and hit another that's lying around. If you don't hit them it is the other person's go. He aims at the hole. He must get in the kell before he can aim at the others.   (Wall Heath and other Staffordshire villages, 1970. Nine other places, England and Scotland, 1950s–1970s)

The second way is simply a further deterioration:

How to play Holesy. Say 'Twosie' has been agreed to. Each player stands and throws a marble at a hole about 4 inches wide at the side of the road. Then they aim another marble in the same way. Usually they don't get it in first go. The one whose marble is nearest the hole has the first turn and flicks the marble into the hole if he can. If he gets it in he has another go with the marble next nearest the hole, no matter who it belongs to, and so on till he fails. The next player takes over—the one whose marble was second-nearest the hole in the first place. He gets as many marbles in the hole as he can and when he can't the next player has a turn (or the first again, if only two people are playing). Whenever you get a marble in the hole you pocket it.   (Alton, Hampshire, and eleven other places in England and Scotland, 1950s–1970s)

The lesser-known way (which is more like 'Chuckie' (p. 110)) is:

A game called Holesy, played with any amount of marbles. You have a strip of ground with no bumps on. Then you get the amount of marbles you are going to play with and roll them and try to get them in the hole. Then you pick them up and mind out of the way. The other boy rolls his. The one who gets most of their marbles in the hole takes them all and throws them all together at the hole. If he gets two or more in the hole he can roll any other marble to the hole. Any marble he don't get in the hole is the other boy's, but the marbles he does get in are his.   (Norwich, 1961; similar at Penrith, 1957, where it is called 'Ben Holey'; and Netley, 1960)

In Guernsey the second method is known under the name of 'Cat's Eye', and is 'not played very much'. The more popular way is to aim at

the hole and 'when at last a person gets a marble in the pot he is allowed
to collect all the marbles that missed the hole, including his own', the
game then being known as 'Potsie' or 'Pots'.

Very often the game has no name but 'Marbles', it being the only
marbles game known. During the 1950s–1970s the name 'Holey' obtained in Ballingry, Fife; Birmingham; Bungay, Suffolk ('Hole');
Gloucester; Langholm, Dumfriesshire; Market Rasen; Parson Cross,
near Sheffield; Penrith; Salford; Sale; Thirsk; and Wilmslow; and
'Holesie' was the name in Alton; Bishop Auckland; Brentwood;
Lechlade, Gloucestershire; Norwich; and Peterborough. Other names
were 'Holes' in Stornoway, 'Ben Holey' in Penrith, and 'Aggie [marble]
in the Hole' in Oxford. 'Muggie Hole' has been the name in Liss from the
1890s to the present day (strangely, 'Muggie' is the name in Alloa,
Clackmannanshire, and 'mug' the name for the hole in Paisley). 'Dumpy',
in Forfar, 1954, has an echo of the old game of pitching lead 'dumps'.

Names for the hole itself are: 'bibby' (Beccles, Suffolk, 1956–60);
'bulley-hole' ('on top of the Pennines', c.1945, *Guardian*, 16 May 1964,
p. 7); 'condy', from *cundy*, a conduit, a drain (Perth, 1975); 'hob' (Ponder's End, Enfield, 1954, and Knighton, 1960); 'kell' or 'kellie'
(Kingswinford, Wall Heath, and Wordsley, villages in Staffordshire);
'kipie' (Aberdeen, 1960);[15] 'kit' (Lambourn, Berkshire); 'mug' (Paisley,
1975); 'pot' (Ponder's End, Enfield, 1954; Langholm, 1960; Guernsey,
1961); 'stank', a grating in a gutter (Glasgow, 1975).

Local rules govern the exact size of the 'pot', and the distance from
which aim is taken; but the same finger-work prevails throughout the
country. The marbles must be rolled up to the 'pot' (after the manner of
bowls), and thereafter the style of propulsion is shoving with the crooked
forefinger ('like the handle of an umbrella', said one boy). In Bungay this
is known as 'shugging'; in Oxford, 'shoving'; in Pontefract, 'codging'; in
Middleton Cheney, near Banbury, 'fingering'. The Hampshire word for
it used to be 'hucking', as used at Twyford, c.1900, Liss and Fareham,
c.1923. However, the marbles were propelled with the thumb in Ponder's End, Rowledge (Surrey), and Enfield; in Aberdeen, in the 1950s,
this was known as 'nipping'.

Historical evidence is scrappy. Correspondents, recalling the game from
c.1900 onwards, describe the first way of playing, above (and most add,

---

[15] 'Kypie', 'Kypie-hole', or 'Capie-hole' are Scottish games of long-standing (see *EDD*, under
'Capie-hole'). '*Kype*, a small scooped-out hollow in the ground, chiefly for use in the game of
marbles' (*Concise Scots Dictionary*), is probably from OE *cype*, Low German *kipe*, a basket. 'Capey-
dykey', the 'game with marbles that is only known in Thrums', in Barrie's *Sentimental Tommy*
(1896), 143, can thus be translated as 'hole by the wall', a *dyke* being a wall.

'It was a girls' game, despised by the boys'); so does Major Barzillai Lowsley in his *Berkshire Glossary* (1888): '*Shuvvy-Hawle*. A boys' game at marbles. A small hole is made in the ground, and marbles are pushed in turn with the side of the first finger; these are won by the player pushing them into the shuvvy-hawle.' Contemporaneously, Alfred Easter, in his *Dialect of Almondsbury and Huddersfield* (1883), describes the other method of play (the second way, above) in a game called 'Hundred', in which every successful hit scores ten, and the first to score 100 'goes in for his pizings' (killings or hittings). He aims at the hole, and if the marble goes in he wins and the game is finished. If he misses, however, the other boy tries for the hole, if successful scores ten, tries to hit his opponent's marble, and if successful adds the opponent's 100 onto his own score, thus triumphantly ending up as the victor—a wonderful reversal of fortune. Maclagan (1901) also notes this method, under the name 'Stealing Numbers' (p. 156). It seems that a hundred-odd years ago there were also two ways of playing 'at the hole'.

FOLLOWS

This very basic one-rolls-a-marble-the-other-tries-to-hit-it game can be—and has been for centuries—played on any piece of smoothish ground. (Only when it is played progressively, along a road on the way to school, or along a gutter, can it be called 'Follows'.)

Its ancestors appear to be 'Span-counter' and 'Span-farthing'. *OED*'s earliest reference to 'Span-counter' is in Thomas Drant's *Horace his Satyres* (1566). 'Span-farthing', the same game but played with farthings, is not heard of until 1688, in Randle Holme's *Armory*. The game was common in the seventeenth century, as seen, for example, in William Hawkins's schoolboys' play *Apollo Shroving*, 1627: 'wee have leave to play, and we play at our best game. What . . . Span-counter?'; and *Poor Robin's Almanack* for 1670, where the prediction for April is the negative one: 'I do not find by the Position of the Heavens, that men shall get great Estates by playing at Span-farthing.' Strutt (1801), 287, provides the link. He describes 'Boss Out, or Boss and Span':[16]

One bowls a marble to any distance that he pleases, which serves as a mark for his antagonist to bowl at, whose business it is to hit the marble first bowled, or lay his own near enough to it for him to span the space between them and touch

[16] A 'boss' was a large stone or iron ball, used in marble playing (*EDD*). A shooting marble in the Battle marbles championship is called a 'bosser' (*Sussex Archeological Collections*, 89 (1950), 90). 'Bost-about' appeared in *The Boy's Own Book* (2nd edn., 1828), 9; if the marbles were shot with forefinger and thumb instead of being pitched the game was called 'Spans and Snops' (ibid.).

both the marbles; in either case he wins, if not his marble remains where it lay and becomes a mark for the first player, and so alternately until the game be won.

and then says that 'Span-counter' is a similar pastime, but played with counters instead of marbles, adding, 'I have frequently seen the boys for want of both perform it with stones.'

Strutt's marbles game is what might be called the 'static' form of 'Follows', and is now the usual way of playing. It appeared as 'Spans and Snops', or 'Snops and Spans', in nineteenth-century books of games for boys who would predominantly be at boarding school, or at home in the holidays, and would not have a daily walk to school; in books such as *The Boy's Own Book* (1828), 9, publishers could scarcely recommend that nicely brought-up boys should set off across country. However, the 'travelling' form is undoubtedly the most fun, though it has been all but eliminated by the increasing difficulties of playing along a road or along a gutter. A contributor to *N & Q*, 9th ser., 3 (1899), 97–8, remembered 'chuck-taw' as a 'goo ter skowl gam'', in Matlock Bath, Derbyshire, *c*.1835: 'It got us over the ground quickly. One lad chucked his taw two or three yards in front, and the next chucked at the first taw with his taw, and so on . . . When played with boulders it was called "chuck-bowder".' A London bus-driver remembered playing 'Leg-a-longs', *c*.1908, as a way of getting home from school 'without losing any sport'. Other names for the game have been: 'Follow my Taw' (*Holiday Sports* (1848), 71); 'Folly-tar' (i.e. 'Follow-taw') (*Provincial Words Used in Teesdale . . . Durham*, 1849); 'Splits' or 'Follows', Farnham, *c*.1867, also a version played by hurling pebbles along the country roads, 'You went fast and far that way' (G. Sturt, *Small Boy in the Sixties*, 148); 'Buckalong', Battersea, *c*.1885 (E. Thomas, *Childhood of Edward Thomas* (1938), 30); 'Bob and Hit', South Lancashire, *EDD*, 1897, cf. 'Bobs Along', Sale, Manchester, 1960; 'Langie Spangie', Aberdeenshire, played with 'muckle bools', straight out along a road, maybe for miles (*EDD*, 1904); 'Hits and Spans', Smethwick, near Warley, Worcestershire, *c*.1905, *London Street Games* (1916), 111, and Ipswich, 1953; 'Chuck Spans', Derby, *c*.1910; 'Plunks', using horse studs (Jim Gott, *Memories of Ripon . . . 1910–1950*, 45–6);[17] 'Penky Follow', Ponteland, *c*.1910, 'Penky', north-west Durham, *c*.1925 (*Times Educational Supplement*, 28 September 1956, p. 1166); 'Long Johns', Bramfield, Suffolk, *c*.1915 (G. E. Evans, *Where Beards Wag All* (1970), 219); 'Follow On' (*London Street Games*, p. 111); 'Long Pops',

---

[17] In the adaptable world of childhood any game can be played with substitute equipment. Travelling 'Follows' was played with buttons, under the name of 'Spawnie' (Revd W. Gregor, *Dialect of Banffshire* (1866)); with toy cars, in Dudley, Worcestershire, 1969; with footballs, Dundee, 1975.

Lincoln, *c*.1920; 'Spits and Spans', Petersfield, Hampshire, *c*.1915; 'Spick and Span' (*London Street Games*, p. 111), Petersfield, *c*.1920, and Alton, 1954; 'Five-Ten', scoring five points a hit, Scottish borders, *c*.1925 (*Southern Annual* (1957), 29); 'Penky', north-west Durham, *c*.1925 (*Times Educational Supplement*, 28 September 1956, p. 1166). Any mention of 'Spans' meant that 'spanning' was part of the game, i.e. that a marble within a handspan of its target was as good as a hit.

Boys' games books, through the nineteenth century, often called the game something like 'Bonce'. This name, like Strutt's 'Boss Out, or Boss and Span', derives from the large marbles used in playing: thus *Juvenile Trials* (T. Carnan, 1772), 'I have won of him, since he came here, two hundred and fifty-nine common marbles, seventy-three allies, and eighteen baunces, all by fair play' (p. 98); *Juvenile Sports* (T. Carnan, 1776), 'The baunce, which is distinguished by that name only on account of its size, being a good deal larger than what we call marbles' (p. 97); *Book of Games, or Schoolboys' Manual* (*c*.1837), 33, 'Bonces, or very large stone marbles, made chiefly in Holland'; H. Mayhew, *London Labour* (1851), i. 39, 'I never learned anything but playing buttons and making leaden "bonces"' (in 1971 a Gloucester boy made the point that 'both Kerbsie and Chasie are best played with a large ball-bearing called a "Big Leady"'); *Cassell's Book of Sports and Pastimes* (1888), 245, 'Each player should know his own "bouncer". "Bouncers", like genuine "allies", are never to be forfeited, they are too scarce.' The game itself is called 'Bonce' in *School Boys' Diversions* (2nd edn., 1823), 37: 'played by one boy placing a marble down, and the other trying to hit it; if he succeed, he wins it; if not, his marble must remain down, for the first player to try at. They thus play alternately, in this manner, as long as they please.' In Paul Percival's *The Youth's Own Book of Healthful Amusements* (1845), 175, it is 'Bounce'; in *The Boy's Own Book* (1855), 12, 'Bonce About'; in *Games and Sports for Young Boys* (1859), 44, 'Bonce Along'.

The modern game seems less interesting (although apparently no less popular), partly because it is usually in the static form, and partly because it often has no name but 'Marbles'. We had descriptions of the static game from thirty-seven places, in the 1950s–1970s: its name was 'Bobs Along' or 'Dobby' or 'Itsey' or 'Sticky Follow' in Sale, Manchester; 'Chasie' in Edinburgh, Langholm, Gloucester, Ferryhill, and Kinlochleven, Argyllshire; 'Clicksy' or 'Follows On' in Market Rasen; 'Dobs' in Oxford; 'Follow the Olly' or 'Knock Along' in Liverpool; 'Follows' in Garforth, West Yorkshire; 'Follows' or 'Straightsies' in Guernsey; 'Follows On' in Dudley, Worcestershire; 'Hitty Three

Times' in Bishop Auckland (the rule of hitting a marble three times to win it is sometimes imposed in other places); 'Keggy' in Forfar; 'Longy' in Perth; 'Ordinaries' in Stornoway, Isle of Lewis (also an alternative name in Langholm).

The travelling game was known only in six places: 'Follows' was played in the gutter at Lydney, Gloucestershire, and 'Follows On' in Swansea; 'Gutterie' in Edinburgh; 'Rolls Along' in Birmingham. In Peterborough they said: 'You mostly play marbles along the gutter.' The gutter game ('Gutter-golf') has been superseded for other reasons beside the obvious one that children no longer walk to school; 'parked cars are a nuisance', for instance, and 'the allies'd go down the drains'.[18] There must, as well, have been considerable parental pressure to discontinue a dangerous pastime—although the children never mentioned this.

Car traffic undoubtedly caused the decline of 'Road Bowls', the adult equivalent of 'Follows', although it may well continue in some parts of Ireland. Roundly condemned by Nonconformists, along with pigeon-flying, the game was traditionally played from pub to pub, involved betting, and was preferred to attendance at divine service. The bowls were iron balls, or round or disc-shaped stones, about 3–4 inches in diameter, found in a stream bed. Before the First World War, especially in Yorkshire and Lancashire, men could be seen shaping the discs with 12-inch flat metal files, as they walked around the streets talking to each other or going shopping.

MINOR GAMES AND VARIATIONS

Like other traditional games, marbles is capable of infinite variation. In Cumnock, 1961, marbles knocked out of a ring must land against a wall, and the game was called 'Jaries'. (Maclagan (1901), 152, defined 'jaries' as brown-glazed earthenware marbles used for shooting.) In Avoch, Ross-shire, in 1975, a ring of stones was placed round a hole, and 'you try to get your marble in the hole without touching any of the stones'. In Stornoway, Isle of Lewis, in 1961, 'Cannons' was played by arranging marbles in parallel rows and shooting between them so as not to disturb the rows. In Birmingham, in 1970, the usual way of playing marbles was

---

[18] Boys in Salford had a heated discussion about the fate of marbles that fall down drains: 'The men come round with a lorry and clean the drains out, and sometimes they get 'em and put 'em on the side.' 'They won't give 'em you though.' 'They will.' 'No they won't, 'cos they got germs on 'em.' 'Well, you can wash 'em in bleach.'

to aim at a large target marble (there known as 'the gobbie'), 'and when you hit it you keep the others that have missed'; the gobbie itself is never in danger, it is returned to its owner. In St Giles High Street, London, in 1955, boys were seen playing a wild roulette game, using an upside-down dustbin lid. Each boy put some marbles into the lid, it was spun on its handle, and when it stopped, the boys grabbed as many marbles as they could. In fact, when one considers the versatile nature of marbles, the possibilities are almost unlimited.

BEGINNING A GAME

First it must be decided whether the game is to be played 'Keepsies' ('Keepies', 'Keeps') or 'Lendsies' ('Lends'—or, in Scotland, 'Funsie', 'Funs', 'Funny', or 'In Funny'); that is, whether the winner may keep the marbles he has won, or whether he must give them back to the loser. Throughout the nineteenth century boys' games books, reacting against the gambling excesses of the previous century, emphasized that keeping marbles was wrong. *The Book of Games* (1805; 1810 edn., p. 156), says, 'He is winner who possesses three marbles, which, however, he is to return to the ring, and not keep them, for that would be gambling.'

It is the general opinion that the second player has the advantage, hence 'there is a race to shout "Bags no throw off"' (Exeter). In other places the cry was: 'Himmie' (East Lothian, c.1940; Cumnock, 1960); 'Laggy', pronounced 'Leggy' in Scotland; 'Lardy' (London and environs). However, 'if someone wants first aim at the hole, they have to shout "first"' (Birmingham), or 'Begs [sic] first shottie' (Knighton), or 'Furry' (Canonbie, Dumfriesshire), or 'First licks' (Langholm). Or 'someone shouts "First Dobs", and someone else shouts "Second Dobs", and so on' (Plymouth).

CALLS AND COUNTER-CALLS

To the uninitiated, a game of marbles seems anarchic. The leaping and shouting is partly caused by excited partisans cheering the players on, and partly by the calls and counter-calls which decide whether various rules can be brought into play or not. This is instant legislation. If a boy finds his opponent's marble is obscured by a stone, stick, or lump, a shout of 'Clears' (or 'Clearsies') permits him to clear them away. His opponent can protect his marble by shouting 'No clears'. ('Shifts' claims the same right; and in parts of Scotland, e.g. Angus, Fife, the word is

'Cleesh'). Thus the game needs anticipation and quick wits, as well as dexterity.[19]

It has to be decided, before a game starts, whether 'you have nothing in the game or everything in the game'. 'When you have nothing in the game', explained a 10-year-old in Vale, Guernsey, 'there is no "strides" or "boots" or anything like that, and when you have everything in the game that means you can have "boots" and "strides".' Or, as a boy in Cumnock put it: 'Before anyone plunks they try to shout "everys" before their opponent can shout "nowts".' In Wingate, County Durham, and Mauchline, Ayrshire, the call is shortened to 'ivs' ('Shout "ivs", if possible, before another person shouts "noughts"'). The Market Rasen phrase was: 'All the rules can be stopped if someone calls "No nothings".'

The calls are legion. The following are the most common. Each was collected in a number of places throughout the country, and may be considered general. Many others, too many to print here, were collected only once, and may have been local.

Backs. '"Backs" allows a boy to move his marble back as far as he likes.'

Barricades. 'If you say that, you can put obstructions, like stones, in front of your marble.'

Bombs, Bombsie, Dobs, Droppy, Dropsies, High Bombs, High Dives (also *Low Dives*), *Highums, High Upsies.* 'Involves dropping one's own marble from about chest height onto the opponent's. A good heavy "dollicker" or "bullicker" is preferred, as one may then have the pleasure of chipping the opponent's marble, or driving it into the ground' (student, Newcastle upon Tyne). Widely used, especially 'if your marble gets in an awkward place'. Sometimes a game in itself ('Let's have a game of Bombs'). If the rule is to drop the bombing marble from eye-level, the call is 'Eye Drop', or if a complete game, 'Eye Dobs' ('He puts a marble to his eye and tries to hit another marble on the floor. Then there is low eye dobs where a person goes on his knees and does the same thing'). *EDD* records

[19] W. H. Babcock found 'the use of "fen" in the old sense of defend, for prohibiting certain actions in playing at marbles' (*American Anthropologist* (1888), i. 276). Edward Moor, in *Suffolk Words* (1823), defined 'fen' as 'a preventive exclamation'. In Alton, *c.*1906, 'Fen ons' prevented a boy from moving his marble further away, and 'Fen uprights' prevented him from aiming from above. In Cinderford, Forest of Dean, *c.*1930, when playing at the ring, 'If you couldn't get a straight shot at your opponent's marble you shouted "Squews, fen doublets". That meant you could move your marble to the side, and your opponent could not shout "Doublets", which meant you would have to move your marble back double the distance away from his.' 'Fen' lingered on in Headington, in the 1950s, where 'Fen clacks' meant 'You cannot be hit', and 'Fen dobs' meant 'The shot must be made without hitting any other marble.'

instances of the aerial approach under *Brentin*, and *Heist* (shooting from the knee), *Belly-marks* (shooting from stomach-level), and *Drop-eye* (from eye-level). The illustration to the German broadsheet version of Cats's moral verses on children's games in *Emblemata*, by Jacob von der Heyden (1632), shows boys aiming marbles from a standing position. The game of 'Bounce Eye' (or 'Bonce Eye'), in which marbles were bombed out of a ring from eye-level, was a regular feature of boys' games books in the nineteenth century, and appears in *London Street Games* (1916), 111.

Bootsie. 'If by chance a marble hits someone's shoe the player calls out "Bootsie", or "Kicks", then he can place his foot beside the marble and bring the other foot hard against the first foot, driving the marble where he wants it to go.' Known as 'Cogs' in Liverpool.

Cannons. 'You hit a marble and your marble keeps rolling on and hits another marble, somebody calls "Cannons" and it means you can't have both the ones you hit.'

Changeys. 'If you lose, when you pay the other boy and one of the marbles is chipped, if you say 'No changeys' he has to keep it.' Cf. *Swops.*

Clicks or *Clinks.* This means that if a marble barely touches the target marble, and does not hit it with the required audible 'crack', a cry of 'Clicks' or 'Clinks' is needed to validate the hit. Other calls with the same function are 'Kisses', 'Tips', and 'Touches'. Otherwise, a scoring shot must be audible; as they said in Salford, 'You've got to 'ear it click before they can 'ave it.'

Dacks or *Dackers.* Means players can put their foot in front of the hole to prevent their opponent getting the last (or earlier) marble in and thus winning the game. (Only from one place, Great Horton, West Yorkshire, 1991). Cf. *Stops.*

Digsie. 'If you don't want the other person to hit your marble, you put your heel on the marble and knock it down, quite far down, and his marble goes right over the top of it. It has to be on soft ground though, like gravel. But he can say 'No digsie' and you can't do it' (Salford, 1970). The equivalent in Edinburgh is 'Grounders' (Ritchie, *Golden City* (1965), 62).

Follows. 'If you don't want your turn. You can miss your turn, and the other player has to have his. It's when you want his marble nearer the hole.'

Hobs. 'You put your feet in a V-shape so that the alley will rebound.' Known as 'Keppies' in Edinburgh (*kep* = intercept): 'If your opponent's marble rolls far further than yours you can shout "Keppies". This means that your opponent must stop your marble with his feet so there is a chance the marble might rebound and hit his.' Cf. *Stops*.

Kicks. See *Bootsie*.

Moves. 'When the marble is in a difficult position you can move it to another place.' Other terms for this privilege are 'Lifts', 'Outs' ('Ounces' in Bishop Auckland), and 'Picksies'. Cf. *Rounds*.

Rebounds or *Rebounds Counts.* 'If my marble was by a wall and somebody rolled their marble at it and missed, it would hit the wall and would land nearer my marble than it would have done before, so I shout "No rebounds" and he has to have another go.'

Rolls. Can mean either that a marble must be allowed to roll without being stopped, or, 'If we are playing in grass and it is difficult to roll a marble on grass and so he has to bomb [q.v.] but if I say "Rolls" he has to try and roll it.'

Rounds, or *All Round.* 'When you're in a difficult position you say "All Round", that means you can move your marble into a position that you can hit your opponent's.' See *EDD*, *Roonses*. The same rule obtained in America as 'Roundings' (with the counter-call 'Fen roundings'), Newell (1883), no. 141. Cf. *Moves*.

Slips or *Slip-sly.* 'When the player's marble slips out of his hand and another shot is wanted.' In Cumnock, 'When a man who is plunking if his finger slips they shout "Slip-sly".' In Edward Moor's *Suffolk Words* (1823), under *Fen*, 'A boy at marbles, his taw slipping, cries "slips over again!" to authorize another attempt; which his adversary averts by sooner, or more quickly, exclaiming "Fen slips over again".'

Slogs. 'When you have to hit the marble hard.'

Splits. 'When you got a lot of marbles together, you can say "Splits" and you knock 'em all over the place, then you can get at 'em' (Salford and Sale). G. F. Northall, *Warwickshire Word-Book* (1896), 223, 'One player holds one of his own marbles, plus a marble of his opponent, over the back of his head, and then drops both—his object being to separate the marbles as far as possible; for the opponent then shoots with his own marble at that of the first player.'

Squibs or *Squirts.* '"Squibs" means you must shoot the marble between your two biggest fingers' (Berkeley, Gloucestershire). 'When you have "Squirts" you grip the bool between finger and thumb and squeeze, and it squirts out' (Langholm).

Stops or *Stopsies.* A cry of 'Stops' allows a player to stop his own marble 'when it is about to go down a drain' or 'is running fast and getting further away', or allows him to stand behind his own marble and stop his opponent's. Cf. *Hobs*.

Straight(s). 'If a marble goes round a corner the second player must say "Straight" and the first player must put his in line with the second player's marble, to give him a clear shot.'

Swops. 'If a person loses a marble to another and it is a favourite or "lucky" marble he can change it for another out of his bag.' In Liss a special formula was used: 'No swops, double quicks, All the world, heaven too, Seven million white rabbits.'

Twosie. 'If a marble hits two marbles with one shot and he shouts "Twosie" he can keep them both.' Cf. *Cannons*.

Ups. 'If a marble goes over the kerb and the owner calls "Ups" before the other player can call "No ups" he can put the marble up on the pavement at the point it rolled into the gutter.' A call of 'Downsie' is also allowed if needed.

Rather different are the calls legitimizing a raid on other people's marbles. It must be said for the mid-century generation of marbles-players, that never, in the course of our surveys, did we have any complaints or accounts of stealing. In previous generations ritualized stealing seems to have been quite common, but may once have been sanctioned as enforcing a quasi-religious Lenten prohibition.[20] F. Austin Hyde, remembering his Yorkshire childhood in the *Scarborough Mercury*, 22 March 1957, recalled that 'a few birds of prey would wait until a ring or hole in the ground was well filled with other players' marbles, swoop down with grasping claws, and with one word, "Bungs!", or two, still more final, "Bungs, Amen!", would scoot away with their ill-gotten gains.' The word is all. In Edinburgh, in the 1960s, 'if . . . ye steal a handfie [of bools] and ye cry "Puggie", that means ye're no a thief' (Ritchie, *Golden City*, p. 61); and 'No Puggies' was the protective cry in Wrecclesham *c*.1892. 'Gobs' was the word of power in Yorkshire (*EDD*), and 'A-rant, a-rant' in Warwickshire (ibid., *Rant*). In Hyndburn, Lancashire, *c*.1900, 'Big lads would swoop on a marble game and pirate the marbles, saying "John Tow Row".' There used to be a strong tradition that marbles should be played only in Lent, and that any marbles in play after Good Friday should be confiscated. 'According to a lady who spent her childhood at Battle over sixty years ago, if the boys began to play after Good Friday the girls used to confiscate their marbles with the cry

---

[20] A teacher at Cumnock Academy, enquiring in 1962 about 'Nunk', which appears in the *Scottish National Dictionary* as a Lanarkshire word meaning a stake in marbles, found, instead, that ' "Snunksy" is used as a battle-cry before a raid on the marbles in the ring, i.e. the stake. There was an influx of Lanarkshire miners to the Cumnock area within living memory.' Cf. 'the dread cry of "Stunks" ' with which some 'big yin' made off with marbles in the Scottish Border country *c*.1925 (*Southern Annual* (1957), 29).

"Goblins after Good Friday!" . . . In Hove about thirty years ago similar measures were taken by the boys . . . but here the slogan was "Lobbing Day"' (R. Merrifield, *Sussex Archaeological Collections*, 89 (1950), 92). In Blaxhall, Suffolk, *c.*1890, the confiscation of out-of-season marbles was accompanied by a shout of 'Tom Fobble's Day!'; and in Petersfield, *c.*1915, by 'Smugs after Good Friday!' 'Smugs' is, indeed, the oldest recorded term to be used in connection with the removal of out-of-season playthings. William Hone, in *The Every-Day Book* (1826), i, col. 253, says that games

among the boys of London, had their particular times or seasons; and when any game was out, as it was termed, it was lawful to steal the thing played with; this was called smugging and it was expressed by the boys in a doggrel: viz.

> Tops are in. Spin 'em agin.
> Tops are out. Smuggin about.

> or

> Tops are in. Spin 'em agin,
> Dumps are out, &c.

'The rhyme is still in use,' says Lady Gomme (*Traditional Games*, ii (1898), 302), 'and may occasionally be heard in the top season.' It is entirely adjustable, appearing in *Games and Sports for Young Boys* (1859), 71, as:

> Marbles are out, smugging is about,
> Tops are in, smugging does begin.

The modern game of marbles, as played by junior schoolchildren, is only a shadow of the game of the past. The decline in skill and the desuetude of many rules has left the game without its former excitement. The polarities of winning and losing have disappeared now that, in an affluent society, it is no longer 'nice' for children to play for material gain; and the loser no longer suffers a physical punishment. In the nineteenth century, and undoubtedly before, the loser (usually in a game of 'Three Holes') had to place his fist upon the ground and allow the winner to shoot at his knuckles. Before a game, it could be decided whether or not to 'Play for *beaks*', as it was called in Northumberland (see *EDD*). It was a sign of manliness if a boy could 'Stand his *rackups*' or *rumps* (Cumberland, Westmorland, *EDD*), since, with the power-shooting of those days, a loser's knuckles could be 'fair skinned'. The punishment, 'Knucklie', gave its name to a complicated version of 'Three Holes' in Maclagan's *Argyleshire*, p. 155; and Jamieson, Supplement (1825), says 'to picket' is 'to project a marble or taw with a smart stroke against the knuckles of the losers' (Roxburghshire), and adds that a loser at tennis had, similarly, to offer his hand for punishment, holding it against the wall to be thrown at. *Blackwood's Magazine*, August 1821, p. 36, reports unsuccessful players having 'dumps' (blows) inflicted on their knuckles.

KINDS OF MARBLES, AND THEIR MANUFACTURE

Alleys. The first mention of 'Marbles' in *OED* is in 1681, and Randle Holme (*Armory* (1688), iii, ch. 16, para. 91) refers to 'Marble Bullets called Marvels'. It must be supposed that, because of their name, the superior sort of marbles were originally made of marble, or were believed to be made of real marble. However, alabaster—closely related to true marble, being sulphate of lime instead of limestone in a crystalline state—was presumably cheaper, and certainly more easily worked.

The 'alley' (first noted in 1720, *OED*) is firmly stated in *Juvenile Sports and Pastimes* (1776), to be 'made of alabaster'; it 'looks exceedingly neat and pretty; the value . . . is proportioned to the variety of its colours; however, this is not so much in use as the common marble, it being much dearer, and more likely to break'. Alabaster is indeed breakable, being the same substance as plaster of Paris; and this fragility may account for the fact that, as far as I know, not one alabaster marble has survived. The 'alleys', like many toys of their time, were imported from Holland, although they were probably not made there but in Germany. In a reading book of *c.*1780 (British Library C.40.a.78, lacking its title-page), a 7-year-old boy rejoices in his 'bag of marbles—fund of pleasure! Here speckled taws, dutch allees—what a treasure!' The charm of alleys ('alley-taws' when used for shooting) was that they were believed to be of 'the purest marble' (*Every Boy's Book* (1856), 17) and sometimes 'of white marble striped and clouded with red', when they were called 'blood-alleys' and considered especially potent (*Games and Sports for Young Boys* (1859), 43).

'Alley' lingered on as a name for a superior kind of marble, usually in the form 'glass alleys'; often the word 'alley' was reserved for 'the big ones, for shooting'. In many places 'alleys' came to denote marbles in general; and in the mid-twentieth century, throughout England, 'alleys'—pronounced 'ollies' in Liverpool—was a general name for the universal imported glass marbles then prevailing.

The alabaster marbles must have been made by the same method as the 'Stonies' (see below).

Ball-bearings (often pronounced 'Bull-bearians'), Big Leadies, Bollies,

Steel Bollies, Steelies, Ironies. Ball-bearings have been used as killer-marbles *par excellence* ever since they have been available. During the Depression in the 1930s grown men played marbles with steel bollies in the streets, for money, especially in Lancashire. 'Steelies are better than glass marbles,' said a boy in St Peter Port, Guernsey, 1961; 'if they are knocked into the hole they don't come out so easily. A steelie is worth about five to eight glass marbles. They come in all sizes, from ball-bearings of cycles to ball-bearings of cars, lorries, and even steam-rollers.' 'Bull-bearian' and 'Ironie' are the names used in England; 'Steelie' in Scotland and the north country. The envied child is the one with easy access to ball-bearings, like the 10-year-old girl in Lambourn, Berkshire, 1973, who told us, 'My Dad's a taxi man. He's got eight cars, and he's given me fifty ball-bearings.' 'But they give 'em you from garages,' said boys in Salford, 1970, 'they give 'em you *free*—that's if they've got any.'

China Alleys. Paul Baumann, *Collecting Antique Marbles* (1970), 33, says that in South Thuringen, in the south-west of the former East Germany, china factories had begun specializing in the production of china marbles by 1800. Hand-worked porcelain was driven through pipes, and cut into equal-sized pieces as it came out. 'These pieces were then laid in plaster-of-Paris forms with long narrow oval grooves. Through a rapid rotary motion these forms moulded the porcelain pieces into spheres.' They were then dried, painted, and fired. The marbles could be glazed or unglazed, and were decorated with floral designs or, frequently, with painted rings of different colours, parallel or at different angles equatorially. Before porcelain, earthenware and stoneware were used, sometimes with plain glazing, sometimes with coloured slip decoration. The London Museum has examples from the sixteenth, seventeenth, and eighteenth centuries.[21] The earlier marbles of glazed earthenware and stoneware were made in the same way.

The glass 'chinas' of the present day are made of opaque glass, often white and looking much like china, with slashes of colour on the surface.

Clays, Commoneys. The much-despised 'clays' probably have the longest history of all. Red-brown clay marbles survive from the Minoan period, in Crete, 2000–1700 BC. A 9-year-old boy in Liss was making himself marbles from the clay in the recreation ground in 1959 ('I don't bother to colour them'). P. H. Gosse, describing the marbles of his

---

[21] e.g. 16th-c. stoneware marble, 0.6 in. diam., glazed dark brown, found in excavation in Angel Court, London EC (A.4956). Three 17th-c. pottery marbles, 0.6 in. diam., found under the floor in 'Jane Shore's house', Wandsworth; two are glazed yellow with brown pattern, and the third white with red marbled effect (A.8434–A.8436). Eleven 18th-c. marbles, stoneware etc., various sizes, marbled decoration, found in a well on the site of Shaftesbury House, Aldersgate Street (A.26020). Four 18th-c. clay marbles, decorated with coloured slip, 'feathered' or marbled (A.3519–22).

boyhood in Poole, *c*.1820 (*Longman's Magazine*, 13, p. 516), said 'There was also an inferior sort, rudely moulded out of red and white clay, which were named "clayers".' A hundred years later, in County Durham, *c*.1925, 'A clay, light in weight and sometimes not quite spherical, was of good colour when new, bought in a little white cardboard carton of 10 . . . each marble of a single colour, red, blue, or green', but 'a few days of play reduced them all to the same grimy brown-ness. No matter, a jam-jar full of "codneys" was a fine rattling possession' (*Times Educational Supplement*, 28 September, 1956).[22] The demise of the fragile clay marble, chiefly used as currency, was only to be expected once cheap glass marbles started flooding onto the market from the Far East, *c*.1960.

According to *The Boy's Own Paper*, 10 (1887), 175, 'Common marbles made of clay [are] rolled up into balls and baked in a kiln, each on a little tripod stand. On every marble so made you will find three little marks where the soft clay has rested on the wire.' But William Bavin, *Marbles* (1991), 8, describes a more efficient method: 'The mass of clay . . . would be forced through a pipe to form long cylinders which were then cut into equal-sized pieces [which were] pressed or rolled into spheres using wooden drums/rollers. They would then be dyed or painted with patterns and colours and placed in a kiln to be fired.' 'Commoneys' appear in Dickens's *Pickwick Papers* (1837), ch. 34, when Mr Pickwick pats a boy on the head and enquires 'whether he had won any alley tors or commoneys lately'. A 13-year-old boy in Forfar, 1954, defined a 'commoney' as 'made of clay or chalk'. 'Potties' seems to have been another name for 'Clays', though these are sometimes said to be of a harder substance, such as stoneware (*EDD*).

Glass marbles: Glass Alleys, Glassies, Glarnies. A contributor to *N & Q*, 9th ser., 3 (1899), 65, giving the values of five varieties of marbles from his Belfast schooldays, *c*.1854, remarked, 'I do not recollect the worth of the glass marbles; they were the newest inventions and not so much used.' Although glass marbles were apparently quite common in the Roman Empire, and a late fifteenth-century Stuttgart manuscript mentions 'the yellow glass used for the little yellow balls with which schoolboys play, and which are very cheap',[23] it was not possible to make the latter-day glass marble, with its magical coloured swirls enclosed in clear glass, until the invention of marble scissors in 1846. A glass-maker makes a 'swirl' marble, as Bavin explains in *Marbles*, pp. 8–9, by gathering a

---

[22] A 'Sunningdale' 'Mammoth Box of Marbles', believed to date from 1950 or a little before, contains twenty-three clay marbles, about 1 cm in diameter, coloured in various subfusc hues of blue, beige, purple, and pink, with several a brilliant silver.

[23] F. J. Mone, *Anzeiger für des deutsches Mittelalters* (1838), 605–6.

glob of clear glass on the end of an iron rod, building coloured glass canes round it to the desired size, rolling his iron on the arms of his chair until a long cylindrical cane is produced, reheating, shaping the end in a mould, and eventually severing the marble with the shears. For machine-made marbles (ibid. 10–11), 'the glass is melted in a large furnace, and when it is the right consistency it pours out through an opening. Shears set at the opening cut the descending glass into equal-sized pieces. These pieces drop into moving, mechanical rollers which have spiralling grooves running down their length. The globs of glass travel down . . . gradually cooling and forming into the round marbles.' Different-coloured inserts, such as those for 'cat's-eye' marbles, are injected into the flowing glass before it meets the shears.

'All the bools now seem to be "glessies", the gleaming coloured glass ones we prized. "Leadies" and the clay type seem to have disappeared', said a writer in the *Hamilton Advertiser*, 15 May 1964. The names 'Glass alleys' and 'Glessies' ('Glassies') gradually died out in the 1960s, in recognition of the fact that all marbles were now the same and could simply be called 'marbles'. 'Glarnies' were glass marbles 'of special beauty and value' around the London area during the first three decades of the twentieth century; the term was also used in Liverpool, *c*.1910, and as a general word for marbles in and around Wolverhampton, 1940–60.

Another journalist (*Guardian*, 23 March 1937) asked some contemporary practitioners about the 'prosaic, earthenware marble of my youth'. 'It was still in circulation, I learned, but was regarded somewhat disdainfully. It had been supplanted by the glass alley. This, in the old days, would have been regarded as sheer hedonism, for glass alleys, with their beautifully coloured interior twirls, were scarce and treasured more as objects of art than as things to be played with.' Machine-made glass marbles were indeed on sale in the 1930s, available in long white cardboard boxes with a viewing slit along the top, and wording such as 'Glass Marbles—7 Pieces—Foreign', or '12 Glass Marbles—Foreign' and the maker's name, Codeg. Codeg's imported ('Foreign') glass marbles, in cardboard boxes, were still on sale in the early 1950s ('Monster Size. 6 Giant Dazzlers'), and in the 1960s in the contemporary small transparent plastic bags, fastened with cardboard fold-overs. Around 1950, IMG Ltd. were selling ten half-inch champion marbles for 10½*d*., in a cardboard box, which were 'Made in England'; these must have been some of the last of the English-made glass marbles. In about 1960, foreign glass marbles were being specified as 'Made in Hong Kong' (twenty marbles for 4*d*., with slogan, 'Marbles to children like water to fish'), or 'Empire Made' (thirty pieces, 6*d*.); in 1966, 'Made in India' and sold by S & K; in

1967, 'Made in China'; in 1974, 'Made in Mexico'; in 1976, 'Made in Korea'; and in Woolworth's, 1978, 'Made in Taiwan' ('Bag of Milky Marbles Our Price 30p Bumper Pack).

Pop Alleys. These are the glass marbles from the necks of the old-fashioned fizzy lemonade or ginger-beer bottles invented by Hiram Codd of Camberwell in the 1870s. The pressure of the gas kept the marble in place in the neck of the bottle; a wooden cap with a peg in the centre pushed the marble down so that the drink could be poured. Boys smashed the bottles to get at the marbles.[24] These were the ordinary glass marbles of the day: 'Ordinary "glassies" were from ginger-beer bottles' (Rochdale, *c.*1900); ' "gingies" were used by everyone for flipping' (east Devon, *c.*1915); 'glass-eyes', 'those you got out of lemonade bottles' (*London Street Games* (1916), 110). 'Pop alleys, out of pop bottles' (Spennymoor; Knighton) were, in the 1950s and 1960s, also called 'bottle alleys' (Lydney), 'bottle stoppers' (Sheffield; Peterborough); 'bottle tops' (Manchester); 'pops' (Bishop Auckland; Wigan).

Stonies. Paul Baumann has a detailed chapter on stone marbles in his book *Collecting Antique Marbles* (Iowa, 1970). Stone marbles were known in antiquity, but the stone marbles of the seventeenth century onwards were made in Germany. Calcareous stone was broken into 1-inch-square blocks with small hammers. The grinding mill, worked by water, consisted of a fixed lower stone with concentric grooves cut into it, into which the cubes were put, and an upper slab of the same diameter, made of oak, partly resting on the lower slab. Water ran between the two blocks while the upper block was kept rotating, and within fifteen minutes the marbles were ready to be finished in a polishing barrel. A mill could turn out 60,000 marbles a week.

P. H. Gosse played with 'stoners', 'made of a compact blue or grey limestone', in his Dorset schooldays *c.*1820,[25] and 'stonies' continued to be appreciated, especially as shooting marbles, throughout the century until, about 1900, competition from the clay-marbles industry put the stone mills out of business.

The most attractive of the stone marbles were the agates, not ground but shaped on a grindstone. They were produced on the Nahe River at Oberstein, in Germany, where they were made in great numbers in the later nineteenth century, chiefly for the American market. Unlike the

---

[24] On 9 Jan. 1897 J. Connor patented an internal stopper for aerated-water bottles that was egg-shaped 'to prevent the wilful breakage of the bottle to obtain the stopper for use as a marble', and J. Meredith followed, 7 Feb. 1899, with an elongated stopper for the same purpose, but these do not seem to have caught on. However reprehensible, the breaking of the bottles was accepted wastage, and the Codd bottles still being made in India in the early 1970s had a choice of marbles, red or blue.

[25] P. H. Gosse, *Longman's Magazine*, 13, p. 516.

'marbles' made of other semi-precious stones—rose quartz and tiger eye—for adults to collect and use as finger pieces, the agates were serious playing marbles. Some must have found their way to Britain, for the word lingered on: all over Oxford, for instance, in 1960, 'aggies' was a general name for marbles—though some children specified 'a big one'.

THE NAMES OF MARBLES

The regional names for marbles in general, during the mid-century, were variations of the word 'marbles' itself: 'mabs' in the west and south of Yorkshire; 'maggies' in Market Rasen, Lincolnshire; 'marbs' in Southam, Warwickshire; 'marlies' in the Midlands;[26] 'marries' in St Andrews and Jedburgh; 'mebs' in Dublin and Langholm, Dumfriesshire; 'merps', found only in Castleton, an industrial suburb of Rochdale, Lancashire, 1969 ('merps'—occasionally 'merkies'—was, until about 1930, the common generic name in Lancashire); 'miggies' in London N22, and Enfield, Middlesex ('miggies' in *London Street Games* (1916), 110); 'muggles' in County Durham and Northumberland. A glance at *EDD*'s *Marble* entry will show that dialect variations have long existed; the most common is Randle Holme's 'marvels' of 1688, which is perhaps the most natural corruption of 'marble'. It seems to have been in general use in the late seventeenth century; in *A Play-Book for Children* (1694), by 'J.G.', p. 39, is the reading sentence, 'Mar-vils are pretty play-things for Chil-dren.'

In Scotland, the generic name for marbles is 'bools'. This is simply the Scottish pronounciation of 'bowls', and is an adoption of the French 'boules'.

In 1951 a man born seventy years before, in Bethnal Green, wrote ecstatically to *The Times* about the games of his boyhood: 'Marbles was a great game in my time. How lovely do the names of them trip off the tongue—alleys, tors, mivvies, glarnies, boncers, bloods.'[27] And truly (for those who like the names of things), the sound of the old names for marbles falls pleasantly on the ear: balsers, bullockers, doggles, dolledgers, dummocks, jaries,[28] moral-leggers (which were 'ring-streaked

---

[26] Thus George Eliot, who spent her childhood in Warwickshire *c.*1830, has Tom saying to Maggie, in *The Mill on the Floss*, ch. 5, 'I've swopped all my marls with the little fellows.' 'Marloes' was the form Husenbeth knew at Sedgley Park School, Staffordshire, *c.*1803–10 (*History of Sedgley Park*, p. 105).

[27] 'Mivvies' was quite a widespread name for marbles at one time, e.g. in Bembridge, Isle of Wight, *c.*1905 (where sheep's droppings were also known as 'mivvies'), and in Petersfield, *c.*1915.

[28] *Jarie*: Scottish, from the diminutive of English *jar*, an earthenware container. 'Brown earthenware glazed and burned . . . The plunker or marble used to strike with is always a "jary", if not . . . of more valuable material' (*Maclagan* (1901), 152). Pronounced 'jorries', apparently the generic name for marbles in Glasgow.

and sometimes piebald'), stonnackerools, stubbers, and tonses. Modern children, having only machine-made glass marbles, are restricted to names describing their size, or the names under which they are sold, or fanciful names of their own inventing. Thus big marbles are big 'uns, bossers, bulls or bullies, bumpers, bunties, cannons, chuckies, coshers or cosses, dabbers, dobbers (an old word for iron or earthenware marbles sometimes 3 or 4 inches in circumference), fighters, fobblers (large ball-bearings are iron fobbers, in Pensnett, Staffordshire), gob stoppers or gobbies, kings (thence queens, princes, and princesses for the lesser sizes), slammers, smashers, stumpers, tattiemashers, and yogis; and small ones are babies, peas, pee wees, peedies, tichies, or tiddlers. Manufacturers' names are difficult to disentangle from the children's own names. The most common glass marbles are of clear glass with a single twist of colour in the middle; who was it first called them 'cat's eyes'? They are sold as 'cat's eyes', but perhaps the manufacturers are using the name children use. More esoterically, about 1990 some children called these marbles 'toothpastes', in reference to 'Aquafresh' toothpaste, which emerged from the tube in stripes of white and green or pink, and was much advertised on television. Children use such names as are suggested by colours, textures, and patterns: 'chinas' look like china; 'clearies' have no colours, 'coloureds' do; 'Frenchies' have three or more colours (Edinburgh, 1975; in Lothian, in the 1890s, a Frenchie was 'of greenish colour, but with strata of lighter colour through it', *EDD*); 'milky ones' look as if milk has been mixed into the glass; 'misties' are of ground glass; 'oilies' are 'like oil in puddles'; 'pearlies' are, naturally, pearly; 'rainbows' are like 'cat's eyes' but have two colours; 'spiders' have a white smoke-effect inside (in Virginia Water these were 'white ghosts'); 'spaghettis' have got white threads inside the glass; 'tods', in Staffordshire, 'have several different colours in them', and the word is rapidly becoming the generic name for marbles; 'water bellies' are plain glass.

Manufacturers are constantly inventing delicious new varieties of marbles, the collectibles of future years. Nowadays the names are less likely to be traditional, or the name on the manufacturer's label, than a child's own invention, and they argue amongst themselves: '*I* call that one a "white shadow"', '*I* don't, I call it a "white knight".' 'Spotted dicks', 'hot dogs', 'black moons', 'blue cannon balls', and 'Coca-Colas', being individual fancies, are not likely to find a place in any dictionary.

The much-prized alabaster marbles with red or pink streaks, called 'blood-alleys' since at least the late eighteenth century,[29] have always

[29] See John Shipp's memoirs, in G. Avery's *The Echoing Green* (1974), 38: 'One autumn morning, in the year 1797, while I was playing marbles in Love Lane, and was in the act of having a shot with my blood-alley . . .'.

been prized for their supposedly magic power, and the same faith is put in their modern equivalents, the marbles of white opaque glass with red streaks in them. Black—'black-eyed'—marbles are also thought to be lucky; they were called 'black-witchers' in Castle Carey, *c*.1912. This was the kind of magic marble Southey was taught to make, at the age of 8, by wrapping an ordinary marble in brown paper, with some suet or dripping round it, and putting it in the fire, 'when you take it out it is as black as jet' (letter, July 1822).[30]

Charms are another way to enlist the aid of the supernatural. 'When losing at marbles you can say "Black cat follow me, not you"' (Swansea). 'If you want to win you put a cross in front of the hole with red chalk and shout "Bad luck" and the person who is playing you gets confused and misses the hole' (Birmingham). 'You jab a circle around the [opponent's] marble and say "Darkies all around it" or make a cross between the two marbles and say "Criss-cross make him loss"' (Newcastle upon Tyne). 'When a person is shooting at a ball we want him to miss we say "Abracadabra wall come up!" and somehow he seems to miss' (Annesley, Nottinghamshire).[31] 'If I am near the hole and someone is in the hole and they go to flirt at my marble I would say "1, 2, 3, lucky, lucky, 4, 5, 6"', and they miss it (Stoke-on-Trent).

---

[30] *Life and Correspondence of Robert Southey* (1849), i. 65.
[31] The same honest wish was being expressed in the west of Scotland in the 19th c.: 'Crecs cross, I wish you may lose, Fine fun for me to win' (*Folklore* (1881), 176).

3

Fivestones

EARLY in the history of the world the knucklebone, or astragalus, which is the ankle-bone of cloven-hoofed animals, was seen to be an excellent plaything. Its neat oblong shape, about 1 inch long in the case of sheep, and four distinctive sides, make it a natural dice: one side is convex, one concave, one almost flat, and one sinuous, ending with a little horn. These bones were used to determine the moves in board games as far back as the time of the New Kingdom in Egypt, *c*.1600–1000 BC, when they were used in the games of *senet* and *tjan*; and for dice-shooting gambling games, like the American game of crap, throughout the Greek and Roman periods.[1] However, the game that concerns us here, which is still played today, is 'Pentalitha' ('Fivestones'). Providentially, one description survives from classical times, in Pollux's *Onomasticon*, written in the second century AD:

In Pentalitha, five little stones or pebbles or knucklebones were thrown up, so that the thrower, turning his hand over, received them on the back of it; if all of them did not land on it, while keeping those that did land lying there he picked up the rest with his fingers.

'The game', Pollux added, 'is rather for women'; and in a famous painting from Pompeii by Alexander of Athens, which is a copy of an original painted in the fifth century BC, two goddesses can be seen playing it. In reddish-brown outline, on white marble, Aglaia, one of the Graces, plays knucklebones with Hilaeira, daughter of Leucippus; and three other goddesses form a sort of chorus-line at the back, each of them being named beneath. Hilaeira has just caught some of the bones on the back of her right hand; more than five bones are present, for two are falling and three lie on the ground. Another game of 'Pentalitha' is taking place on an Attic red-figured toilet-box, now in the Metropolitan Museum of Art, New York. The knob on the cover is in the shape of a knucklebone. One of the two unpleasantly muscular little players is depicted with a knucklebone balanced on her wrist. The lady who owned the box was undoubtedly a knucklebone-player herself. Again, a vase of 420–400 BC

---

[1] Also for divination, which—even more than their use as dice—is beyond the boundaries of this book.

Right: Winged Eros absorbed in a game of knucklebones, on a bell-crater painted by the Amykos Painter, 420–400 BC (British Museum E 501).

Below: The two goddesses in the foreground are playing knucklebones. The three in the background may be playing another game, now forgotten. Painting from Pompeii, by Alexander of Athens, copied from an original of the fifth century BC. National Museum of Naples, No. 9562.

in the British Museum shows winged Eros, seated, absorbed in the game
of knucklebones—a game that it is perfectly possible to enjoy on one's
own. He does not stick to Pollux's rules either, but has about eight bones
on the back of his hand, and about six on the ground.[2] The ancients liked
to make themselves comfortable while playing knucklebones. The girl
knucklebone-players depicted on an Arretine bowl of $c.5$ BC, found at
Haltern, in the valley of the Lippe in Gaul, are seated on stools around
a low table;[3] and the French still considered it civilized to sit down to
one's game of knucklebones centuries later, for in the eighteenth century
Chardin's teenage 'Joueuse d'Osselets' is seated at a table, and so are the
young knucklebone-players in *200 Jeux d'enfants* ($c.1892$; pp. 164–5).

From time to time literature provides a glimpse of knucklebones or
fivestones. In the second part of the *Roman de la rose*, written in the late
thirteenth century, Pygmalion provides his statue with a purse adorned
with five beautiful round stones, 'such as children might pick up from
the seashore as playthings'.[4] Rabelais gave Gargantua an appropriately
large number of games to play; among them are 'pingres' (knucklebones),
and 'martres', added in the 1542 edition, which Urquhart, in 1653,
translated as 'ball and huckle-bones'.[5] A ball is easier to throw up and
catch than a knucklebone or pebble, especially if it is allowed to bounce
and is caught on the rebound; a pottery ball, or a marble (which is more
bouncy than one might think), has been used in the game right through
to the days of 'Bobber and Kibs' and the latter-day rubber ball that
accompanies metal jacks.

Apparently knucklebones was too quiet a game to have been included
in *Les Trente-six figures* (1587), or in *Jeux et plaisirs de l'enfance* (1657).
It is first noticed in a games book, *School Boys' Diversions* (1820), 52; but
the stilted description of 'Hucklebones' there is suspiciously Gallic,
containing such un-English moves as 'the Kisses—kiss the bone while
the other is in the air', and is probably a translation from *Les Jeux des
quatre saisons* (1812), a copy of which I have failed to find. The French
game, as described in books such as Madame Celnart's *Nouveau manuel
des jeux de société* (1827; 1846 edn., pp. 318–20) and Madame de
Chabreul's *Jeux et exercises des jeunes filles* (1856, p. 84), was highly

---

[2] Vase no. E501. Cf. a fragment of a poem by Anacreon, born $c.570$ BC: 'Astragals are Eros'
passion and battleground'. Several statuettes have survived of young women playing with knucklebones (e.g. British Museum terracotta D161, Italian Greek, $c.340$–330 BC). They are not playing
'Pentalitha', but probably 'Omilla', the game comparable to ring marbles.

[3] F. Haverfield, *The Roman Occupation of Britain*, rev. G. Macdonald (1924), 179.

[4] The important lines are 21781–4. See P. Huot, *Étude sur le roman de la rose*, pub. Société
Archéologique de l'Orléanais (1853).

[5] See M. Psichari, *Revue des études rabelaisiennes* (1908), vi. p. 157.

Jean-Baptiste-Simeon Chardin (1699-1779), 'Les Osselets', *c.*1734.
Chardin's homely teenage girl is seated at a table, in the French manner,
and plays knucklebones with four bones and a pottery ball.

developed, and included a long sequence of variations such as 'passes-passes', in which the bones are pushed through an arch made by the thumb and index finger of the left hand, and moves that use the shape of the bones, such as 'rafles', in which the left hand turns the bones one by one onto their flat sides—thus making it even more of a two-handed game.

   The English game must have been just as elaborate. Most of the nineteenth-century games books found it too complex to describe, and dodged the issue by saying that 'all the requisite manœuvres . . . can readily be learned by any person acquainted with the game'.[6] However, a Captain A. S. Harrison wrote 'a full description of the game of Knuckle Bones' for *The Boy's Own Paper*, 28 May 1881, in reply to an enquiry, describing the fourteen stages known to him in his boyhood: 'Beginnings, Ones, Twos, Threes, Fours, Short spans, Long spans, Creek mouse, Second Creek mouse, Bridges, Cracks, No cracks, Exchanges, and Everlastings'; and 'Toodles' outdid him, in the issue of 2 July, by sending his own 'Order of Knuckle-bones' of twenty-nine stages: 'Practice, ones, twos, threes, fours, creepmouse, chuck-ups, dux 1, dux 2, dux 3, dux 4, clicks, non-clicks, postman's knock, double postman's knock, fingers, inches, spances [spansies], short arms, long arms, and triangles; squares, daggers, swords; pockets 1, pockets 2, pockets 3, pockets 4, everlastings.' 'Everlastings', at least, deserves to be described here, since, as Captain Harrison said, it is a task that 'approaches the everlasting'. 'The whole of the bones are tossed from the palm, and any number caught on the back. These are tossed from the back and caught in the palm; and any that have fallen in the first toss have to be picked up while the whole of the others are in the air, so that at one moment there may be four dabs in the air and one to pick up.'

   The advantage of the game is that it can be played in idle moments, and with objects that lie close at hand. Five suitable pebbles can easily be found.[7] Animals' bones were once part of ordinary household refuse. G. H. Kinahan (*Folklore*, 2 (1884), 266) conveys the intimacy and antiquity of the game in Connemara: 'Jack-stones . . . have been found in the crannogs or lake-dwellings in some hole near the fire-places, similar to where they are found in a cabin in the present day. An old woman, or other player, at the present time, puts them in a place near the hob, when

----

   [6] *Boy's Own Book* (1855), 63.
   [7] They are, however, selected with care. A girl in Street, Somerset, said, 'With Cocky Fivestones you need five medium stones with like three corners on. You can buy wooden stones for playing but I think you can get on just as well with ordinary stones'; boys in Glastonbury stressed that the 'cocky fivestones must be flat so that they stay on the back of your hand'.

they stop their game, and go to do something else.' A correspondent described the homely use of meat bones: 'My mother says that 'Chink-stones' was a very favourite game in Suffolk in the 1870s. They played it with the round bones from pigs' trotters. Everybody kept pigs so there were always plenty of those.' Sheep's knucklebones could be saved from joints of mutton; but bones from a sheep's spine were used as well—and, in sheep-farming country, both kinds of bone were salvaged from the skeletons of sheep lost on the hillsides during a bad winter. Plum- or peach-stones were often used (as they are frequently in Italy) and some present-day children use them on their fathers' recommendation. It has also been reported that for the game of 'Snobs' in Enderby, Leicestershine, c.1935, 'Shoe buttons tied into five separate bundles were considered best.' Almost any small object that is heavy enough can be tried out: 'You can play with halfpennies, or marbles, or acorns, or shells.'

The manufacture of knucklebones or 'stones' has been a considerable trade. In the ancient world artificial knucklebones were made from terracotta, glass, lead, bronze, ivory, onyx, rock crystal, precious stones, and even gold. Madame Celnart said (1827) that although knucklebones could be obtained from the shops, carved in ivory or ebony, the natural bones were best. *The Boy's Handy Book* (1863) mentioned that 'artificial dibs may be obtained at the ivory turners, and at many toy-shops'. C. H. Rolph, harking back to his childhood, c.1905, said, 'You could buy a set of fivestones for a few pence and usually they were shaped to resemble the knucklebones of a sheep' (*London Particulars*, p. 37). On the Continent, imitation knucklebones were being sold until the late 1960s, and possibly thereafter.[8] The manufactured substitutes for real stones were cubes about the size of a sugar lump, in boxed sets of either five 'stones', or four 'stones' and a large pottery marble, and each stone was a different colour—such as red, blue, orange, yellow, and green. (In an effort to suit as many customers as possible, one enterprising manufacturer printed a different name on each side of the box: '5 STONES 5, CHINA GLAZED BOBBERS AND KIBS, CHINA GLAZED DIBS, CHINA GLAZED SNOBS'.) They were 1½ centimetres square, had fluted sides and flat ends, and could be made of glazed china, china clay, or wood. They were in production from around the 1880s to at least the late 1970s.

There is probably no country in the world where fivestones is not played. The game would serve as an ice-breaker at any international

---

[8] Examples from the 1950s–1960s in my possession are full-size, four red bones and one white, in a long box made of clear plastic. Metal 'bones', also dating from the 1960s, are smaller, and the set is silver, with one red bone used for throwing-up.

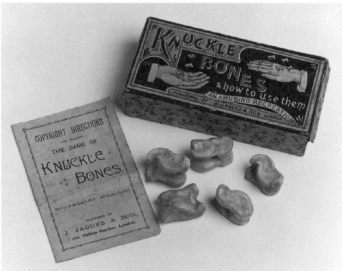

Above: Girls playing 'Jacks', *c.*1885, outside David Storry's grocery shop in Whitby, at the foot of the 199 steps leading up to St Mary's Church.

Left: J. Jaques & Son's boxed set of knucklebones, with 'Copyright Directions' for playing, *c.*1900.

gathering. The following are some exceptions to the usual style of play, noted or reported to us by people from foreign parts: in China, girls sometimes employ little bags filled with beans, or, in Malaysia, with rice or tiny stones ('The girls say bare stones hurt your hands');[9] in Singapore, children play with imitation knucklebones carved in wood by their fathers; in Greece, sometimes, with the cones from the cyprus trees; in Capri, in days gone by, with small squares of mosaic picked up from the tessellated floors in Tiberius' palace; in Mongolia, in the 1920s, with the ankle-bones of reindeer; in the Philippines, with shells.

'Jacks' is, nowadays, the universal English-language name for fivestones, a word which has been 'laundered' in its transition from Ireland to the United States to refer to metal 'jacks', those commercially produced, six-knobbed metal stars (or, according to an 8-year-old in Stepney, 1976, 'funny things made of spikes with bobbles on the end') which are easier to manipulate than stones, cubes, or bones. They have been made in plastic since around 1960, though plastic is too light a material. The standard equipment is five jacks and a small hard rubber ball, but, since ten jacks are often provided in a packet, children have frequently played with all ten. According to all the evidence, they are an American invention. W. W. Newell, *Games and Songs of American Children* (1883; 1903 edn., pp. 190–3) describes 'Jack-stones' and its variations as played with pebbles, and adds 'Instead of pebbles, little double tripods of iron . . . are generally in use.' W. H. Babcock, 'Games of Washington Children', *American Anthropologist*, 1 (1888), 268–9, called them 'a set of oddly shaped bits of metal'. They made a regular appearance in American mail-order catalogues: for instance, in the Montgomerie Ward & Co. mail-order catalogue for 1889, p. 33: 'Jack Stones Nicely Coppered, good sizes. Per dozen $0 04 Per gross $0 30.' Exactly when the metal stars first appeared in Great Britain is not known, but innumerable people remember playing them in the mid-1930s, especially in girls' boarding schools.

Boarding schools—both boys' and, later, girls'—have in fact been a natural habitat for the game. A glossy set of knucklebones, polished by use, is companionable in the pocket, and as soothing as worry-beads. The

---

[9] See S. Culin, *Games of the Orient* (1958), 58–9 (originally published as *Korean Games, with notes on the corresponding games of China and Japan* (1895)). He says that 'Kong-keui' ('Jackstones') is played by boys with five or seven stones or pieces of brick, and that 'Tja-ssei', a similar game, is played by girls, using coins. However, he reproduces a picture from the Japanese book of amusements, *Wa Kan san sai dzue* (1714), of a young woman playing a classic game of fivestones; and Chinese labourers in the USA told him that jackstones is regarded as a girls' game in Kwangtung. Pl. xiii shows a Korean village girl sitting cross-legged in front of a small bolster to play 'Tja-ssei', while two others wait their turn.

bones do not cause the problems associated with pseudo-currency, since they are not won or lost. It is known that the boys of Christ's Hospital played knucklebones *c*.1825, and undoubtedly long before (*N & Q*, 8th ser., 4 (1893), 201). However, the best example is Cheam School (the Duke of Edinburgh's preparatory school), which was founded in Cheam, Surrey, in 1646, and is now at Headley, near Newbury; there the tradition of playing 'Dibs' with sheeps' knucklebones appears to have continued unbroken since the seventeenth century, with one unfortunate hiatus when F. B. Beck and M. Wheeler, taking over the school in 1947, found a large number of unsavoury-looking small objects lying around and consigned them to the incinerator. The losses were slowly recouped, and the *Cheam School Chronicle* (1960–1), was able to report that 'Dibs is regaining its popularity among the junior boys, Mr Munir having returned from France with 90 sets of cast metal dibs, which are as alike bone dibs in size and weight and bounce as makes no odds.' The junior competition that year was 'won quite easily' by Mary Beck, the headmaster's daughter.[10]

Although most players have been female there have been plenty of male addicts, as has been seen, and not only schoolboys but youths as well. An irascible writer to *Notes & Queries* (1894), 256, complained that 'This game is much played in this place [Coventry] by young men and others who have nothing better to do, and much to the detriment of the lawn mower, as they pick the five stones out of the gravel-walk, and when done with leave them on the grass.' Although present-day boys tend to dismiss 'Jacks' as having 'not enough fun in it, not enough action', whenever there has been a strong 'Jacks' craze the boys seem to have joined in too.

THE NAMES OF THE GAME

The names of the game are also the names of the objects used.

Bobber and Kibs. Played with a pottery ball and small cubes. Predominantly a Lancashire name (earliest *EDD*, 1887), with most recordings from Manchester and Salford. The most recent recording, however, was from Marple, Cheshire, 1954.

---

[10] Even in modern times the bones were preferred to stones as being cleaner and more regular in shape, and not likely to scratch a table or other flat surface. The Cheam dibs sequence has thirty-nine movements: sums, ones, twos, threes, fours, back-handers (three times), arches, pigs, pots, horses, cracks, non-cracks, eggs, change 'ems, turn 'ems, Kiss 'ems, elbow's length, arm's length, hand's breadth, dib's breadth, finger's breadth, hair's breadth, one sweep, two sweeps, three sweeps, four sweeps, double sweeps, cross sweeps, one dob, two dobs, three dobs, four dobs, cross dobs, rats, creep-mouse, tea-pots, triangles, everlastings.

Checkstone, Chackstone, Checkers, Checks. The earliest recorded English name for the game. Arthur Golding, translating a French religious work in 1587, talked of 'yoong children, which set al their felicitie in Checkstones and pins'. Cotgrave, *Dictionarie* (1611), defined 'Cailleteau' as 'a chackstone, or little flintstone'. 'Checkers' seems to be peculiar to Lincolnshire (Peacock, *N. W. Lincolnshire. Glossary* (1877); children, Scunthorpe and Caistor, 1952). 'Checks' (or 'Checks and Bobbers') was the name in Yorkshire and Lancashire, the equipment being four or six earthenware cubes and 'a large earthenware marble, almost the size of a golf ball'; and was still sometimes the name in the 1950s, as an alternative to 'Jacks' (Pontefract, Holmfirth, and Ecclesfield, all in Yorkshire, and Ulverston, Lancashire).

Chucks, Chuckies, Chuckie Stones, Chuckers, Chucks and Handies, Chucks and Marvels. Once widespread, but later, as often happens, known only in Scotland and the far north of England. First recorded, as 'chuckers', in a 'remarkable occurrence' in Avon, Devonshire, in 1760 (*Annual Register*, p. 82): 'As one John Wilson, an old labouring man . . . was laying on a bench fast asleep, some boys being at play with chuckers, and the old man's mouth being open, one of them chuck'd one directly into his mouth, which waking him, and he not being aware of it, sticking in his throat, choaked him before any assistance could be procured. He was upwards of ninety years of age.' Jamieson (1808), has: 'A game used by girls, in tossing up, and catching pebbles as they fall, is called the *Chuckie-stanes.*' *Blackwood's Magazine*, August 1821, referring to girls playing *chucks* in contemporary Edinburgh, says the game is 'played with a *bowl* [marble] and *chucks*, a species of shell (*Buccinum lapillus*) found on the sea-shore'. In R. O. Heslop, *Northumberland Words*, (1892–4), 'chucks an marvels' (apparently the county-wide name) is played with five sea shells and a 'marvel' (marble), and 'chucks and handies' is given as the name in South Shields. During the 1950s and 1960s the game with stones or cubes was declining in favour of 'Jacks' with metal stars, but we had 'Chucks' from Langholm, Findhorn, and Airdrie (all in Scotland); Newcastle upon Tyne, Annitsford, and Hexham (all in Northumberland); and Spennymoor, Bishop Auckland, Thorpe Thewles, and Low Fell, Gateshead (all in County Durham); 'chuckies' from Stirling, Forfar, Aberdeen, Perth, Glasgow, Kilmarnock, and Ayr, and 'chuckie stanes' from Kirkcaldy. 'Chuckies' was rare south of the border—only from Berwick, and as an alternative in Newcastle upon Tyne.

Clinks. 'Clink' was one of the 'joys eternal' John Clare remembered from his childhood in Helpstone, Northamptonshire, *c*.1800

('Remembrances', in *Poems*, ed. J. W. Tibble). The name has a sparse and random distribution. *EDD* has 'clinks' from correspondents in Cumberland and Westmorland. *Farmer's Weekly* readers (November 1954) recalled 'clinkers' from Dumfriesshire, *c.*1900, and 'clinks' from Penrith, Cumberland, *c.*1915; and children in Penrith and around Ecclesfield, Yorkshire, knew the name in 1954.

Cocky Fivestones. Probably throughout Somerset in the 1960s, certainly the only name known in Street and Glastonbury; 'Cocky Fivers' reported from Plymouth, Devon, 1960.

Dabstone, Dabs, Dabbers. John Donne (son of the poet), in the Epistle Dedicatory to *Paradoxes* (1652), writes that 'Lelius and Scipio are presented to us as playing at Dabstone before they fought against Hannibal'. Donne was born in London in 1604, and educated at Westminster School; 'Dabstone' was still known to young Londoners in 1960 (Putney), as well as being the word in Norwich (1961). 'Dabs' is the more usual term: recordings from Richmond, Surrey, *c.*1890 (*Traditional Games*) and 1936; Teddington and Hanworth, Middlesex, 1946 and 1966; Watford, Hertfordshire, 1948; Worcester Park and Tolworth, Surrey, 1948 and 1950; London, 1952; Croydon, 1953. 'Dabbers' was the form entrenched in Oxfordshire until about 1950, and was the alternative form in Watford, Hertfordshire, in 1948 (does this represent a boundary?), and correspondents to *EDD* reported the name from Warwickshire and as far away as Northumberland.

Dandies. Entirely Welsh. J. A. St John, born in Carmarthen in 1801, describing 'Pentalitha' in his *Customs of Ancient Greece*, p. 160, said it 'is still played by girls in some remote provinces of our island, where it is called "Dandies"'. *EDD* has an 1888 quote from Pembrokeshire. D. Parry Jones played 'dandies' at Capel Mair on the Carmarthenshire banks of the Teifi, *c.*1905 (*Welsh Children's Games*, p. 99). In Swansea, in 1953, 'Dandies' was 'the most widely used term for fivestones', and a child there advised, 'The best source for dandy stones is the gravel path round a park bowling green.'

Dobber and Jacks, Dobs. 'I used to play this game at the Wesleyan day school at Old Glossop [Derbyshire] between 1890 and 1893. It was known as 'dobber and jacks' there, and was played on the steps leading up to an unused entrance. Steps or flagstones were necessary, because the dobber would only bounce on stone' (*Guardian*, 3 March 1936). 'Dobber and Jacks', the name in Blackburn, Lancashire, *c.*1905, and Hyde, Cheshire, 1930s, was also known as 'Jacks and Dobbers', for instance by the singer Gracie Fields when a child in Rochdale, *c.*1905, and in Manchester 1910–15 (*Journal of Lancashire Dialect Society*, January 1976, p.

15). The name, shortened to 'Dobs', survived in Manchester until at least the mid-1950s, when it was still being played with four pot cubes and a large pot or glass marble.

Dibs. The most prevalent name in the south. Many recordings (1950–70) in Dorset, Hampshire, Surrey, Sussex, Kent, and also in Gloucestershire, Oxfordshire, Wiltshire, and Berkshire. First noted, as *Dibstone*, in Locke's *Thoughts Concerning Education* (1693), para. 152, 'I have seen little Girls exercise whole Hours together and take abundance of Pains to be expert at Dibstones as they call it.' Bailey, *Dictionary* (1730–6), has '*Dibbs*, a play among children'.

Fivestones, Fivies. Played with real stones or cubes of wood or pottery. Sometimes the game is called 'Fivestones' even when metal spikes are used. The name is so obvious it has been ignored and does not appear in *OED*. 'Fivestones' was listed by Edward Moor, *Suffolk Words*, from his childhood, *c*.1780. J. M. McBain, *Arbroath [Angus]*, p. 346, played 'fivies' as a boy *c*.1840, as Barrie did in Kirriemuir (*Sentimental Tommy* (1896), ch. 37). It could be considered generic, and is found wherever a local name is not dominant. It occurs evenly across the south from Cornwall to Kent, and into mid-Wales; but not further north than Humberside, and Skipton in Yorkshire; and not in the heart of the Midlands, where 'Snobs' holds sway.

Gobs. The older form was 'Gob-stones'. Charles Vallancey, writing about jack stones in *Collectanea de Rebus Hibernicus* (1784), iv. 65, adds, 'It has another name among the vulgar, viz. gob stones, because one part of the ceremony is, to convey them into the gob or mouth.'[11] Later, 'Gob-stones' was almost exclusively the London name. 'In London it is called *Gob-stones*, and the taw is the *Buck*, but the stones are five, and are anything that can be found in the street: stones, coal, or broken-pot' (S. Dyer, *W. Riding of Yorkshire* (1891), 108). Shortened to 'Gobs', it appeared in phrases which included a word for the marble thrown up, as well as the stones. The Revd Allen Moxon (*St Anne's Soho Monthly*, July 1907), said that 'Buck and gobs' was 'certainly known to every girl and most boys in all this neighbourhood'. Norman Douglas, in *London Street Games* (1916), 69–70, said, 'Gobs and bonsers are used by girls. "Gobs" . . . are shaped like dice . . . and a bonser or bonk or buck or bonster is a large marble that bounces . . . With these things you play BUCK AND FOUR [and] ALLEY GOBS'. Girls interviewed in Stepney in 1976

---

[11] Was he thinking of the variation, part of the French sequence, 'kiss the bone while the other is in the air'? I suspect that 'Gob-stones' or 'Gobs' continued to be current in Ireland. Unfortunately our surveys did not cover Ireland, but 'Gobs' is noted in County Cork in 1938 (Irish Folklore Commission, MS vol. 353).

said, 'Gobs are five little squares with lines down the side. You get them in a little packet, and they cost 5p. They're made of china clay. You buy them in toyshops or sweetshops.' All the recordings in our 1950s–1970s surveys came from the London area or adjoining parts of Essex.

Jacks, Jack Stones, Jacky Fivestones. 'Jacks' was played with real stones long before the metal spiked 'jacks' were invented. Vallancey, *Collectanea de Rebus Hibernicis* (1784), iv. 65, says '"Jack Stones" is the name used by the English in Ireland.' Like most of the old names for the game, it is not neatly regional, although it has always been common in Ireland. However, 'jacks' were 'square pieces of stone' in Hull in the 1890s; and in *N & Q*, 8th ser., 5 (1894), 256, were described as 'five smooth stones'. *EDD* has locations for 'Jack Stones' in Ireland, East Yorkshire, Lancashire, Cheshire, Warwickshire, North Staffordshire; and for 'Jacky Stones' in Cumberland, Lakeland, and Westmorland. The thirty or so Opie recordings of 'Jacks' (less often 'Jack Stones' or 'Jacky Fivestones'), using stones or cubes, from the mid-1930s to the mid-1960s, are widely scattered and show no strong geographical pattern.

Jinks. A name once common in Cambridgeshire, Hertfordshire and Derbyshire (e.g. J. Cussan, *History of Hertfordshire* (1881), iii. 320, 'smooth, water-worn pebbles'; E. Porter, *Cambridgeshire Customs* (1969), 211–12; correspondents). Students at Bedford Training College said they were playing 'Jinks' in these counties c.1945–50, but only one child, in Glossop, Derbyshire, produced this name for our 1950s survey.

Snobs. Very strong in the Midlands (Derbyshire, Nottinghamshire, and Leicestershire), both in the past and up to at least the mid-1960s. Also in Northamptonshire, Lincolnshire, Ipswich, and occasionally in South Yorkshire. The earliest mention is in *N & Q* 8th ser., 4 (1893), 273, 'In South Notts we . . . played it with small stones or marbles. Our "feats of legerdemain" were, I believe, nine in number, of which I only distinctly remember "one-ers", "two-ers", "three-ers", "four-squares", "magic", and "fly-catchers".'

THE BASIC GAME

The basic game has been the same ever since Pollux's day. The standard routine of 'onesy, twosey', up to 'fives' is performed before any variations are tried; and in places where the tradition is weak, this routine constitutes the whole of the game. The method is much the same all over the country, so only two accounts are offered. The following is from an expert who perfected her game at Bicester Grammar School about 1950:

Start. Every game starts with throwing the stones up and catching them upon the back of your hand. The secret is to hold your hand low to the ground, gather the five pebbles close together in your palm, and throw them up only a little way, just enough to turn your hand over. Your hand should not just be flat, but with the fingers pushed back as far as possible, so that it is slightly concave. In this way you are certain to catch a good many of the stones. The stones that are upon the back of your hand must be thrown up again and caught in the palm. If you fail to catch any of them you are out. You may, however, drop off any stones you think are difficult to catch before you throw them up. From the stones in your palm choose one to be the 'jack'.[12] This is the stone that you throw up in the air, and it must not be dropped.

Onesy. The stones are thrown out on the ground, the jack is thrown up, one of the stones is picked up, and the jack caught in the same hand. The rest of the stones are picked up singly, in the same way, keeping all the stones in the catching hand.

Twosy, threesy, foursy. The stones are picked up in pairs, threes, etc. You may knock them into position by throwing up the jack, moving a stone, and catching the jack again. In 'foursy', if you have caught three stones and only have two on the ground, you need only pick up those two together.

Fivesy. When you have caught as many stones as you can on the back of your hand, your opponent places the remainder on, and then you must throw them up and catch them all. A friend will put them close together, an enemy make it as difficult as possible (the extreme back of the hand by the wrist is a very nasty place).

And this, from Fulham, London, *c*.1960:

One of us throws the fivestones up and sees how many they can get on the back of their hand, then the next one does it, and the next. The one who catches the most on the back of their hand starts the game. You throw out the stones on the ground, quite close together, choose one stone to throw up, throw it up and pick up one stone from the ground. Go on till you have picked up each of them one at a time. Throw the stones out again, throw up the one and pick the stones up in pairs—that is two pairs. Throw the stones out, throw up the one stone, pick up one stone, then three together. Now you are on fours, throw up the one and pick up all the other four together. When you get to fives you throw them all up in the air and catch them or you are out.

The basic routine is often repeated in the form of 'double bounce'; 'it is exactly the same but you bounce the ball twice instead of once'.

VARIATIONS

Ever since a girl in Banbury assured us that 'this game is played to different rules by nearly every child', we realized that it would be impos-

[12] In the Birmingham game of 'Jack Stones', 1952, the marked stone used for throwing-up was also called the 'jack'. Other special names for the stone thrown up were 'king' (Pontefract), 'bonker' (Enfield), and 'master' (Ipswich).

sible to give the order of play for each locality. Even within one school,
the players may know different variations and play them in a different
order. It must be understood that each variation is embedded in a
throw-up-and-catch ritual: 'throw up the stones and catch one on the
back of your hand, throw it up and catch it, throw it up again and
while it is in the air push one of the stones under the bridge' (or whatever
movement is required by the variation). A player says where she has got
to in the long sequence of variations with the phrase, 'I'm on ———'; thus
if play is interrupted by the bell, when the girls settle down to continue
the game in the next playtime one will say, 'What are you on?' and the
other reply 'I'm on "Trams"', or whatever variation she had been performing. The following were the main variations in mid-twentieth-
century Britain:

Bridge, or Dogs in the Kennel, or Marble Arch. 'You put your thumb
and first finger of the left hand into the shape of a bridge, and knock your
stones under the bridge one at a time.'

Bunny in the Hole, Rabbits in the Hole, Bunny Hole. 'Put the first finger
and thumb of your left hand on the floor, slightly apart, to make the
burrow. Push the stones into the burrow one at a time.' When the finger
and thumb touch, and 'the hole is above the floor and horizontal', the
name is 'Dustbins', 'Netball', 'Manhole', or 'Buckets'.

Butterfly, Flycatcher, Swipes, Grabs. This is a different way of catching. The basic game can be repeated, but instead of catching the thrown-
up stone in the usual way you pounce on it from above, 'with a downward
swoop of the hand'.

Cavesie. 'A cave shape is made on the ground with the left hand and
the stones are pushed in one by one.'

Crab or Crabsie. 'When you have caught the stones on the back of
your hand then you pick up the ones you have dropped between your
fingers, keeping the ones you have caught on the back of your hand. (It
is easier to drop them all but one, even though it means more to pick up.)
When you have got a stone between each of your fingers, throw the one
on the back of your hand into the air, and catch it, and work the stones
between your fingers into the palm of your hand without dropping any.
Double Crabsie is like *Crabsie*, but two stones between each finger. *Lobster*
is the same, but you gather the stones under the palm of your hand
instead of between your fingers.'

Ducks in the Pond, or Hearts. 'Two people are needed, one to play and
one to make the pond by putting her hands on the ground in a pond
shape. Flick the stones into the pond one by one.'

Eggs in a Basket. 'Your left hand is the basket and you put the stones in one by one.'

Flydob Scatsie, Shiggle, No Disturbs. 'Flydob Scatsie is very hard. You get the fivestones, throw them up and catch them on the back of your hand. If you catch them you throw them on the floor, throw up the master, get a snob and catch the master by hitting it as you catch it. If you disturb another one you are out' (Ipswich, 1960).

Horses in the Stable, Pigeon Holes, Under the Arches. 'Finger tips of left hand on the ground, with arches between them and palm raised. Stones must be knocked through each arch.'

House that Jack Built, or other nursery rhyme, or phrase. An old variation. The four stones are arranged in a square, or occasionally a row, and, tapping them with the fifth, a rhyme or phrase is used to count round. The old rule was that the stone the rhyme finished on was taken away, and the player was out if he finished at the place where a stone had been, instead of on a stone itself. The modern rule is that when the last word is reached, that stone has to be picked up while the fifth stone is in the air—but in some places the old rule still obtains, or 'you go on round till all the stones are picked up'. A common phrase was 'My little pussy-cat likes sweet milk' (Perth, *c.*1910) or 'My little pussy-cat sells pipe-clay, My little pussy-cat gets no pay' (Scottish Highlands, *c.*1920, *Farmer's Weekly*, November 1952) which, remarkably, was collected from a 10-year-old boy in West Ham, 1960, as 'My cat likes hot milk'.

Ladies, Little Tich. 'After you have finished the first part of the game you toss again and toss again to the front part of your hand and keep the jacks in the hand that throws the ball up.'

Nelson's Column, Lamp-post, Peaser, Big Tich, Big Ben. Build a column, then pick or knock off the stones one by one. 'Nelson's Bed' is the same, but there are three stones on the ground for the bed, with one sticking up at the end for the bed-end. 'Lady in the Bath' is the same. *London Street Games* (1916), has 'Lamp-post', p. 128.

Over the Garden Wall. 'The wall is your left hand, held sideways on the ground, fingers together, thumb at the top, then you chuck each of the jacks over it to the other side.' In *London Street Games* (1916), as 'Over the wall, one, two, three', p. 128.

Snakes. Four stones are spread out in line with a short space between. The jack is placed on the back of the hand. One finger of this hand traces a snake in and out of the stones, along the length and back. Then the jack is thrown up and one of the stones on the floor picked up. Another snake is made along the remaining stones and so on till all are picked up.

Sweep the Floor. A two-handed game. Four stones are placed at corners of a square, representing chairs. 'You say, "Sweep the floor, Lift the chair, Sweep below it, Put it there," and you say the rhyme for each stone. You are throwing up the fifth stone of course, in between actions.' Each chair is addressed in turn. 'Your four actions are: make general sweeping motion; pick up the first "chair"; sweep the place where it was; replace it.' A long-time favourite, since about 1890. The 1960s schoolchildren were vague in their instructions, and the full glory of the original game can only be appreciated from the long description in J. T. R. Ritchie's *Golden City* (Edinburgh), pp. 71–2.

Sweeps, Magic, Black Magic. 'Toss up five stones. If you catch all of them on the back of your hand and toss them and catch them in your palm, you win the game.'

Tapsie. 'That means that while picking up a jack you must tap it on the table or on the ground before you catch the ball.'

Trams, Long Swipe, Short Swipe, Big Medal and Little Medal. Set the four stones out in a line, with about 3-inch gaps between them. Throw the bonker up, and scoop all the stones up together. 'Long Swipe' is the same as 'Trams'; in 'Short Swipe', arrange the five stones along the back of the right hand, one finger-width apart, toss and catch, then arrange them two fingers-width apart, then three fingers-width, and continue until the stones extend up the forearm. In 'Big Medal' two stones are placed a hand-span apart on the ground, picked up and caught; in 'Little Medal', the stones are only the width of a closed hand apart.

Then, of course, there are the various calls for different contingencies, such as 'Split Jacks', when 'if the ball bounces on a jack and is impossible to catch you may have it again', and 'No drops', when 'you cannot drop any stone off the back of your hand', and 'Cushions', which means 'if the ball bounces on your clothes or your leg or anything else and goes cockeyed you can have it again', and 'Sweeps', which means 'if the stones are impossibly scattered you can throw one up and get the others together again', and 'Double Jacks', when 'if one comes down on top of another and you shout Double Jacks you can knock it off, but if the others see it and shout first you are out', and some others, too—but that is probably enough about fivestones.

4

Throwing and Catching

GAMES of throwing and catching are as old as recorded history, and probably older. Already, before 2000 BC, symbolic ball games were depicted on the walls of Egyptian tombs,[1] and we know from Homer that ball games were played in early Greece, for it was a lost ball that roused the shipwrecked Odysseus from his sleep and led to his discovery by Nausicaa.[2] It was as natural to the ancient Greeks to stand in a circle on the sea-shore tossing a ball to each other, as it is to families today taking their summer holiday at Bournemouth. The Romans, it seems, were more strenuous. The fast and skilful game of 'Trigon', in vogue during the Empire, was played by three players standing at the points of a triangle drawn on the ground, with servants acting as ball-boys. In the statement that the players 'probably sent the balls without warning, trying to surprise each other', we recognize an element of the latter-day game of 'Broken Bottles'.[3]

BROKEN BOTTLES

'Broken Bottles' is a catching game, with 'forfeits' for missed catches. It can be played by two people opposite each other, or by three people in a triangle, or by four people in a square; but the usual way is by six people in a circle. The ball is thrown to any other player, suddenly, 'and if they are not properly awake they don't catch it and they're called "Butter-fingers"', but if they catch it, they throw it to someone else. In a more sedate form of the game, the ball is thrown from one to another round the circle.

According to our older correspondents (*c*.1903 until the 1940s) names then varied as much as they did during the 1950s–1970s. 'Broken Bottles' is possibly the oldest and most popular name; but 'Crazed [or Chipped], Cracked, and Broken', 'Sick, Ill, Dying, Dead', 'Wounded

---

[1] See e.g. Egyptian Exploration Society, *Survey of Egypt*, Memoir 2 (1893), 16.
[2] *Odyssey*, book vi.
[3] See L. Becq de Fouquières, *Jeux des anciens* (1869), 179, and Daremberg and Saglio, 'Trigon', under 'Pila'.

Soldier' (later 'Nelson'), and 'Donkey' have also had long innings. The order of the forfeits varies, too, but the basic idea is that if a player drops the ball, he is disabled in some way. He may have to hold one hand behind his back, and then, after dropping the ball for the second time, stand on one leg, then go down on one knee, then two knees, and finally go out of the game. ('In Nelson you have to go on one knee and then two knees, one eye, one hand, then no eyes.') However, as one mature player put it, 'If you catch a ball after you have missed one it cancels out the miss and you go back to the position you were previously in.' 'Donkey' does not involve any loss of limb. 'You stand or sit in a circle and throw the ball to each person in succession and if any one drops it they are known as "D" and the first person to get the word "Donkey" completed is called the Donkey.' In Sale, Manchester, 1960, the 'donkey' had to give everybody else a ride on his back. In Galashiels, 1975, the game of 'Bomb' was simpler and more dramatic: 'Six or seven people stand in a circle. If you drop the ball you have to explode.'[4]

PIG IN THE MIDDLE

'Pig in the Middle' is a game so elementary and universal that it has become a metaphor. Three players stand in a line. The outside two throw the ball to each other; the one in the middle tries to intercept the ball, and then the last thrower is pig. Who can forget the ignominy of being 'pig', leaping in vain as the ball flies over one's head, while the spectators jeer? 'Piggy in the Middle' was the more common form of the name in the mid-century. Variations are: 'Fool in the Middle' (Bacup, 1960), 'Fool in the Midst' (Whalsay, Shetland, 1960), 'Mug in t' Middle' (Rossendale and Accrington, Lancashire, 1953 and 1960). A version with several people in the middle was called 'Hot Rice' in Dovenby, Cumberland, 1952.

The name may have been adopted from a captive-in-a-ring game, 'Pig in the middle and can't get out' (*Folklore* (1887), 50). The game itself was perhaps thought too simple to be included in any of the usual sources (except *London Street Games*, p. 3, as 'Catch' or 'Teaser'). The only reassurance of its antiquity comes from Rabelais; scholars have equated it with Gargantua's 'J'en suis' (see M. Psichari, *Revue des études rabelaisiennes*, vi. 321).

---

[4] Would it be far-fetched to see the ancestor of 'Broken Bottles' in the French 'casse-pot', defined in Cotgrave's *Dictionarie* (1611), as a game in which 'one tosseth an earthen pot at another, who, if he catcheth it not, falling, it breaks, and he forfeiteth'?

AIMING

The emphasis in the following games is more on aiming. Boys take pride in having a good aim, and scorn is poured on the boy who cannot throw straight. A Bishop Auckland boy wrote, 'When someone throws a brick at a person and misses the other shouts "You couldn't hit a pig in a passage" or "Good shot for a bad miss".' Others say, 'Good shot for a donkey', 'Good shot for a blind man', 'Good shot but bad aim', 'Three times for a Welshman' (proverbial in Ruthin), 'All monkeys miss', 'Yah, missed, go home to your mummy', 'You couldn't hit an elephant off a rice pudding', 'You couldn't hit a barn door with a bulldozer', 'You couldn't hit a barn door if you were inside', 'Better luck next time', 'First time he hit it he missed it, second time he hit it in the same place'.

It seems to be instinctive for boys to throw missiles, and they boast that they are better at throwing than the girls. 'We skittle 'em,' said one boy, 'in other words we beat 'em easy.' (It is apparently a scientific fact that males have better visuospatial abilities than females.) A multitude of aiming games have been recorded, which are no less satisfying for being simple and often improvised on the spur of the moment. Sixteenth-century children, as Thomas More knew, made 'castelles of tile shards, & then make them their pastime in the throwyng down agayne'.[5] Three centuries later the pastime of knocking stones off a pile was called 'Cogs', 'Cogs-off', or 'Cog-stone' (see *EDD*). In East Finchley, in 1962, a 9-year-old wrote that his 'very best game outside is called Knock-it-down. You stand five tins on top of each other and in your hand you have two tennis balls. You take aim carefully and fire one tennis ball. And it is smashing fun.'

Knocking down or off is different from throwing at to dislodge or throwing at to break. 'At home', said a girl in Tunstall, Staffordshire, 'we have a very big beech tree and to pass our time away we throw an old mop into the tree and we throw stones and try and get it down.' 'A popular game in Presteigne', said a 12-year-old boy, 'is to take about five bottles down to the river and throw them in about 30 yards up river from where your pals are standing. When the bottles come down it is the first one to break a bottle.' Of course they try to break bottles lined up against a wall, too.

The ball as extended body-power can be used in other ways. It can move small objects out of a circle or square drawn on the ground, though

---

[5] *The answer to the first part of the poysoned booke whych a nameles heretike hath named the Supper of the Lord, the fourth booke, the xxii chapter*, in *Workes*, ed. Wyllyam Rastell (1557).

in the first sighting of this game, in *Chums*, 31 January, 1894, p. 375, the street boys were 'jinking' coins out of a circle with spinning tops. In the early years of the twentieth century pennies or ha'pennies, or small prizes such as buttons, marbles, or cigarette cards, were knocked out of a circle or square drawn about a foot away from a wall; a ball was bounced so that it hit the prize, bounced onto the wall, and was caught on the rebound. By the 1920s (juvenile gambling then being frowned upon) it had become a question of dislodging stones, and the game was called 'Three Stones' or 'Five Stones'. This game was popular in a number of varieties in the 1950s–1970s, usually under the name of 'Five Stones'. In Croydon 'Master out of the Ring' was played in a circle on open ground, with five 'quite large' stones placed round the edge and a smaller stone in the middle. The winner managed to move all the stones out of the ring without moving the small one. Corks or buttons or metal bottle-tops, or, not infrequently, pennies, could be used as targets. In Salford a variation was to throw the ball backwards over one's head, hoping to shift a penny out of a circle: 'You can play it anywhere, but mostly in the house because boys could pick up the penny when you weren't looking.'

LEDGER OR KERBIE

Accuracy in aiming is a necessity for this game, which involves a window ledge, or the ledge that is often found near the foot of school walls, or a gutter at the side of a street—in other words, an angle into which a ball can be thrown so as to bounce back. Players are not required to catch the ball on the rebound, just to make it bounce back. Under the name of 'Ledger', it was 'a very common game around Wolstanton schools' in 1961; a successful throw scored a point. 'Ledgy', at Camberwell in the same year, was rather more complicated. If a player hit the ledge she took a step back. (If she failed it was the next person's turn.) When she reached an agreed distance, she started coming forward again, and 'when you reach the ledge again, you do it all over again, only in dumb, then steel, dumb steel, one leg, the other leg, one hand, the other hand, claps, drop, twists'. 'Gutter Ball', a game 'which could go on for ever if you didn't want it to stop', is played with two people standing on pavements opposite each other, 'and you have to have a large ball and you throw it and try to hit the other kerb and if it bounces back off the kerb back over to the person who threw it, well that is a point to them'; and, in some places, when it bounces back, 'if you catch it without letting it touch the ground then you get two points'. The scoring varies, but typically 'When someone has got ten points it is half time, and you change sides, and twenty is the winner.' The game was a craze in Netherton, Worcester-shire, in 1969; and in the summer of 1971 Plymouth children were 'playing it all over the town'. 'Kerbie' was the name in Scotland, where, in 1974–7, 8- to 10-year-old boys and girls in a number of places de-scribed it as 'a game we play often', and none of them mentioned any danger or interruption from traffic.

SCRUBBY

This is in the nature of a 'garden escape'. It started as an organized game,[6] and was naturalized into streets and playgrounds. It is played on a chalked court—a block of four squares—and is reminiscent of tennis. The following is an amalgamated description from 10-year-old boys and girls at Kilburn, London NW6, 1961:

Scrubby has two other names, and they are Four Square and Fish and Chips. You have four squares and you have four people, one standing in each. You have two people over, in case anyone is out. You draw a ring in the centre and write OUT in it, and if anyone lets the ball bounce in the ring he is out. In square 1 you write PORK, in square 2 PIE, in square 3 FISH, in square 4 CHIPS, and in square 4 you write CHAMP as well. The person in square 4 is the champ. The champ bounces the ball and hits it to someone with both hands, and that person must hit it to someone else. You have to pat the ball with two hands, not one. If the ball bounces twice in anyone's square they are out and one of the spare people goes in and everybody moves round one square except the champ who stays in square 4. If the champ is out then the person in square 3 is the champ.

Of course, the method of play varies. 'Chinese War', in Wolstanton, Staffordshire, 1960, had two sides, of two players each; 'You go in one square and your other man on your side goes in the diagonal square, and you have to get the ball over to your man.' A Salford boy, cock-a-hoop because he knew a game the others did not know, said:

This American that was on the 'olidays where I was, in Norfolk, he says to 'ave a game of Bounce, and I says, 'What's that?' and he shows me. There's four squares in a row. You stand in one square and another boy stands in another square that's not next to you but next door but one, and he has to catch it, then he does the same, and if you don't catch it, or if you miss the square, you change round and the other person gets a point.

The 'official' rules for 'Boxball', in the New York street-game Olympics of 1972, in which genuine street games were played by adults, show it to be the same game that arrived (via Norfolk) in Salford (*New York Times*,

---

[6] A version appears, for instance, in J. B. Pick's *Phoenix Dictionary of Games* (1952), which is for teachers and other adults who need to organize games.

29 August 1972, p. 35). In an account sent from a Cincinnati school, in 1973, 'Four Square' is like 'Scrubby', and the size of the courts varies— 'for small children we have smaller courts'.

The game must be well known in many countries, judging from such news of it as has drifted our way. A New Zealander on a train (1973) told us that she tried to teach the New Zealand game of 'Square Ball' to children in Fulham, and they knew the game already, as 'Champ'. A man from Mölndal, near Göteborg, told us (1995) the game was popular in Sweden and was called—in English—'King'. His 10-year-old son played it with friends on summer evenings.

WALLIE

One person has a ball and tries to hit the others, who run across in front of a wall of the school building, more or less one at a time. The boys hurl themselves across, as if under fire, sometimes turning head over heels on the wet asphalt, sometimes grazing their knees. When one is hit, he changes place with the thrower. The girls scarcely ever play this game. 'We would get our socks dirty,' they say.

In the most popular version, called 'Jumping Beans' or 'Dancing Dollies' or 'Ball He' or 'Slam' or 'Hits' or 'Pelt Ball' or 'Tin Army' or 'Wooden Soldiers' or 'Dodgems' or 'Dodge Ball', the players must dodge about in front of the wall, not run from side to side ('and if the weather is too hot they just stand still'). 'You are supposed to hit people below the knee,' they explained: 'you are not supposed to hit the people in the face.' However, there is a pleasurable sense of danger, and a 9-year-old in Guernsey said, with relish, 'This game is called Wall of Death because when anyone goes against the wall they nearly have their heads blown off.'

ECHO

This great old game was still being played in the north country in the 1950s and 1960s, as 'Echo', or 'Echo Pusho', or 'Willie Wit', or 'Ball over the Wall' in Penrith; as 'Pie Bally' in Dovenby (though 'Pie Ball' is normally a name for 'Rounders'); as 'Throwing over the Water Stone' in Barrow-in-Furness.

It is a game of throwing a ball over a roof top—'a rather long and not too high a building, say an ordinary farm byre'. Teams are chosen and stand each side of the building. One person throws the ball over the building, calling out a warning cry, and if someone catches it, he runs round to the other side of the building and throws it at one of his opponents. If that person is hit he has to return with the thrower as his prisoner; the object is, of course, to go on until all one team are caught. Naturally the game has spawned variations, the chief being that all the catching team run round to the other side of the wall (in Penrith they say 'Standum no flinchums!' before the catcher tries to hit his victim); in Barrow-in-Furness, where the game is played 'over a large square boulder in the middle of a common', 'when the team who have thrown the ball see the others coming they run to where the others were but while they are running the person who caught the ball has to hit as many of them as possible.'

The names of the game reflect the warning cry, for instance in Cross Fell, c.1885, 'Ickly Ackly Aiko' (*JLDS* 7 (1945), 7); and in Cumberland, 'Conny-Co', *EDD*. Across the Atlantic the common name was, and is, 'Ante-over' (Laura Ingalls Wilder remembered this from her schooldays in South Dakota in 1880, *Little Town on the Prairie*, ch. 14), or 'Anti Anti I Over' in mid-twentieth-century Canada (children in Edmonton in 1961 talked of playing the game 'over the house or a low garage' and 'if the ball does not go over the person who threw it shouts "Pig's tail"').

The appeal of this game lies in the excitement of not knowing exactly where the ball will appear, nor who has thrown it—an appeal felt just as strongly five centuries ago, it seems, for it is a sport specifically forbidden in 'Symon's Lesson of Wysedom for all Maner Chyldryn': 'Chyld, cast no stonys ouer men hows'.[7]

BAD EGGS

Hippolyt Guarinoni, in *Greueln der Verwustung* [*The Horrors of the Devastation of the Human Race*] (1610), p. 1212, describes a ball game which 'holds good for the Jesuit College in Prague about 1580':

Small holes of a size for the ball to go into are made in the ground, to the same number as there are players and to each player is assigned a hole. When the ball, bowled pretty well straight at all the holes, falls into one of them, the player to whom that ball belongs must be smartly on the alert, to seize the ball and throw it into the cluster of players scattering before it. But because this player is stooping towards the ball the group have a chance to make off scot free across the field or playground, one here, one there. Nevertheless they all duck because no one is sure of being safe from the blow.

'Nine Holes', 'Ball Hat', or 'Egg in Cap' was one of the major European games, and was doubtless played in many other parts of the world

---

[7] Bodleian MS 832. fo. 174, line 43.

as well. In later days it was played with a row of boys' caps set against a wall; and then, possibly because boys no longer wore caps, died out in favour of the similar game of 'Bad Eggs' (see below). The only mid-twentieth-century recording of a 'cap' variety was 'Cap and Ball' from a girl in St Helier, Jersey, in 1963.[8]

The version known as 'Bad Eggs' is more convenient, needing neither holes nor caps: a ball is thrown up, a name called, and the named player must catch it. Herrick wrote of it, some time between 1620 and 1641: 'I call and I call: who doe ye call? The Maids to catch this Cowslip-ball.' Randle Holme, *Academie of Armory* (1688), listed it amongst country sports: 'Ball play, or call at a call' (in 1922 a Somerset boy sent a description of 'Call Ball' to *Pulman's Weekly News*, hoping to win a prize). The sport may have been as popular in the Regency period as it was in the 1950s–1970s, and was known under as many names. In *Schoolboys' Diversions* (1820), it was called 'Days of the Week', and it was pointed out that 'the number of players should not exceed seven, each of whom is to take the name of a day':

Sunday must take a ball, and making it bound high against a wall, must say, 'Here goes up to Tuesday', (or any other whom he may choose), and then all the rest should run away. The boy who represents Tuesday, must catch the ball before it falls, and if he can throw it so as to hit either [any?] of the others, it reckons one [penalty point] to that person; but if it touch the ground before it is caught, or miss striking the boys, it reckons one to him. When any of the boys reckon seven, the game is finished: and then Sunday takes the ball, and Monday, putting his hand flat against the wall, Sunday must throw the ball against it as many times as Monday has owing to him. When Monday has been paid, he must pay Tuesday, and Tuesday must pay Wednesday, and so on. The distance Sunday should stand from Monday, while throwing the ball, should be about six yards. And each boy may choose whether he will have the ball thrown at his hands or his legs.

The same game is given in *The Boy's Own Book* (1828), under the name of 'Catch Ball'.

John Mactaggart included 'Burly Whush' in his *Scottish Gallovidian Encyclopedia* (1824):

---

[8] See J. C. F. Gutsmuths, *Spiele für die Jugend* (1796; 1802 edn., pp. 116–20); E. L. Rochholz, *Alemannisches Kinderlied und Kinderspiel aus der Schweiz* (1857), 387; T. and F. Vernaleken, *Spiele und Reime der Kinder in Österreich* (1873); F. M. Böhme, *Deutsches Kinderlied und Kinderspiel* (1897), 609–10, 611 ff.; A. de Cock and I. Teirlinck, *Kinderspel und Kinderlust in Zuid-Nederland* (1902–8), iii. 116–26; *N & Q* (1907), 306, *re* 1806 engraving of boys playing the game in Rome; M. M. Lumbroso, *Giochi* (1967), 356, contemporary schoolchildren in Mantua and Sardinia; Newell (1883), 183; *Jeux des jeunes garçons*, c.1810; H. d'Allemagne, *Sports et Jeux* (1904), 153; and Çanakkale, Turkey, 1970, where stones called 'eggs' are placed in the holes of the losers, and the game is hence called 'Mother Hen, where is her chick?'

'Bob-in-the-Cap', a name for 'Bad Eggs' when the ball was thrown into a line of caps, in *Phil May's Gutter-Snipes*, 1896.

The ball is thrown up by one of the players on a house or wall, who cries on the instant it is thrown to another to catch or *kep* it before it falls to the ground; they all run off but this one to a little distance, and if he fails in *kepping* it, he bawls out *burly whush*; then the party be arrested in their flight, and must run away no farther. He singles out one of them then, and throws the ball at him, which often is directed so fair as to strike; then this one at which the ball has been thrown is he who gives *burly whush* with the ball to any he chooses. If the corner of a house be at hand, as is mostly the case, and any of the players escape behind it, they must still shew one of their hands past its edge to the *burly whush man*, who sometimes hits it such a *whack* with the ball, as leaves it *dirling* for an hour afterwards.

The appeal of this game largely lies in the savagery at the end, and this is so even in the present day, when games have become less robust. A cluster of 8-year-olds in Salford, in 1970, were telling me about 'Three Bad Eggs', played by throwing a ball against the end wall of a house. One of them began, 'Say we were all playing it and I was throwing up the ball, and say they were "One", "Two", "Three", "Four", and "Five", and I said "Four" and it was Jacqueline, well Jacqueline would run and get the ball, and while she was running for the ball all the others would run away as far as they could, and then she'd say "Stop" and take so many steps, and throw the ball, and if it doesn't hit the person she's aiming for they all run to the wall . . .'. Another, unable to contain herself for excitement, broke in, 'And they all put their hands up on the wall, and Jacqueline has to run through and you got to smack her bum.' 'What do you like specially about that game?' I asked, and there was a chorus of, 'Miss, where you smack their bums!' This punishment is, of course, the old one of 'running the gauntlet'; it is called 'through the mill' in Cardiff, and Sleaford, Lincolnshire, and 'through the rag-tag' in Forfar. In 'under the arch' in Peckham, the loser gets 'pinched, punched, and kicked'; but in West Ham he is allowed to choose between 'hit, pinch, punch or kick'. In Denholme, near Bradford, in 1990, 'anyone who is "three bad eggs" must go through what is called a slapping machine'.

If a player is hit (or, in the softer climate of the 1990s, has the ball rolled at his legs or between his legs) he has a 'bad egg', or has a point against him. (The usual modern name, 'Bad Eggs', is descended from the 'Egg in Cap' of previous generations, when stones, called eggs, were put into the players' caps as penalties.[9]) According to the local rules he is

---

[9] 'Egg-in-Hat' appears in *School Boys' Diversions* (1820), and was much played, as 'Egg in Cap' or 'Ball Hat', up to the 1930s. A correspondent knew a variety called 'Stink-Egg' or 'Addled-Egg' in Bembridge, Isle of Wight, *c*.1905; players took turns at rolling the ball at a row of holes called 'nests', each owned by a different boy. The owner of the hole the ball rolled into had to snatch it out and throw it at the other boys, who were running away. If he was successful, the one who was hit had a pebble put in his hole; if not, the thrower had to put a pebble in his own hole. Anyone accumulating five pebbles was 'stinkered' by the others, who held their noses and cried 'Stink-egg', and he was out of the game.

punished when he has three, or six, bad eggs; or in Wigan, where the game is called 'Sixpence', 'If someone is hit he is a penny, and you keep on till someone is sixpence, and he has to go under the pigsty six times and we all have to hit or thump him.'

A multiplicity of names always seems to have been characteristic of this game, and often there are several different names in one locality. Lady Gomme, in *Traditional Games* (1894–8), recorded the names 'Colley [Calley] Ball', from Hemsby, Norfolk, 'Hommer-the-let', from the Isle of Man, and 'Monday, Tuesday', from London. In the *Journal of the Lake District Society* (1951), local names recalled from the late nineteenth century are 'Conny Co', 'Hoosie', 'Hah-go', 'Hi-co', 'Nickelly nickelly hah-go', 'Ikkelly ikkelly hah-go' and 'I Call'.

The most common present-day name is 'Bad Egg(s)' (forty places, widespread over England and Wales). Variations are 'Three Bad Eggs' (Salford, Denholme near Bradford, Coseley in Staffordshire, and Welshpool), 'Rotten Eggs' (Alton, Basingstoke, and Lydney), 'Egg If You Budge' (Cleethorpes), 'Egg-o' (Scarborough, and Sleaford, Lincolnshire), 'Eggy' (Bu, Aberdeenshire), 'Egg Flip' (Banbury), and 'Eggity Move' (Market Rasen, Lincolnshire). Other names reflect the secret naming of the players, as 'Flowers' (Berry Hill, Gloucestershire, Welshpool, Forfar, and Penrith), 'Fruits' (Spennymoor), or 'Colours' (Flotta and Stenness, Orkney); 'Names', at Forfar Academy, meant that different categories could be chosen ('It could be cars, chocolates, washing powders, or perfumes'), and 'Airys', at Langholm, meant that players were different makes of aeroplane. The old name of 'Monday, Tuesday' was found in Croydon, and 'Days of the Week' in Penkhull, Stoke-on-Trent. Some names are taken from words called out in the game: 'Halt' (Edinburgh), 'One and a Halt' (Stornoway, Isle of Lewis), 'Stand-i-o' (Lincoln), 'Stand-on' (Spennymoor), 'Standers' (Berry Hill, Gloucestershire), 'Stop' (Thurso, Caithness, and Stromness, Orkney), 'Stork' [?Stop] (Aberdeen), 'Flinch' (West Bromwich, and Welshpool— 'You mustn't flinch when someone is trying to hit you'), 'Flinchers' (Liverpool), 'Ditto' (Radcliffe, Lancashire—'What you shout if you catch the ball'), 'Flinnies' (Barrow-in-Furness). Less explicable are 'Aggymoo' (Helston), 'EIO' (Aberdeen), 'Gobbo' (Banbury), 'Piggie Bliggie' (Blackburn: 'One person throws a ball in the air and shouts "Here I come, here I crawl into the hands of——"and they call a name'), and 'Shorey' (Buckie, Aberdeenshire). 'Throw up and Call' was the name in Grimsby and Cleethorpes. In Garndiffaith and Glastonbury the game was 'Cabbage': 'You all have numbers and as soon as the person with that number has caught the ball he shouts "Stop" and then looks for the nearest person and takes as big strides as possible for every letter of the word "Cabbage" and then tries to hit him.' The same idea was expressed in the more mundane form of 'Hop, Skip, Stride, Jump' in Accrington.

The permutations and combinations of children's games are endless. In some places, especially in boarding schools, 'if you hit someone it's a letter against them', and when a word such as 'Spuds' or 'Donkey' has been spelt, 'you give them a biff on the bottom'. In other places the throwing-up and calling has disappeared and the ball is dapped ten times on the ground before the thrower takes aim. Leslie Paul used to play a form of 'Bad Eggs', called 'Kingy', in a London suburb before the First World War (*The Boy from Kitchener Street*, pp. 42–3). When a boy was hit he joined the boy who had hit him, retrieved the ball, and hit another player who then joined them. It seems that this game, spawned from 'Bad Eggs', is the missing ancestor of the immensely popular game of 'Kingy' (Opie, *Games*, pp. 95–9).

DUCKSTONE AND CANNON

'Duckstone', 'Duck', 'Ducker', 'Ducky', 'Duck on a Rock', or 'Ducks and Drakes' is a game that combines throwing, tigging, and a considerable element of danger. 'Ducking stone' was a joy John Clare remembered from his boyhood in Helpstone, Northamptonshire, *c.*1800 ('Remembrances', in *Poems*, ed. J. W. Tibble). *Blackwood's Magazine*, August 1821, p. 32, mentioned that 'The duck is a small stone placed on a larger, and attempted to be hit off by the players.' *The Boy's Own Book* (4th edn., 1829), 36, has detailed directions for playing 'Duck', and naturally the later boys' games books followed suit. The attraction of the game, however, is shown in the flood of reminiscences that poured into newspapers, local journals, and our own files. It was clearly a memorable ingredient of many lads' lives in the nineteenth century and up until the late 1930s, and filled whole summer evenings with contest and excitement not just in the rugged fell country of Lancashire and Yorkshire, but in the towns of the Black Country, the quiet villages of Wiltshire, and the streets of London—a nationwide sport, one might say, for the tougher kind of teenager. An account from a 16-year-old who had played 'Ducky' at Cartmel Fell, in Lancashire, in 1933, gives the flavour of the game:

Two suitable stones are found, one stood flat down and one propped up behind. The next job is to find, for each person who is going to play, a suitable stone about the size of a grapefruit. Pick three yards from the Ducky Stone, and that's the boundary. The last to get to the ducky stone is the Ducky Man. He sits his ducky on the ducky stone, and the first player throws his stone and tries to knock it off. If he succeeds the other players throw their stones down a respectable distance from the boundary and the ducky man rushes and grabs his ducky and places it on the stone and runs to tig one of the players. Whichever he tigs is ducky man. Then all the players have to pick up their stones. You can't run unless you've picked your stone up, and he can't tig you unless his ducky is on the ducky stone and unless you have your stone in your hand. If you can get past the boundary with your stone in your hand you are clear. After tigging, the ducky man has to throw his stone down like the others, pick it up again and then run to the boundary before he is tigged back. You can also pick up other people's stones and then there is a great mix-up.

It may be that *The Boy's Own Book* can clarify the central part:

Each player, as soon as he has cast his duck, watches for an opportunity of carrying it back to the offing [boundary], so as to cast again. If the player who is duck, can touch him after he has taken up his pebble, and before he reaches the offing, provided his own pebble remain on the large stone, then the player so touched becomes duck.

There had, perhaps, been a growing realization that throwing stones is dangerous; and yet the principle of the game, a struggle between those intent on destruction and those intent on restoration, was too fundamental to abandon. Here is an interesting intermediate stage between 'Duckstone' and the widespread modern game of 'Cannon', from Maclagen's additions to the *Games of Argyleshire*, in *Folklore* (1906), 223–4:

Dulges. Played in the Orkneys . . . is played by two chosen sides. [One side, the 'ins'] takes possession of what is called the 'Hales', while the 'outs', at a distance of eighteen or twenty feet . . . make the 'Dulges', that is, a pile of their stones. They also appoint a keeper of the Dulges, whose duty it is to rebuild it when knocked down. The side in the Hales try in succession to knock the Dulges down by throwing at it, those missing it leaving their stones lying. When it is knocked down, all those who have thrown rush to pick up their stones and regain the Hales, before the keeper of the Dulges can rebuild it and any of the 'outs' tig them after this has been done. Those tigged stand aside till this has happened to all of one party.

In the Isle of Dogs, East London, c.1905, this team game was called 'Piles', and a ball was thrown to knock down a pile of five stones; in Bethnal Green, also in the East End, c.1914, a ball and three flat stones were used, and the name was 'Tiles' (a name still used in Liverpool in 1960).

In the 1950s–1970s stones were still sometimes used ('In the game of Weasel you put five stones on top of one another, against a wall', West Ham, 1960); but much more often three or four sticks were set up against a wall, 'like a cricket wicket', or were balanced on a tin which was set in a circle, and a rubber ball or tennis ball was thrown to knock them down. Firewood was often specified, or lollipop sticks. The game is recognizably the same as in Orkney (1906), but the use of a ball means that members of the throwing team can be caught out, and the second part of the game becomes very like 'Kingy'. A Pontefract boy describes 'Sticks':

You have an 'in' team and an 'out' team. The 'in' team take turns throwing at the sticks to knock them down. The 'out' (or fielding) team spread themselves out and try to catch the ball when it rebounds off the wall. If the ball is caught the player who threw the ball is out till all his team is out or they run off. If all the 'in' team are caught out before the sticks are knocked down then the fielding team is 'in'. If the sticks are knocked down and the ball is caught all the team is out. When the sticks are knocked down all the throwing team must run off, and the catching team must try to hit them with the ball and then they are out of the game. The throwers cannot run with the ball, they have to pass it to another player of their own team. Meanwhile the team which has run off (the 'in' team) are trying to get the sticks built up again. If they do, they go in again.

A 14-year-old from Spennymoor, County Durham, said, 1960: '"Cannon" is usually played in a street. It is unpopular with passing motorists or people who have gardens nearby.' Spennymoor provided the only account of 'Duckstone' sent in during our surveys—it was played in the proper old way ('Each player in turn tries to knock the small stone off the large flat stone by throwing his own small stone at it . . .'). By far the most common name was 'Cannon' (twenty-seven places, five of them districts of London: but widespread, and including Scotland). Other names were: 'Sticks' (London, Pontefract, Denholme); 'Three Stones' (Oxford, Ipswich); 'Tin Can Alley O' (Stoke-on-Trent, played with a pile of tin cans); 'Tin Can Topper' (Swansea, played with two crossed sticks on a tin can); 'Tins' (Peterborough); 'Weasel' or 'Whistle' (West Ham).

In the game of 'Seven Stones', played in Denholme, near Bradford, in 1990, the target was a pile of seven stones built inside a circle drawn on the ground. The opposing team stood on the far side of the stones, and three throws were allowed for each player; otherwise the rules were as in 'Cannon'. Knowing the perpetual circulation of games round the world, one wonders whether this version might have arrived with immigrants, for 'seven flat stones', 'three throws each player', and the same disposal of the teams is customary in present-day Amritsar, in the Punjab, under the name of 'Pitou'. The same rules distinguished the game of 'Seven Stones' played in Ethiopia in the 1930s (M. Griaule, *Jeux abyssins*, (1935), 82–4); and the familiar-seeming game being played in a dusty square in Rhodes in 1969, where a few girls were trying to rebuild a column of seven pieces of tile while being pelted by the boys' team was, naturally, called 'Seven Stones'.

DONKEY

The first person throws the ball against the wall and jumps over it on the rebound, shouting the letter 'D'. If she misses, she is 'D'. The next in line repeats this, and so on. The second failure to jump over the ball gives 'O'. In some places, when someone has failed six times and gets to 'Donkey' the rest make an arch and the donkey has to go 'through the mill' where they are whacked six times; in other places, the loser has to 'give everybody a donkey ride'.

The game is immensely popular (versions from forty-five places); and at Liss Junior School it seemed to go on continuously throughout the year (until at least 1984, when weekly observations ceased). Only occasionally does the name vary, such as when children have the bright idea of spelling 'Elephant' or 'Duck' instead of 'Donkey'; or when they borrow a skipping term and call it 'Jump the Moon', a propensity in Scotland (Cumnock, Langholm, Edinburgh, Tighnabruaich). In Fortrose, Ross-shire, it was 'Jump the Dyke'; in Welshpool, 'Wally, Wally, Bashit' ('As the ball came down and bounced on the ground you jumped over it, and as you ran to the back of the line you would cry "Wally, wally, bashit", and if you missed you were out, and the winner was the last one in and when she was given the bumps they cried, "Wally, wally, bashit, smashit, crashit."')

'Donkey' has no recorded history, but a number of correspondents recalled playing it in the 1920s, including one who remembered it being considered '"not quite nice", since enthusiastic players tucked their skirts into their knicker elastics and "showed their shapes" which, if there were boys in the school, was very dreadful'.

5

Gambling

Behold, how quick the itch of gaming advanceth upon youth! The boy which yesterday thought himself as rich as Croesus, with a pouch of marbles, to-day longs for something more like to money. Invention is set to work, and dumps are coined from lead and pewter,[1] and become the current money, which he lavishly hazards in a way, that tends to corrupt the morals of youth.

(*The Instructor and Guide for Little Masters* (1772))

Dice is another game usually played at street corners (when there are no stakes) but in someone's back yard if money is involved. The stakes are small, one penny entrance into the game and a halfpenny a round.

(Boy, 14, Spennymoor, County Durham, 1960)

No amusements are older, simpler, or more infectious, than the little games of chance in which clusters of children may be seen absorbed, on sunlit evenings, in the alleyways alike of London, New York, and Beijing. Similar groups, it seems, could have been observed in classical Athens and Rome. Plato describes a knot of boys in the undressing-room of a gymnasium, engaged in 'Odd or Even'; Horace looked upon 'Par impar' as one of the first games of childhood; and in a mime written for an Alexandrian audience about 270 BC, a mother drags in her son, complaining to the schoolmaster: 'His writing tablet lies neglected . . . But the knucklebones in his bag are a lot glossier than our oil-flask which we use every day.'[2]

'Gambling games', suggests a present-day youngster, 'are games for pleasure from which one hopes to make a profit.' The games that follow (usually played out of doors) are not all pure gambling games, since chance is not their only element, yet they are games which would be little played if it were not for the excitement of being able to win or lose something while playing them. The two major wealth-acquiring games

---

[1] Dumps were lead discs used instead of money in pitching games. Their manufacture occupied schoolboys from the late 18th c. until at least the beginning of the 20th.

[2] *Herodas, The Schoolmaster*, mime iii, cited by F. A. Wright, *Greek Social Life* (1925), 231–2. The knucklebones referred to here would have been used for dicing, rather than in the essentially feminine game of 'Pentalitha' ('Fivestones').

in which skill predominates over chance appear elsewhere in this book: marbles (Ch. 2) and cigarette cards (Ch. 7).

Boys are of course ready to gamble on anything: the runners on Sports Day, the length of the headmaster's oration on Speech Day (a popular sweepstake at one school), and the sequence of advertisements to appear in the next break on television. They bet on beetle races, spider races, and snail races. ('If the snail is reluctant to get going,' advises a 10-year-old, 'one must put a little water in front of him, then he'll go.') In Guernsey, boys were found to be playing the old game of 'Prick the Garter' or 'Fast and Loose', a fairground swindle familiar to Shake-speare, and one which, in the previous millennium, was minutely described by Pollux under the name *Himanteligmos*. We found, too, that the games 'Deuce', 'Brag', 'Pontoon', 'Poker', 'Rummy', 'Dice', 'Crown and Anchor' (with either spinner or dice), 'Put and Take', and 'Bingo-Roulette', were not unknown to British schoolchildren; but would emphasize that we had no evidence of a vast amount of gambling amongst children, nor of play for high stakes. Rather, it is commendable that, in quite affluent times, children should continue to be content to play for marbles, flick cards, sweets, nuts, buttons, badges, bottle-tops and comics, just as in previous centuries children played for the trifles that were closest at hand: pins, points,[3] shells, or cherry-stones.

BIRDS IN THE BUSH

In the words of a Birmingham 10-year-old: 'A child turns his or her backs on the other child, and that child gets some marbles and rattles them in his or her hand and shouts "How many birds in the bush?" and the first child turns round and shouts a number.' 'If he guesses rightly,' explains a Spennymoor boy, 'he receives all the marbles in the other's hand. If, however, he guesses wrongly, he has to pay the other the amount he is out. Some play that the guessing changes around in turns, others that the same person always guesses until he guesses correctly. In some games a limit is set at ten, in others twenty.' 'Supposing a player had seven,' instanced a Stornoway boy, 'and one boy guessed five and

---

[3] Points were tagged laces used to fasten clothes. After the laces wore out, the metal tags were used in games such as 'Spurn-point', 'Dust-point', and 'Blow-point', games whose rules have not survived. Panurge lost all the points of his breeches when playing at 'Primus et Secundus' and 'Peck Point' (Urquhart's translation of Rabelais (1653), ii, ch. 18, just as in the 19th c. boys in north-east Scotland gambled away their trouser buttons so that 'it was marvellous by what means our little breekies were suspended' (J. M. M'Bain, *Arbroath: Past and Present* (1888), 341). In the *History of Tom Thumb* (1630), Tom, playing with children in the street, gambles away his whole stock of 'counters, pins and points, and cherry stones' (Hazlitt, *Dictionary of Faiths and Folk-Lore*, 'Pins').

another sixteen, then the one who guessed five would give him two (5 + 2), and the one that said sixteen would give him nine (7 + 9). If you guess correctly then it is up to the other person who told you to guess to give you the number in his hand.' 'When we get to the interval at one o'clock,' stated a 12-year-old girl, writing in Gaelic from Kershader, Lewis, 1962, 'everyone tries to be first out on the road to get the best stones for playing "Twenty Stones" ("*Fichead Clach*").' They then ask the others how many stones they have in their hands, trying to make the hollow of their hands look big if they are holding only a few, or small if they are holding many. Thus children in the mid-twentieth century provide a scholium on the ancient authors.[4]

Names: 'Bird in the Bush' (Farnham, Surrey, *c*.1870, George Sturt, *Small Boy in the Sixties* (1927), 149); 'Birds in the Bush, how many?' (*Warwickshire Word-Book* (1896), and Derby, 1900); 'Buy, an' buy 'em!', ''Ow does ta sell 'em?', 'Penny bunches, a penny', 'Rattle 'em', played with cherry-stones, Ripon, *c*.1910 (J. Gott, *Bits and Blots* (1987)); 'Eggs in the Bush' (*The Boy's Own Book* (1855); the 1872 edition adds 'This amusing game may be played with nuts, buttons, etc.'); 'How many eggs in your basket?' (London, *c*.1900); 'Ship sail, sail fast. How many men on board?' (Cornwall, *Folklore*, 5 (1887), 59); 'Hairry my Bossie' (Keith, Banffshire, *Traditional Games*, i (1894); 'Ho-go' (*Berkshire Words* (1888)); 'Over and Under' (*Book of Games or Schoolboy's Manual* (*c*.1840), 34).

In the United States: 'Hull-Gull' (Asa Greene, *Dr Dodimus Duckworth* (1833), i. 30; and Newell (1883), 147–8); 'Hully-Gully' (*Midwest Folklore* (1951), 248); 'Jack in the Bush' (*North Carolina Folklore*, 1 (1952), 60).

In France: 'A l'ingent' (*Jeux de l'enfance* (1883), 174–5); 'Grelot, grelot, combien de sous dans mon sabot?' (current Albi, Tarn, 1955).

In Italy: 'Galota, galota' (see Newell (1883)); 'A Sivaleri' (Pitrè (1883), 72); 'Marcia quando?', played with buttons, Capri, *c*.1925. Compare the game 'Egg or wind?' (i.e. 'Something or nothing?'), played by the royal couple on the fourth day in *Il pentamerone* (1634). In one of the games in Ovid's poem on a nut tree, *Nux*, the player makes his opponent guess the exact number of nuts held in the fist (lines 79–80).

In Japan: 'Telaga-tari' (E. Hillyer Giglioli, *Viaggio intorno al globo* (1875), 177).

[4] Cf. Xenophon in his treatise *Hipparchicus*: 'As boys, when they play at "How Many?", hold out their hands in such a way that, having few, they pretend to have many, and having many, they make believe to have few.' Xenophon, *Minor Works* (1884 edn.), p. 318 and n.

PITCH AND TOSS

'As many people as you like can play this game as long as they have plenty of pennies.' Each person in turn throws his penny at a stick stuck in the ground, trying to get it to land as near the stick as possible; or, if the game is played in the street, they stand in a line in front of a wall, and each in turn throws a penny to the wall, 'remembering the penny they threw'. When everyone has thrown, they look to see whose penny is nearest the stick or the wall, and the person who has made the best throw gathers up the pennies, holds them in the palms of his hands, and tosses them in the air (or, what amounts to the same thing, shakes them and lets them drop), and the pennies which come down 'heads' are his to keep. The player whose throw was next best picks up the rest and tosses them in the same fashion, taking possession, in his turn, of the pennies which land 'heads' up. The third-best thrower then tries his luck with the remaining coins, and so on until there are no coins left. If there are only two players, they take turn about until the coins have been divided out. 'This is a very good game if you win,' remarked one informant. 'I often see the game played behind the shelter in the playground,' remarked another. In Walworth the game is known as 'Penny up the Wall'. In Dublin little huddles of newsboys may be seen engrossed in the game as soon as they have sold the majority of their evening papers. In England it is said that boys have sometimes gambled away their school-dinner money.

A game which was virtually indistinguishable from present-day 'Pitch and Toss' was 'Pitch and Hussle', listed by Randle Holme in 1688 as one of the sports of 'our countrey Boys and Girls'. According to Strutt (1801), who described 'Pitch and Hustle' as commonly played in his day, the player whose halfpenny was pitched nearest to a mark claimed the privilege to hustle first. This meant merely that 'all the halfpence pitched at the mark are thrown into a hat held by a player who claims the first chance' and 'after shaking them together, he turns the hat upon the ground; and as many of them as lie with the impression of the head upwards belong to him; the remainder are then put into the hat a second time, and the second claimant performs the same kind of operation.' It can be assumed that 'Hussle-cap', mentioned in Smollett's *Peregrine Pickle* (1751), ch. 16; Sterne's *Tristram Shandy* (1759), i, ch. 10; and 'Jingle-the-bonnet' in Jamieson's *Scottish Dictionary Supplement* (1825), are either the same game or the second, hat-shaking, part on its own.

Francis Place, as a boy in 1783, gambled at 'tossing up' with the young

The little h Play.

PITCH *and* HUSSEL.

POISE your Hand fairly,
  And pitch plump your Slat;
Then ſhake for all Heads,
And turn down the Hat.

MORAL.

How ſickle's this Game!
  So Fortune or Fate
Decrees our Repentance,
When oft 'tis too late.

CRICKET.

Above: 'The little h Play—PITCH and HUSSEL', from John Newbery's *A Little Pretty Pocket-Book*, 1744 (1767). The game was played by pitching coins to a mark, the nearest to the mark claiming the right to 'hustle' the coins in a hat and turn them out on the ground. Those that lay heads up belonged to that player; then the second nearest did likewise, and so on.

Right: 'Playing at Hustle-Cap', one of the pictures on a writing-sheet, 'Coming out of school', published by John Marshall & Co., 1786. Writing-sheets had pictorial borders, and in the centre a boy would write samples of his best hand-writing, hoping for rewards from family and neighbours during the Christmas holidays.

Playing at Hustle-Cap.

blackguards around Drury Lane (*Autobiography*, ed. M. Thrale, 48–50) in exactly the same way as children do today, and it may be that they, too, believe that the result can be decided by the way the coins were placed in the hands before 'a peculiar twist of the wrist of the right hand sent them up into the air'. In *Blackwood's Magazine*, August 1821, pt. 2, p. 35, Edinburgh boys play 'Pitch and Toss' by throwing halfpence or buttons to 'a mark, or gog' before tossing up for heads or tails; and in mid-nineteenth-century London, Mayhew noted that one of the chief amusements of the street lads was 'gambling at pitch and toss, for halfpennies or farthings'.

PICTURE OR BLANK

In the East End of London the memory remains of the days when boys regularly played pitch and toss with cigarette, or 'tab', cards, under the name 'Pict or Blank'. Juvenile accounts from the mid-twentieth century are apt to start, 'Me Dad told me'. The cards were 'skated' to the wall, and the owner of the one that landed nearest was entitled to first toss. Coins are more satisfactory for this game, and since they are no longer in short supply, the card version is dying out. However, the name 'Pictures or Blanks' has been adopted for a slightly different game:

You and another person decide how many cards you are going to play for. Then the person who is going first gets the picture card lengthways in his hand and with a twist of his wrist it spins round and you shout 'Picture' or 'Blank'. Then it is the other person's turn, and so on. If there are more pictures than blanks the person who shouted 'Blanks' is the winner. If there are as many blanks as pictures it is a draw.   (Boy, 11, West Ham)

This, in turn, is not dissimilar to 'Odds and Evens' as now played (particularly in Scotland), where each boy tosses a coin simultaneously, one of them calling either 'Odds' or 'Evens'. If both coins fall the same side upwards, 'Evens' takes them; if different, 'Odds' has them.[5]

   In County Durham a modified form of pitch and toss with cards is played under the name 'Wallie'. The term 'dabbing' in the following

[5] The game of 'Odd or Even' played in classical times was more akin to 'Birds in the Bush'. One player held a number of small objects in a clenched fist, and challenged his opponent to guess whether their number was odd or even. If the guesser was wrong he had to pay one to the challenger to make it correct. This amusement is still known on the Continent and in the United States, but has not been found today among British children. It was much played in England up to the early 19th c. (see e.g. *The Old Curiosity Shop*, where boys play it 'even under the master's eye', ch. 25). The game of 'Odd Man', formerly 'Tommy Dodd', in which three players toss, and he whose coin is at odds with the other two pays for the next round, apparently continues among some drinkers in rural places.

description by a 14-year-old Spennymoor boy means 'throwing with force', and the *dabber* is the article so thrown, in this case a picture card:

Each player chooses a card out of his pack to be used as a dabber, and these dabbers are placed together with the pictures facing each other and dabbed against a wall. The card with the picture facing up is the winner. If there are two pictures facing up then the person that is dabbing gets another go. When the picture comes up the persons whose cards were turned down each pays him a card. There cannot be more than four people playing because there will be too many cards.

In Bishop Auckland they speak of 'casting'. 'If someone wants to play he asks if he can cast in, and if someone wants to stop playing he asks if he can cast out.' Further, according to some 11-year-olds: 'If one of the fagcards falls upright against the wall you say "flicks" (or "falls") which means that you hit the fagcard so that it will fall flat on the ground. If both cards show the back you say, Ha, ha, the cat and the fiddle, The one to touch black gets the riddle, and you get the card.'

'Picture or Blank' was popular in Edwardian London (correspondents), and Norman Douglas lists 'Picture or Blank', 'Skate Them', and 'Wallie' (or 'Up the Wall') in *London Street Games* (1916), 123. The name 'Picture or Blank' probably dates back to the days when the backs of cigarette cards were indeed blank, with no wording on them.

6

Hopscotch

It is almost impossible to walk along the streets of any ordinary-sized town without coming across some version of the familiar Hop Scotch diagram drawn on the pavement, and indeed it would be very difficult to find a ground more suitable for playing the game than is provided by the ordinary granite or asphalt paving.

(*Cassell's Book of Sports and Pastimes* (1888))

First the hopscotch must be drawn on the pavement, and as long as it is not in front of your house it doesn't matter where it is.

(Girl, 10, Peterborough, 1952)

NOT many children's games leave traces behind them (who can tell where a game of tig has exploded, whirled, and faded away?); but hopscotch diagrams on pavements and in the parks are welcome signs that an old tradition has not died out. 'I saw a hopscotch last week,' people will say; 'I didn't think they played hopscotch any more.'

The history of hopscotch can be told through its diagrams, which have changed over the centuries. From time to time some writer has made the statement that hopscotch was played in ancient Rome, either because he felt that such a widespread game must have been played in antiquity, or because he knew that the paving stones of the Roman Forum are covered with incised diagrams and thought that hopscotch might be among them. However, the diagrams in the Roman Forum, when their uses are known at all, are for adult 'board' games and gambling games; nor is it easy to visualize the Romans kicking stones along with sandalled feet.[1]

The earliest known hopscotch pattern is the straightforward ladder pattern of the seventeenth century. Jacques Stella, in *Les Jeux et plaisirs de l'enfance* (1657), shows six little cherubs playing 'La Marelle a Cloche-pié'.[2] The performer is hopping between a row of six parallel lines (i.e. a ladder, without the sides) and has kicked a stone into the space ahead of him. A verse offers the rather forced moral that 'cet enfant' who hops

---

[1] See J. P. Toner, *Leisure and Ancient Rome* (1995), 90; and Nicholas Purcell, 'Literate Games: Roman Urban Society and the Game of *Alea*', *Past and Present*, 147 (May 1995), 18–21.

[2] *La marelle* is the French word for hopscotch in general, and may be related to the Low Latin *merellus*, a token, coin, or counter. *Sauter à cloche-pied* is to hop; *clocher*, to limp.

across the diagram, imitating a cripple, is perhaps working in vain, because if he goes outside the lines his stake will change hands. The eighteenth century saw a change from this simple ladder-rung pattern to one with diagonal lines in one or more of the compartments, or beds, creating four small diamond-shaped sections, though how early in the century this occurred is difficult to find out. The eighteenth century displayed an unhelpful levity where children's games were concerned. In his satirical *Useful Transactions in Philosophy* (1709), Dr William King complains that a classical author 'has omitted the delineation of a pair of Hopscotches, with the names of their several Apartments, which I hope to retrieve and publish'. Ebenezer Forrest, writing up his diary of a jaunt in May 1732 as a burlesque on 'historical writers recalling a series of insignificant events entirely uninteresting to the reader', recorded merely that 'Hogarth and Scott stopped and played at hop-scotch in the colonnade under the Town-hall [at Rochester]' (*Peregrination* (1782)). Henry d'Allemagne, in *Sports et jeux d'adresse* (1904), 313, reproduces 'La Marelle a Cloche-Pied d'après le *Kinderspeel* de Katz, XVIIIe siècle', an engraving of a youth kicking a round, flat stone towards the 'exit' of a hopscotch figure; this has five oblong compartments and, at the top, a semicircular compartment containing a circle, which presumably served the same purpose as later semicircles—to allow the hopper to put both feet down and turn around ready for the return journey.[3] The use of the circle within the semicircle is explained in *Cassell's Book of Sports and Pastimes* (1888), 60: 'The order of the players . . . is decided by pitching the "clipper" . . . towards the top compartment . . . He who pitches nearest to the small circle [within the semicircle] plays first.' In *A Little Pretty Pocket Book* (1744), under 'Hop-Scotch', the simple version is still in fashion; the slots are numbered 1 to 9, and the semicircle at the top has 'PT' written in it (see below). Instructions are given in verse: 'First make with Chalk an oblong Square, With wide partitions here and there; Then to the first a *Tile* convey; Hop in—then kick the *Tile* away.' Joseph Strutt, playing 'hopscotch' at Chelmsford, *c.*1760, also knew a 'parallelogram about four or five feet wide, and ten or twelve feet in length . . . divided laterally into eighteen or twenty compartments . . . called *beds*; some of them being larger than others'. The Revd Joseph Hunter, however, as a boy in Sheffield, *c.*1790, used a diagram in which three compartments are each divided into four with diagonal lines. He annotates Brand's *Antiquities* (1813), ii. 304, thus: 'This [i.e. "Scotch Hoppers"] is, I suppose, the game with which I am very familiar under

---

[3] Cats's poem 'Kinder-spel' does not mention hopscotch, and I have not been able to find any 18th-c. edition of the work with these illustrations.

The little p Play.

HOP-SCOTCH.

FIRST make with Chalk an oblong
    Square,
With wide Partitions here and there;
Then to the first a *Tile* convey;
Hop in—then kick the *Tile* away.

RULE of LIFE.

Strive with good Senfe to ftock your
    Mind,
And to that Senfe be Virtue join'd.
                                        Who

Above: 'La Marelle a Cloche-pié' ('hopscotch') in Jacques
Stella, *Les Jeux et plaisirs de L'enfance*, 1657, showing the
early 'ladder' diagram.

Left: 'The little p Play—HOP-SCOTCH', in John Newbery's
A Little Pretty Pocket-Book, 1744 (1767). The 'ladder'
shape is still in fashion, and in the semi-circle at the top
'PT' probably stands for 'POT'.

Below:'Hop-scotch' in *Phil May's Gutter-Snipes*, 1896,
with double-column diagram.

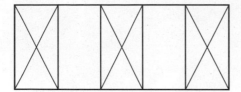

the name of Hop-Scotch. A number of partitions are chalked on the ground thus . . . A piece of broken pottery is thrown upon it, and the player is to kick it from one compartment to another with the toe.' A sheet of woodcuts of children's games published by H. van Munster en Zoon in the Netherlands, *c.*1780–90, has the first compartment divided similarly, and also has a semicircular top (L. de Vries, *Bloempjes der Vreugd* (1958), 29); and *The Boy's Own Book* (2nd edn., 1828), 33–4, has the same diagram (without the semicircle) for 'Hopscotch' ('in some parts of England . . . called Pottle . . . played with an oyster shell').

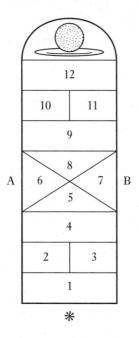

Thereafter, probably because they are described more often, and are included in books of games for boys throughout the nineteenth century, hopscotch diagrams seem, at first sight, bewilderingly various; but this is chiefly because the square with diagonal lines can be placed almost anywhere in relation to the plain squares. In Dean and Munday's *Every Boy's Book* (1841), 30, is the first appearance of the complicated twelve-division pattern which appears to have been the most popular through the nineteenth century.[4] The semicircle at the top is called the 'pudding', and has a plum pudding on a dish depicted therein. The players 'pink' for first go, that is, they try to pitch their tile into the 'pudding' (the descendant of the eighteenth-century circle-within-the-semicircle—see the description of the 'Kinderspeel' illustration, above). The winner throws his piece of tile or lead into division 1, hops into the space, and kicks the tile out of the diagram the way it came. He continues thus through successive numbers until he comes to 8. He is then allowed to put his feet in the beds marked 6 and 7, and rests himself on both feet in 8 before hopping into beds 9, 10, and 11 and putting both

---

[4] This may be because it appeared in subsequent games books, either identically or redrawn, e.g. Routledge's *Every Boy's Book* (1855 and 1868), and their *Games and Sports for Young Boys* (1859); T. Nelson's *Out-of-Doors*, by Alfred Elliott (1872); Kate Greenaway's *Book of Games* (1889). However, in Ward and Lock's independent publication, *The Boy's Handy Book* (1863), a very similar diagram appears, missing only the long slot beneath the diagonals and the picture of the plum pudding in the semicircle (which is numbered 12).

feet down again in 12—the second resting-bed. When he comes to plum pudding he must kick the tile with such force that it passes out through all the other beds with one kick.

The semicircular resting-place at the top of the diagram has a history of its own. The PT written in the semicircle in *A Little Pretty Pocket Book* very probably stands for 'Pot', the usual word for this space until semicircles were dispensed with altogether. In Mactaggart's 'Hap the Beds', 1824, the two farthest divisions are called 'kail pots'. *Punch* (1845), 204, dismisses writing 'POT' in the semicircle as 'mere shallow pedantry'. Heslop, in *Northumberland Words* (1892–4), says, 'The top bed is marked "pot", and the player counts by getting the "dabber" safely into this bed and calls it "one-a-pot", "two-a-pot", and so on.' 'Potsy' was written in the semicircle at the top of the six-square 'Beddies' diagram in Aberdeen, in 1960; and 'Pot' in Forgue, by Huntley, Aberdeenshire, in 1975. Other names have been written in this dome-like space: for instance, 'BOX', Pontypool, 1954; 'CAT', Market Rasen, 1960; 'DEN', Isle of Skye, 1950; 'HOME', Perth, 1954, and Swansea, 1960 (and in Brooklyn, 1891 (*JAFL* 4 (1891), 229); 'LONDON' (*Traditional Games*, i (1894), 224, and Newcastle upon Tyne, 1959); 'MOONSIE', Kinlochleven, 1961; 'OXO', Norwich and East Dulwich, 1961.

As for the circle within the semicircle, first seen in the eighteenth-century Continental engraving (above), that was eventually turned (by whom, and when?) into the Victorian 'plum pudding'; and occasionally into a cat's head. The statement in Dean and Munday's *Every Boy's Book* (above), that players try to 'pink' their tile into the 'pudding' for first go, is reinforced by *Cassell's Book of Sports and Pastimes*, p. 260, where the top compartment is named 'pudding' (though having no picture of such), and 'he who pitches nearest to the small circle in that compartment plays first.'

The Revd J. G. Wood's *Modern Playmate* (1870), *Cassell's Book of Sports and Pastimes*, and Maclagan's *Arglyeshire* (1901), give diagrams in which slots, squares, and crossed squares appear in varying combinations. The variety of design is indeed bewildering in these years before the First World War, when street-play was at its most prolific, and when the Revd Moxon enthused, in his parish magazine, August 1907, 'I have in my collection no less than twenty varieties of game and diagram as played in Soho.' A late nineteenth-century elaboration of the game was, after the usual hopping part, to place the stone on successive parts of the body and walk through the diagram without letting it fall. For example, in *Traditional Games*, i. 224, in one of several similar games from Crockham Hill, Kent, a five-slot diagram is hopscotched normally, then, starting from the fifth slot, the player returns with the stone on her thumb, then with the stone on her eye (head held well back), then with the stone in her palm, then with the stone on her head, then with the stone on her back. In each case, after completing the course, the player throws up the stone and catches it. The idea that there should be extra trials when the main game was over was still extant in the mid-twentieth century, chiefly in connection with 'Six Beds' (see below).

AEROPLANE HOPSCOTCH

The most modern pattern is often called 'Aeroplane Hopscotch', because the diagram looks something like a flattened biplane. The pattern probably came into fashion with the arrival of city pavements, for the diagram simply follows the edges of the paving stones.

There's a square, and then two squares side by side, then another square, then two side by side, then one, then two side by side. You number them 1 to 9 with chalk. You can have more than two people playing, but the more you have the longer you have to wait your turn. You get a flat stone and throw it in the first square, and hop over square 1 and land with one foot in square 2 and the other in square 3, hop into square 4, astride into 5 and 6, hop into 7, astride into 8 and 9. Jump round to face the other way and come back down again. Then stay on squares 2 and 3 and pick up your stone and jump right off. If you make a mistake it's someone else's turn. Then you throw into number 2, but you step into number 1 and go up the beds just the same way as you did the first time. Then you throw to 3, and so on. When you get to number 9 you can claim any square as your own and you write your name in it. This means that no one can tread in it except you.

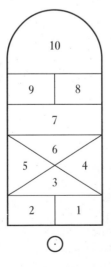

Sometimes, instead of choosing a square, the girl turns her back to the diagram and throws the stone over her shoulder; whichever square the stone lands in is hers, and she writes her initials in it. The general effect is of jumping, not hopping; after landing split-legged into two side-by-side squares, the player hops into a single square and immediately jumps into the next side-by-side squares, one foot in each (or as a 7-year-old explained in Edinburgh, 'How you play hopscotch is hop, split, hop, split, hop, split, hop'). In aeroplane hopscotch continuous hopping is not required; neither is the skill of hopping while driving a stone from bed to bed. This split-leg type of game was first noted by Newell, in 1888 (see diagram). The young Americans were managing to kick the stone along as well as straddling between squares 1 and 2, triangles 4 and 5, and squares 8 and 9. In England, it was first sighted by Lady Gomme's correspondent, the excellent Miss Chase, in Crockham, Kent. Here was the orthodox 'aeroplane' shape, achieved by chalking round six paving stones. A pebble was 'put' into no. 1 and picked up, the hopping and jumping followed exactly as in the present day, and the game finished with the requisite trials of walking through the figure with the stone balanced on various parts of the player's body, in this case on her shoe, and then on her eyelid. When the feat had been successfully accomplished the stone was kicked or thrown up in the air and caught.

SIX BEDS

'Peever is played on beds which are chalked on a piece of hard ground. There are at least two kinds of beds used in our district, "aeroplane" and "sixy beds"' (girl, 12, Helensburgh, 1960). Six-bed hopscotch is 'proper hopscotch'; in other words, the stone is kicked from square to square as it has been for the past three centuries. Although 'Aeroplane Hopscotch' is played almost exclusively by girls, the six-box game in which the stone is kicked is often played by boys; in fact in Workington, 1961, it was called 'Boys' Oppy Beds', and was a boys' game only. The diagram is simple: a rectangle of, say, 8 feet by 4 feet is divided by crossed lines into six smaller rectangles. The rectangle of the late 1890s to the 1920s more often contained eight squares (*Traditional Games*, i. 224; correspondent, Dundee, *c*.1915; Stockland, Somerset, Macmillan Collection, 1922). Six squares was the most popular number in the 1930s and until at least the mid-1970s; but, said the children, 'you can have eight or nine or ten or twelve, or thirty if you want'. Local rules varied, but after the main game was over, the player often had to walk through the diagram in different ways, thus maintaining the late Victorian tradition:

Slide a slate into number 1, then hop into the square you slid it. Then kick it into number 2, then 3. In number 4 you can have a rest. When you have had your rest you kick it into number 5, then 6 and out. Then you throw into number 2 and go round like before. You go like that till you have thrown into number 6. Then you go back. When you have finished going backwards you do 'handy', which is you walk right round with the slate on your hand. Then 'footy', with the slate on your foot. 'Heady', with the slate on your head. Then you walk right round without standing on any lines with your eyes closed.   (Girl, 10, Scarborough, 1960)

A subspecies of 'Six Beds' is named 'London to York', because the hopscotcher kicks a stone through nine or eleven oblongs from a semicircle called 'London' to a semicircle called 'York', 'with a rest-place half-way along'.

The general rule is that children's games are found in a better and better state of preservation the further north one goes; thus the most varied and intricate hopscotch is (or was) to be found in the north of England and in Scotland, especially in Edinburgh (a marvellous variety of diagrams was recorded by J. T. R. Ritchie in *Golden City* (1965), 96–110).

Another way of recounting the history of hopscotch is by means of its name. The word 'Hopscotch' is simply a description of what happens in the game: a player hops over the 'scotches', or lines of the diagrams, driving a flat stone before him. ('Scotch' is an old word meaning a score or incision). Randle Holme listed 'Tip stones or Hop skotches' amongst the 'Playes with Instruments' in his *Academie of Armory* (1688). But in the seventeenth century, and after, the word seems to have been equally acceptable the other way round. *Poor Robin's Almanack* frequently used 'Scotch-hoppers' in its burlesque prognostications: in April 1668, for instance, 'Mercury being Lord of the nineth house, presages that there shall be much playing at Scotch-hoppers among children.'[5] Grose, *Dictionary*, 1788, under *Tray Trip*, uses the term 'Scotch hop', of which Moor's 'Scotch-hob', in *Suffolk Words* (1823), may be a corruption. Local forms of the word were 'Hop-crease' (East Anglia), and 'Hop-score' (Shropshire, Yorkshire).

Frequently the game was called by the name of the squares or oblongs in the diagram, which, once uniformly throughout Britain and now chiefly in Scotland, were known as 'beds'. John Ray recorded, in his *Collection of* [North Country] *Words* (1691), that *Beds* was 'A game of children, in which they hop on one foot through different spaces, chalked out, called beds.' Strutt played 'hop-scotch' in his schooldays in Chelmsford, *c*.1760, and referred to the compartments in the diagram as 'beds'. Husenbeth, at school at Sedgley Park, Staffordshire, *c*.1805, says 'There were numerous games of "beds", played with a tile, or a piece of wood . . . such as "pudding and beef", "running beds", "back beds",

---

[5] Another 17th-c. name for hopscotch was, apparently, 'Scotch Morris', as in 'Morris dance'. The following conversation takes place in T. Duffett's play *The Mock Tempest* (1675), IV. ii: *Miranda*: 'Now let's leave this idle talk, and play the *Scotch* Morice.' *Dorinda*: 'Then I'le play forward, and backward, for that's the way now.' *Miranda*: 'No I won't play Boyes play.' Hopscotch began to be a girls' rather than a boys' game around 1820 (cf. Ch. 9, 'Boys and Girls'). When T. Bachelor printed a little chap-book entitled *King Pippen's Delight* in *c*.1830, he used, for the game of 'Step Scotch', the verse that originally appeared in *A Little Pretty Pocket-Book* (1744), but felt it necessary to redraw the woodcut so that girls were shown playing instead of boys (he also changed the diagram from the 18th-c. ladder pattern to the contemporary pattern with diagonal lines).

"cross beds" and "funny beds".' Nowadays 'Beddies' is a common alternative to 'Beds'. Sometimes the name can be compound, as in Mactaggart's 'Hap [hop] the beds', Kirkcudbright, 1824; 'Hip the beds' or 'Hippy-beds', Northumberland, 1892, *EDD*; 'Hoppy beds' (e.g. Workington, 1961, Dumfries, 1975).

The name could vary according to the local dialect word for 'hop': see, in *EDD*, 'hackct' (Somerset), 'hick' (Yorkshire, East Anglia, and the West Country) and 'heck' (Pembroke and Somerset). 'Heck-a-Beds' was noted in Wexford in 1862 (*Dublin University Magazine*, November issue), and 'hecky' was the name for hopscotch in Crickhowell, Breconshire, 1961. *EDD* gives 'hitch', to hop on one leg, from Northumberland, Cumberland, Westmorland, Yorkshire, and Suffolk, and this spawned the hopscotch names of 'Hitchy-bay' (in the north-east, 'bay' is the word for a 'bed', or for the whole diagram, and the game is sometimes just 'Bays'); 'Hitchy-dabber' (the 'dabber' is the square piece of wood, flat stone, shard of earthenware, or glass-jar stopper, which is kicked round the diagram); and 'Hitchy-pot' (the 'pot' is the piece of broken crockery or stone).[6] This nomenclature was still in place in the 1960s. A Bishop Auckland girl wrote: 'The game I like best is called Hopscotch, better known as Itchy Bay. You get a stone but we call it a dabber. You have to stand in a bay and throw the dabber'; and from Spennymoor came the statement: 'Hopscotch is known as "Hitchy-dabbers". It is played with a tin filled with soil.' A 14-year-old girl at Penrith, describing the pattern known in the south as 'Aeroplane Hopscotch', said: 'Two to three people can play "Hitchy-pot". You have eight squares and go up to the end and jump round without going off the "Hitchy-pot".'

The two famous Scottish names for the object thrown, and hence for the game itself, are 'Pallie' and 'Peever'. The well-known connection between the Scottish and French languages is not, as often stated, a consequence of the Auld Alliance, but is due to the survival in Scotland of Norman French. Thus 'pallie' is possibly from the French *palet*, the flat disc long used in France for various pitching games;[7] and 'peever' is possibly from *pierre*, a stone. Jamieson (1808), defines *Pallall, Pallalls*, as 'a game of children, in which they hop on one foot through different triangular spaces chalked out, driving a piece of slate or broken crockery before them. From the figures made, it is also called *the beds*.' 'Pallal' (or 'Palaulays', or 'Pauldies') was a well-documented term in southern

---

[6] Another 'hitchy' formation, 'Hitchy-beds', formerly in use in Cumberland, Westmorland, and Yorkshire (see *EDD*), does not seem to have survived into the 20th-c.

[7] See e.g. the numerous quotations in Michel Psichari, 'Les jeux de Gargantua', *Revue des études rabelaisiennes*, vi. 319.

Scotland and as far south as Yorkshire during the nineteenth century (see *EDD*), and was still very much in the games vocabulary in the mid-twentieth century. Thus, a 13-year-old girl in Kirkcaldy, 1952: '"Beds" are used for peevers, or paldies . . . The stone is known as pall-all, and children cherish a special one—for luck or because it slides easily'; and a 14-year-old in Forfar, 1954: 'Pally is a well known game. Some people use a stone to play it, but it is better with a real "pally", which is a piece of old tile from a fireplace'; and a 9-year-old in Lundin, Fife, 1975: '"Pauldis" is a game that we play at our school. You mark out six squares, and kick the stone.'

'Peever' is not as well recorded, and John Strang, City Chamberlain of Glasgow, did not provide the most helpful first account when he described 'the young misses' of the City 'scoring the flagstones with their peevers, for the purpose of playing at pal-lall' (*Glasgow & its Clubs* (1856), 218). In the Addenda to the 1887 Supplement to Jamieson's *Scottish Dictionary*, 'Peever' is defined as 'The pitcher or flat stone with which the children's game of *beds* or *pallall* is played; the game is therefore sometimes called "*peever*" or "*the peever*". West of Scotland', and Maclagan confirms the western connection with his 'peaver . . . a smooth flat stone' from Argyllshire. However, 'Peever(s)', as game and stone, turned up from a variety of places in our surveys. It seemed to be the usual name in Edinburgh and Glasgow. In Perth, Kirkcaldy, and Lairg and Helmsdale in Sutherland it was subsidiary to 'Hopscotch'. In Helensburgh 'Hopscotch' was the game 'shaped like an aeroplane', and 'Peever' was reserved for the game played 'with a straight diagram of six squares and a stone called a peever that you kick all round'.

A proper peever, with vertical sides and a flat bottom, is desirable for 'real' hopscotch, in which the thrown object is kicked from square to square. (Children nowadays in England, and increasingly in Scotland, can use any irregular-shaped stone, although it should not be round or it rolls; they only have to throw it so that it stays in a certain square—they do not kick it round the diagram.) The best peevers are made of marble,[8] but a good (though more ephemeral) substitute is a shoe-polish tin filled with earth. Correspondents remembered buying peevers: 'A smooth, round piece of granite, about 2 inches wide, costing a penny' (Edinburgh, *c*.1910); 'A marble peever could be purchased for the price of a week's pocket money from a corner shop or a mason's yard' (Glasgow, 1920s); but more often a child's father made her a peever from an odd

---

[8] *The Book of Games and Sports* (pub. T. Nelson, 1856), 15, says hopscotch is played 'with a piece of smooth flat stone about half an inch thick, and twice as large as a penny. Any kind of stone will answer the purpose, but black or white marble is best.'

piece of marble, such as got broken off old-fashioned wash-basin stands, or a friendly monumental mason made one gratis, from a scrap of waste marble, sometimes decorating it with a raised pattern and the child's initials. It may be that some peevers have been passed down through generations from mother to daughter; perhaps it was the owner of an inherited peever who, in a tough area of Glasgow, in 1975, said emphatically, 'Peaver is a game. It is called peaver because you use this thing called a peaver. You need a bit of chalk. Then you need a peaver. If you do not have a peaver a polish tin will do.'

Peevers could be made from the base of a broken bowl, or the bottom of a glass bottle. In fact, since children are very resourceful, a wide variety of objects have been used in the game of hopscotch: an oyster shell, in the nineteenth century when oysters were cheap; the head of a flax-mill bobbin (Tayside, 1920s); a piece of fireplace tile, 'with the white side downwards, which soon gets smooth and polished' (Blackburn, 1950); a cake of pipe clay (Glasgow, 1950); 'a piece of wood, stone, canvas, or linoleum', 'a round piece of wood with a hole in the middle to put your finger in' (Aberdeen, 1960); 'You get an old tin for a slider' (Wigan, 1960); 'The dabber can be a stone or a safety-pin with several buttons on it' (Spennymoor, 1960); 'You have a piece of slate for a shyer' (Headington, Oxford, 1960); 'A piece of glass or a brick, a little bit of brick flat on the bottom' (a teacher commented, '"Brick" is their word for a stone') (Oldbury, Worcestershire, 1970); 'We throw the chalk we drew it with' (Aberdeen, 1976).

'Pickie' and 'pickiestone', as the names for the game and stone respectively, seem to be almost exclusively Irish, although the word 'pick', to pitch or throw, had dialect uses all over the British Isles. 'Pickie' has often, in the past half-century, degenerated into 'piggy'.

SPIRAL HOPSCOTCH

A stone is not used in this game, but 'you are a lot longer on your one leg than in ordinary hopscotch'. A 12-year-old from Tiverton, Devon, described it thus in 1954:

Hopscotch starts as a round circle. Then a road is made round the circle getting wider and wider as it goes round. The road is about one foot wide. Then the road is made up into lots of squares. In the middle circle the word HOME is written, because that is where you have to get. Any amount of children can play, except one. A person starts to try to hop round until she gets to HOME. If she falls over or hops on a line she is out. Then the other person will start to go round. She might successfully get home but then she has got to hop back. When someone gets to home and back she can mark in any square with her initials, and when the other person gets to that square she will have to jump over it. It goes on till all the squares are filled with initials, and the one with the most with their initials in wins.

Spiral hopscotch has had various names, for example: 'Circular Hopscotch', Aberdeen, 1952; 'Cockleshell', Enfield, 1954; 'French Hopscotch', Stratford, east London, Selborne and Liss, Hampshire, 1960; 'London' [LONDON is written in the centre], Swansea, 1952; 'Round Hopscotch', Market Rasen, 1952; 'Snail', Plympton St Mary, Devon, 1957; 'Spider's Web', Kinlochleven, Ayr, 1961.

It seems, though, that in older days a stone *was* kicked from square to square. Thus in Grimsby, *c.*1915: 'First you went round just plain shooing. The second round was "Killing ants", after each hop you tapped the ground once with your free foot. The next round was "Killing beetles", when you tapped the ground twice with the free foot.' Some of the children in our surveys said they had learned the kicking method from American visitors, confirming that the USA is tenacious of tradition. In the first notice of the game, 'Snail, or French Hop', *Boy's Own Book* (1872), 34, a chip or shell is kicked round the spiral, and the French attribution may be correct. 'La Marelle Ronde' (with 'Enfer' along the way and 'Paradis' at the centre) appears in *200 Jeux d'Enfants* (*c.*1892), 115. H. Carrington Bolton, writing to *JAFL* (1901), says:

In a review of Paul Sébillot's 'Le Folklore des Pêcheurs' it is mentioned that the author states that the children on the seacoast in Upper Brittany, in playing hopscotch, make use of a diagram resembling 'the circumvolutions of the helix of a snail,' and he regards this as the result of the environment of the children. My criticism is that the children of Washington, D.C., employ the same diagram with no thought of ichthyological surroundings; all over the sidewalks of this city one sees the helicoidal hop-scotch diagrams, chalked on the surface of the flags. (p. 314)

Spiral hopscotch must have been flowing about the world for quite some time.

BALL HOPSCOTCH

This is played all over Great Britain and can be played on any kind of diagram; it does not seem to matter whether it is the 'aeroplane' diagram, or (more often) a straight diagram of six or eight or ten or twelve squares. A 10-year-old girl explained how it was played in Crickhowell, Breconshire, in 1962:

Hugh Thomson's girl playing
spiral hopscotch, from
*Highways and Byways in
London*, Emily C. Cook, 1902.

Father Damian Webb's
girl playing spiral hopscotch,
in Workington, 1962.

You first get a chalk and draw an oblong on the path. Then you draw a line down the middle, then draw four lines across and there you have eight squares and you put numbers in them, One, two, three and four up one side and five, six, seven, and eight down the other side. Now you can begin the game. You roll the ball into square number one and catch it before it rolls into number two. You then bounce it once in number one, two, three, four, and keep on like that until you have been around all of the squares. Then in number two you bounce it twice all the way around. Then three times, and so on. The player must bounce the ball according to the number on the square, for instance the ball must be bounced eight times when you are doing number eight.

The earliest description of this game is in an article, 'Children's Games of 1900', in the *Journal of the Institute of Education of Durham University*, 4:21 (1953), 19. Called by the local name of 'Bays', and having a semi-circle at the top called 'Pot', it was performed as at Crickhowell except that the ball need only be 'stotted' once in any square. Before the ball was rolled the player had to say 'Welcome Onesy' (or 'Twosey', etc.). There were final rounds called 'Hard Labour', hopping through all the bays while stotting the ball, and 'Blindies', walking through all the bays with eyes shut, while holding the ball. If all was accomplished 'the player began the last ritual' (not described, but probably involving the award of a mark in 'Pot'), saying: 'Tid, Mid, Miserere, Carlin, Palm, Paste Egg Day; Set a cup upon a rock, and mark me one-a-pot, and for my Welcome Onesy.' The rhyme names the Sundays in Lent, and was also used—exclusively in County Durham—for keeping the score in ball-bouncing (from at least *c*.1880), shuttlecock, and skipping.

BROOM JUMP (OR BROOMSTICK, OR CHECKS)

In this hopscotch game no stone is used at all. 'You need a huge square divided into about twenty smaller squares. The numbers are not put in evenly, they are jumbled up, but they must be within reasonable reach of each other. Then one person with the help of a broomstick has to hop or jump from 1 to 20 in the right order.' In some places a few squares are left blank, 'and you are allowed to stand in them'. Correspondents have mentioned this game as current about 1925: 'a variation using a sweeping brush—but this caused trouble on the home front'.

Hopscotch is one of the truly international games. In Moscow, one of the ways of knowing spring is on the way is to see children marking designs for hopscotch on pavements newly clear of snow; in South Africa, the subject is big enough for an honours thesis: 'Hopscotch, or Hinkspel, a

Description of the Game as it appears in South Africa', Dr Connie Mulder, Wits University, 1951. In China it is called 'T'ao Fang Tze'; in Denmark 'Hopskok'; in Germany 'Hopse'; in Turkey 'Sek-sek'; in Italy 'Campana'; in Spain 'Rayuela'; in Albania 'Kambëxinkthi'; in India the Hindi name is 'Ikarh dukarh'; in the former Yugoslavia (Belgrade) 'Shkolitse'; in all these countries, and everywhere else in the world, the names and patterns of hopscotch undoubtedly vary as much as they do in Britain. Jan de Vries produced a detailed study, *Untersuchung über da Hüpfspiel: Kinderspiel-Kulttanz*, Folklore Communications no. 173 (Helsinki, 1957), which covers Europe and America. Here I have only the space to remark that on the Continent the Church seems to have converted the game into a religious exercise, as it did with some of the singing games; the diagram has assumed the character of a journey through life to death and beyond; in Holland, for example, the hopscotchers hop to 'Dood' (Death), in Germany from 'Erde' (Earth) to 'Hölle' (Hell) and 'Himmel' (Heaven), and in Italy to 'Paradiso'.

7

Chucking and Pitching

THESE games are usually played by boys, who are believed to be better at aiming and throwing than girls are. The primary appeal of such games to boys, however, is that they are the means of making a fortune; that it is a fortune in nuts, counters, bottle caps, buttons, or other near-worthless objects does not matter—these things represent the juvenile currency of the time. Small humans are just as acquisitive as large humans; but adults—rightly—believe that children should not have control over valuable objects or more than negligible sums of money. They sentimentalize over their own childhood, when, they say, 'we made our own amusements, we made do with whatever was at hand', while at the same time they prevent their children from ever being able to make the same boast by giving them expensive manufactured toys. When the game of 'Pogs' swept America in 1992 (see 'Flipping-over', below), some adults—far from welcoming the return of a traditional game—were scandalized that children were playing with a zeal that caused squabbles and recriminations between vanquished and victorious, and some were so carried away that they bet money on the results. The reaction was a symptom of a prevalent malaise: middle-class North Americans and Europeans are addicted to, and disgusted by, their affluent state.

CHUCKIE

The game of throwing small objects into a small hole in the ground has provided instant entertainment over a very long period of time. In ancient Greece, play-goers emerging from a religious performance were permitted the relaxation of playing 'at the hole' (Cratinus, c.450 BC, in *Seriphians*); and a chimney sweep in Alton, Hampshire, 1956, reminisced that as a lad there, c.1906, he had often had a game 'at the hole'. The small objects thrown could be pebbles, counters, coins, marbles, broken stems from clay pipes, buttons, nuts, date-stones, or, most famously, cherry-stones.[1] The game itself was usually named from the implements

---

[1] In *Vulgaria* (1519), a book of sentences for the use of schoolboys, William Horman, vice-provost of Eton, opined, 'Playenge at cheriston is good for children', and this statement held true until after

used, although some names are mysteries; thus, with the date when the name was first recorded: 'Anti' (c.1930, Henfield, Sussex); 'Auntie' (Farnham, Surrey, c.1870, G. Sturt, *Small Boy in the Sixties*, p. 148); 'Buttons' (*Traditional Games*, i. 54–5; R. C. Maclagan, *Games of Argyleshire* (1901), 47); 'Cappie-hole' (*OED*, 1605); 'Cherry-pit' (*OED*, 1522); 'Chock' (*EDD*, from Nottinghamshire, Northamptonshire, Warwickshire; c.1925, Kirkby-in-Ashfield, Nottinghamshire, with marbles); 'Chuck' or 'Chock' (*OED*, 1711); 'Chuck-button' (*OED*, 1863); 'Chuck-farthing' (*OED*, c.1690); 'Chuck-fibs' ('small bones of sheep's feet', *Pegge's Derbicisms* (Hallam & Skeat, 1894)); 'Chucky' (Bredbury, Cheshire, c.1905); 'Counter-hole' (Robert Armin, *Two Maids of More-clacke* (1609)); 'Frips' (with pipe-stems, a 'frip' being something worthless, Burnley, Lancashire, c.1910); 'Gamma' or 'Slosh' (c.1914, Bethnal Green, London, using lead buttons).

Glimpses can be seen of men and children playing this game through the centuries. In the margin of the Luttrell Psalter (BM Add. MS 42130 fo. 193ᵛ), for instance, a man carefully aims a handful of small objects at a depression in the ground. The manuscript was illuminated in East Anglia c.1340. At about the same time, the child Froissart was playing 'fosselettes' at Valenciennes. He says, in 'L'Espinette amoureuse', lines 238–40, 'Et se faisions fosselettes, Là où nous bourlions aux nois; Qui en falloit, c'estoit anois' ('Or small pits in the ground would make, And play at nuts, which he who lost, His pleasure bitterly was crossed'). Rabelais has Gargantua playing 'à la fousette' after dinner (1534; i, ch. 22), which Urquhart translates (1653) 'at cherrie-pit' ('pit' is the pip, not the depression); and Michel Psichari, *Revue des études rabelaisiennes* (1908), vi. 340, gives a number of quotations showing how popular 'la fossette' was in the sixteenth century. Plate 8 in *Les Trente-six figures* (1587), shows the game in action. Six nuts lie around the hole; one of the two players is aiming a handful of nuts at the hole, while the other stands awaiting his turn, his nut-bag in his hand. Jacques Stella's naked boys play 'La Fossette aux Noyaux' in *Les Jeux et plaisirs de l'enfance* (1657), pl. 16; they, also, are using nuts.

---

the Second World War, when cherries were no longer cheap and plentiful. In Mayhew's *London Labour and the London Poor* (1861), i. 85, a costermonger explained, 'No fruit is cheaper than the cherries at 1d a pound . . . Then boys buy, I think, more cherries than other fruit; because, after they have eaten 'em they can play at cherry-stones.' Cherry-stones were known as 'cherry bobs' (Hertfordshire, c.1895, and London, c.1903), 'cherry-cobs' (W. Dickinson, *Dialect of Cumberland* (1878–81)), 'cherry gobs' (Devon, *Reports Provinc.* (1891)), 'cherry odds' (F. T. Elworthy, *West Somerset Word-Book* (1886)), 'cherry oggs' (London, c.1905–c.1920), 'cherry wags' (Liverpool, 1960), and 'cherry wobs' (Birkenhead, c.1905); in Scotland, 'paips' (= English 'pip').

The great A Play.

CHUCK-FARTHING.

AS you value your Pence,
  At the *Hole* take your Aim;
Chuck all safely in,
  And you'll win the Game.

MORAL.

Chuck-Farthing, like Trade,
  Requires great Care;
The more you observe,
  The better you'll fare.

B 3   *Flying*

Left: 'The great A Play—CHUCK-FARTHING', in John Newbery's *A Little Pretty Pocket-Book*, 1744 (1767).

Below: Men playing 'Chuckie', in the Luttrel Psalter, *c*.1340. The manuscript was illuminated in East Anglia.

At foot of page: *Les Trente-six figures*, 1587. The four boys on the left are playing 'à la fossette'.

Jacques Stella, *Les Jeux et plaisirs de L'enfance*, 1657. In 'La Fossette aux Noyaux' the game was played with nuts.

'La Fossette ou le Jeu de Noyaux', by Augustin St Aubin, *c*.1770, in the series 'Differens Jeux des Petits Polissons de Paris'.

Strutt gives the most precise description of a chucking-into-a-hole game in *Sports and Pastimes* (1801), 289:

I have seen a game thus denominated [chuck-farthing] played with halfpence, every one of the competitors having a like number, either two or four, and a hole being made in the ground with a mark at a given distance for the players to stand, they pitch their halfpence singly in succession towards the hole, and he whose halfpenny lies the nearest to it has the privilege of coming first to a second mark much nearer than the former, and all the halfpence are given to him; these he pitches in a mass towards the hole, and as many of them as remain therein are his due; if any fall short or jump out of it, the second player, that is, he whose halfpenny in pitching lay nearest to the first goer's, takes them and performs in like manner; he is followed by the others so long as any of the halfpence remain.

Francis Place, the radical reformer, born 1771, played 'Pitch in the hole' with the same rules in London *c*.1780 (*Autobiography*, ed. Mary Thale (1972), 49–50). The dialogue between the two boys playing 'Chuck Farthing' in William and Cluer Dicey's print, 'Sport upon Sport', *c*.1750, further brings the game to life: 'Next plump'—'No I'le Slide'.

There were of course variations (when are there not?). The chief one was that an agreed number of marbles or nuts were put into the hole by each player and the game was to knock these out. In Dollar, Clackmannanshire, *c*.1890, the game was played with large brown porcelain marbles: 'A marble was stotted against the wall, caught and then pitched into a hole at the foot of the wall. The object was to hit other boys' marbles out and forfeit them.' Exactly the same game was known as 'Hashie-Bashie' in Forfar, *c*.1910. In Manchester, *c*.1910, a boy's cap was used and the game was played 'about Christmas time when the nuts came in. You got the weakest little boy's cap away from him and filled it with nuts and put it on the ground. Then you took turns throwing nuts at the nuts in the cap and trying to knock them out. The nuts you knocked out you kept.' Norman Douglas mentions 'Nuts in Cap' in *London Street Games* (1916), 10.

In Spennymoor, 1960, 'The main [marbles] game was to scrape a shallow dip in the sand and place four marbles each in it. The object was, in your turn, to knock out with another marble as many of the marbles in the dip as you could.'

Although the game has not altogether disappeared, the old rules no longer apply. Rather than simply getting the objects into the hole, the emphasis is now on hitting or overlapping the objects already in the hole:

thus, Dublin, 1953, 'Chuck into a small depression in turn. If your button hits someone else's you win'; Aberdeen, 1960, 'They all try and get a penny into the hole, and if you get your penny on top of his you win'; Knighton, 1960, 'We get a nice bag of nuts and a tall thin tin, with the round part about as big as a penny. The boy with most nuts puts one in the tin, then we start throwing at the tin and the first to get one in the tin has the lot that missed.'

From this sport the most popular of the present-day marbles games possibly evolved: see 'Holie' (p. 36).

KNOCKING DOWN SCREWS

This was a game played especially in London and its environs in the first half of the twentieth century. In 1957 a boy from High Laver, near Ongar in Essex, provided the only post-1950 description. A screw was stood on its head, the missile preferred being a cherry-stone—a marble or pebble being looked on as a poor substitute because it made the game too easy. The screw might itself be the prize, in which case the owner in recompense pocketed the cherry-oggs which were cast at it; alternatively, he who made the successful shot won the stones as well. Long screws were considered more valuable than shorties; and brass screws, known as 'brawnies', more valuable than steel screws or 'milkies'. The boy from High Laver explained that 'the number of threads' on the screw were first counted and 'the same number of paces were taken from the screw' before the cherry-stones were bowled at it. An alternative was to flick cigarette cards at the screw.

This is apparently the sole representative among modern schoolchildren of games like ninepins, in which a standing object is knocked over by a projectile. It is in direct line of descent from a game much played in the late eighteenth century—throwing at leaden cocks, which imitated the sport of throwing at real cocks on Shrove Tuesday (see John Brand, *Popular Antiquities* (1849 edn.), i. 72–82). Francis Place gives all the technicalities of 'Cocks and Dumps' in his *Autobiography*, pp. 47–8: 'They were cast in moulds made either of Fire Stone or Chalk. I soon found that a small quantity of pewter mixed with the lead made the cocks much more tough . . . Each of two boys put a cock upon the ground and having settled who should have the first shy the game commenced, by throwing a piece of lead or a smooth stone at the cock, he who upset the Cock won it.'

LINE OF MARBLES, BUTTONS, ETC.

The basic principle of this game is that the players should put an equal number of marbles, nuts, or buttons in a line, and then take turns at knocking them out of line in order to win them. The game is old, and widespread. 'La rengée' was the 136th game played by Gargantua after dinner (Rabelais, i. ch. 22), a name translated by Cotgrave in his *Dictionarie* (1611), as 'A play with a ball at nuts, placed on a row'. Pitrè, in *Giuochi fanciulleschi siciliani* (1883), 85, described Sicilian boys playing 'A Murari' with a row of nuts, apples, or apricot-stones: 'The players . . . pitch similar objects at this row, and what each knocks out he wins.' Brewster, in *American Non-singing Games* (1953), 139, describes Armenian children playing 'Djanakjagh' in the same way, with knuckle-bones stood on end in several lines, each line belonging to a different player.

The British game is probably long established, although it went unnoticed until Anne Elizabeth Baker was gathering material for her *North-amptonshire Words and Phrases* (1854). 'Plum-pudding', she records, 'is the name of a game at marbles . . . a mark is made at some little distance, called *taw* . . . the first boy bosses at the row . . . so as to strike as many as he can out of the line; all that he strikes out he takes; the rest are put close together again, and the other players take their turns.' A 'boss' was a large stone or iron marble, and 'to boss' was 'to bowl with a boss'; the word also meant 'to bang or throw' (*EDD*). The game appeared in a succession of boys' games books, under the name 'Picking the Plums'; according to the first of these appearances, in *The Boy's Own Book* (1852), 13, a player could 'bounce or shoot' a marble at the line. In other games books, 'bounce' is defined as aiming (dropping) from above. 'Picking the Plums' seems to have been the best-known name for the marbles version of this game, and Norman Douglas lists it as such among marbles games in *London Street Games* (1916), 111. In Midgley near Halifax, 1890–1900, the game was called 'Plonks of the Mark'. In America it was 'Picking Cherries' (*American Boy's Book* (1864), 35).

Maclagan reports an Argyllshire game of knocking marbles away from a line, 'American Tag' or 'Chippy Smash', in which the shooting marble is called 'the knicker' (*Folklore* (1905), 347). 'Knicker', for a shooting marble, goes back at least to the late seventeenth century (see *OED*, *Nicker*, the derivation being from Dutch *knikker*, a marble, via *knikken*, to crack, snap). When cheap brass trouser-buttons came into manufacture in the mid-nineteenth century, and almost immediately became favourite playthings, the term *nicker* was also used for the large metal button, heavy lead disc, piece of slate or stone, or even door-key, which was skimmed towards the line of buttons to knock them away. This game seems to have been a particular craze in London, *c.*1890–1915; or did it just happen to be frequently described? (See *Strand Magazine* (1891), 515; *Living London* (1903), 268–9; *London Street Games* (1916), 15; and several Opie correspondents.)

The game is still played with marbles, but in a nebulous fashion, unadorned with a name: 'Line them up and, using one marble, try to knock them off the line' (Dudley, Worcestershire, 1960); 'I don't play marbles but when I do I put them out in a row and try to knock them out of line' (Newport, Isle of Wight, 1969); 'We each put four of them in a line, and each person has six chances to hit them' (Alloa, Clackmannanshire, 1975).

BUTTONS

'Buttons' or 'Buttony' is played in a variety of ways, although some of them have become vestigial and most extinct. In the basic game a circle is drawn on the ground, about 18 inches in diameter, and the players each throw or flick one of their buttons from about 6 or 8 feet away. If anybody's button rests in the circle, the thrower is entitled without further argument to every button so far thrown, and the game recommences. If nobody's button lands in the circle, the person who is nearest attempts to flick in his button (with a flip of the index finger off the thumb), and is then allowed to withdraw it, if he can, by licking his thumb, pressing it on the button and lifting it away. If successful he or she—the game is perhaps played more by girls than by boys—can attempt to carry off other players' buttons by the same admixture of spittle and pressure, continuing until a button falls from the thumb before it has been removed from the circle. In Stoke-on-Trent the game is played with further circles, or penalty places, known as 'fat'[2] and 'kisses'.

It will be appreciated that this game has a fundamental similarity to 'Pitch and Toss' (see Ch. 5, 'Gambling') and 'Chuckie', in that the player who initially makes the best throw has the greatest opportunity, though not the certainty, of winning the most from his fellows. Although the game has been reported from the south as well as the north, it has been on the wane from *c.*1950. Boys no longer need fly-buttons, nor buttons to suspend braces (both were considered expendable), and metal buttons, always the most prized for this game ('one mite is worth two or three

---

[2] *Fat* was an epithet for a marble or tipcat stranded outside the ring and therefore 'dead', see *EDD* and *N & Q* 3 (1899), 66.

twecks'), are becoming unknown. However, metal bottle caps are a fair substitute, and in Aberdeen the game is played with these under the name of 'Stoppers'.[3]

BANG-UPS

In this game a small object is thrown against a wall, to bounce back and hit, or be within a handspan of, a similar object on the ground. By the mid-twentieth century the object used was always a marble, and was often rolled along the ground, not thrown through the air. Thus at Crawshawbooth in Lancashire, 'The first boy sends his marble against the wall, letting it roll away where it will. Then the second does the same. If the second boy can span the distance between the two marbles then he claims the first marble. If he cannot the next person has his turn and as many marbles as he can span he can claim as his own. When a boy is successful he must start the game off again.' In Knighton, however, where the first marble is placed in position, rather than rolled at random, he who lost his marble, the 'loster', has to put down another to be shot at. At Alton, where the game is known as 'Wallsey', the children point out that it can equally well be played against a kerb. In Bedfordshire the name was 'Spanning the Marble'.

This game has a long history. In the years before the First World War it was much played in the streets of London with metal buttons (correspondents; *N & Q* 3 (1899), 184; and see *Traditional Games*, i (1894), under 'Banger'). In *London Street Games*, Norman Douglas commented:

Everybody knows the old game of 'Buttons' (or 'Bang-Out' or 'Bangings') where you pitch them against a wall and have to measure the intervals between them with the span of your fingers and always end up with a row about cheating distances. You can make a fine gamble out of 'Buttons' if you play the game with halfpennies; you can win quite a lot, when there are no coppers about . . .

Across the Atlantic the same game was known as 'Spans' (Newell (1883), 188); and Norman Douglas also lists 'Spans' (or 'Spanners') as a game with fag cards (p. 124).

In its earliest sighting the game is played with marbles. Richard Johnson, in *Juvenile Sports* (1773 or 1774), 96–8, says, 'The most common of all games at marbles is *nigs*, which some young artists very improperly call *lag-out*, for these are two different games'. In *nigs* two boys take turns to strike a marble gently against a wall, endeavouring to

---

[3] Douglas, *London Street Games* (1916), 151, has 'Buttons in the Ring'. Correspondents, recalling the game from 1900 onwards, knew it simply as 'Buttons'.

send it 'in such a direction as may give him hopes of hitting that of his playmate'. *Lag-out* is a longer game. If the second player fails to hit his opponent's marble on the rebound, 'he then takes up his marble, and shoots it at yours and if he hits yours he has it: but if he misses, you then take up the farthest marble, lag out as before, and then shoot at him. In this manner the game may be carried on as long as you like, or till one of you is a broken merchant.' Jamieson's *Scottish Dictionary* (1808), vol. ii, gives the game as 'Spangie', in which 'A marble or halfpenny is struck against the wall. If the second player can bring his so near that of his antagonist, as to include both within a span, he claims both as his'; *School Boys' Diversions* (1820), has 'Lag' (1823 edn., p. 36); *Boy's Own Book* (2nd edn., 1828), 10, 'Knock Out'; in *The Boy's Own Paper*, 24 December 1887, p. 207, it is 'Cannon'; but the writer adds, 'When "Cannon" is played out of doors it is called "Laggings", and then the rough ground or uneven wall reduces it to a game of chance.'

In *Les Trente-six figures* (1587), pl. 10, two boys are shown playing 'Le claque mur aux iettons' ('Bang-wall, with counters'); one of them is bending down spanning the distance between his own *jetton* and his opponent's, while the other looks on. Except for their costume they might be boys of the twentieth century. In Jacques Stella's *Les Jeux et plaisirs de l'enfance* (1657), a stylized group (naked) are shown playing the same game, but under the name 'La Patte aux Jettons' ('Palm, with counters'). One boy has his hand flat against the wall, and seems about to release his *jetton* in the manner of a London boy playing the cigarette game 'Droppie' 300 years later. In Brueghel's picture of children's games, painted in 1560, the game is shown twice. In the top-left corner, two boys are playing it with marbles, and in the top right two more play, one of whom stands in the same attitude as Jacques Stella's *putto*, with his hand flat against a wall. In *200 jeux d'enfants* (c.1892), 203–4, the game is called 'Le tic-toc', and is played with buttons.

BOTTLE TOPS AND OTHER FLIP-OVER GAMES

In the summer of 1979 a new craze arrived in Liss playground. The children were eager to explain.

It's called Bottle Tops. There are two people and they have a bottle top each. [These 'bottle tops' are the crown caps made of metal, with crimped edges, that are used to close beer and soft-drink bottles.] You flick them in turn until they meet and then with one finger you flick yours so it gets on top of the other person's and the one who gets it on top of the other person's wins the bottle top that's underneath.

Probably the game was being played in many other places; crazes seldom erupt in isolation. We had one other sighting. The game was being played, exactly as in Liss, in West Africa in 1975.

As usual, the game was a hybrid, the result of an old games-principle applied to new games-equipment. The principle is comparable to that of the cigarette-card games of 'Topsy' and 'Droppie', in which cards are thrown to overlap other cards, and the flip-over game 'Pogs' (see below). The equipment, crown caps, have been in existence since 1928. Occasionally they have been mentioned as substitutes for orthodox equipment in established games: flattened on the railway line, they were used in a game similar to 'Droppie' in Swansea, 1928; in a peg-top game called 'Chip 'ems' the top was set spinning and 'chipped down on to the metal cap, whizzing it along the ground' (R. Gamble, *Chelsea Child* (1979), 146, referring to *c.*1930). But the great crown-cap game is 'Skelly' (variously 'Scully', 'Scelsy, or 'Lodies'), which predominantly, if not exclusively, belongs to New York City—and this is only right, since the caps were invented in the USA. 'Skelly' is a game of flicking the upside-down caps into numbered compartments chalked on the sidewalk, and scoring accordingly. An equivalent game, played on a plain chequered diagram, is played in Spain under the name of 'Chapas' (*Times*, 18 July 1963, p. 7).

A craze for 'Pogs' swept the United States in 1992–4. In the 1960s a Hawaiian dairy began marketing POG, Passion fruit—Orange—Guava Juice, in bottles with cardboard caps. Children used the caps, as they used the ordinary milk-bottle tops, for a flip-over game. A teacher in Hawaii remembered the game of 'milk-covers' from her childhood, and taught it to her pupils, and the game spread not only to the mainland but to Canada and Mexico. Inevitably it became commercialized, and manufactured 'Pogs' appeared, bearing a variety of (collectable) designs, and costing between 10 cents and $1 each. Heavier, thicker discs are required for pitching; they are known as 'slammers', cost up to $6, and are made of plastic or metal—they are, in fact, the successor to the old lead 'nicker' (see p. 116).

The basic modern 'Pog' game is played thus. Each child tosses a pog into the arena, face-up or face-down, as agreed. Each player in turn takes his slammer and pitches it hard onto the accumulated pile of pogs. The caps that land opposite-side-up are the slammer's to keep.

As with all games in the ascendant, variations have proliferated ('There are now sixty-seven varieties of Pog playing', said one expert); and, in typically American fashion, there are now organized Pog tournaments. (See *Los Angeles Times*, 20 June 1994, section E; and *The Outlook* (Santa Monica), 4 July 1994, A8–7.)

Above: 'Le franc du carreau' (Bang-ups), in background. *Les Trente-six figures*, 1587.

Left: 'Chuckie', using marbles, *Phil May's Gutter-Snipes*, 1896.

Below: A set of Pogs, six discs and a slammer.

'Pogs' arrived in Britain early in 1995. Discs manufactured by Waddington Games Limited, copyright of the 'World Pog Federation', were sold in packets of six different designs of disc and a plastic slammer called the 'Kini' ('Kini' is Hawaiian for 'king') for 99 pence. Rules are included: 1. Decide who goes first and if you are playing for 'keeps' or not. 2. Each player puts an equal number of Pog milk caps into the stack, face (colour side) up. 3. Grip the Kini between two fingers. 4. Throw the Kini at the stack, making sure it leaves your hand and hits the stack. 5. All Pog milk caps that flip over, face down, go to that player. 6. All Pog milk caps remaining face up are restacked for the next player's attempt to flip the stack. 7. Players take turns until every Pog milk cap has been flipped and won. The player with the most Pog milk caps is declared the overall winner.

However, the journey from Hawaii to Britain is not the whole history of 'Pogs'. A pog-like flipping game was sighted in Jerusalem in 1965, where the play materials were folded chewing-gum wrappers, of varying values according to rarity. They were stacked with the pictures facing up or down, as agreed. The first player hit the ground beside the stack, or a draught was created by a swift downward swoop of the hand, the aim being to win some by dislodging some of the wrappers and turning them over (see Rivka R. Eifermann, *Determinants of Children's Game Styles* (1971), 50–1, 'Chewing Gums'). A similar game played with football cards was seen in Capri in 1971, but the players dispersed and flipped the cards by bringing a hollowed hand down onto the top of the stack. Undoubtedly the Hawaiian milk-cap game was only one manifestation of a world-wide flipping game, played with whatever small, flat, light objects were considered desirable enough to be worth winning. Predecessors of 'Pogs' can be seen in the card-flipping games, ways 2 and 4, played in Japan *c.*1940 (see p. 127). The pleasure of winning, as with cigarette cards, is not only in acquiring a larger number than one's competitors, but in gaining all the different varieties for one's collection.

CIGARETTE CARDS AND OTHER PICTURE CARDS

Games of skimming cigarette cards can only have been played after the invention of the cards themselves in 1880; they were originally inserted in the packets by the manufacturers to keep the cigarettes from being squashed. Perhaps the games started in America. In 1891 an astonished contributor to the *Journal of American Folklore* commented, 'With what rapidity child notions travel today! Cigarette pictures were a craze among street-boys for months before they were used for chance games. I think that flipping of cards struck New York, New Haven, and Baltimore within a single week.' That cigarette-card flipping is a matter of chance, is, however, vigorously contested. A Liverpudlian player, vintage *c.*1910, insisted, 'There's no question of luck. It was a game of skill. They went to the target straight and swift like an arrow.'

Although the major tobacco concerns stopped issuing cards soon after the beginning of the Second World War and never resumed production, picture cards have been continuously obtainable. They have been given away free inside packets of jelly, custard, cereal, and cheese; and, most famously, with Brooke Bond tea. They have been distributed with comics and have been inserted regularly in packets of bubblegum and sweet cigarettes. A wide range of picture cards can be bought, the most popular featuring footballers and cars. However, once our main survey was over, we could not be sure that the skimming games were still going on. Then in 1991 came news from Great Horton Middle School, in West Yorkshire: 'Some of the boys were flicking Euro Football cards [made by Panini], and seeing who could land nearest a wall. They were quite expert, and able to send them floating for 20–30 metres in some cases. The nearest the wall naturally won all the other cards flicked.' 'Nearest the Wall', at any rate, was apparently not forgotten.

NEAREST THE WALL

The boys 'flick', 'flirt', 'flip', 'fling', 'pitch', or 'skim' the cards to a wall, taking the back-right corner between the first and second fingers (the near-right corner resting against the ball of the thumb) and spinning it forward ('it took me six months to learn to do it properly') with a flick of the wrist: 'The idea is not to flick too hard, or the card will bounce back too far, and not to flick too gently, or it will not go far enough. The winner is the one whose card or cards is nearest the wall. He keeps the other cards.' As well as 'Nearest the Wall' the game is known by the generic names 'Flick', 'Flickers', 'Flicksie', 'Flickums', 'Flicks 'ems', 'Flippers', and, in Aberystwyth, 'Flakers'; also as 'Faggies', 'Ciggies', and, in Manchester, 'Dockers'. In West Ham it is 'Shorts, as opposed to 'Longsie' or 'Longs' (see below).[4]

For twenty years or more the game was also played with milk-bottle tops. The cardboard disc tops of the 1950s were flicked in the same way as picture cards, and the game was then known as 'Milkies' or 'Milk Toppers'. Later, when tinfoil caps became general, the tops were either

---

[4] 'Nearest the Wall Takes' is listed amongst games with fag cards in Douglas, *London Street Games* (1916), 123.

carefully smoothed out or, more often, the turn-down edge was held between the tips of the first and second fingers (the cap being held horizontally), the fingers were flicked apart, and the cap went spinning away. 'This action,' said an Ipswich boy, 'is usually accompanied by a sound that vaguely resembles an aeroplane.' In 1951 the game, when played with tinfoil caps, acquired the topical name of 'Flying Saucers', a term which rapidly became general (reported from places as far apart as London, Langholm, and Aberystwyth). In the late summer of 1976 a further contemporary implement came into use. A Dundee local paper announced joyfully: 'That old children's game "Penny up the wall" is having a new lease of life here. But it's not pennies this time, but the rings off can tops. One boy throws a ring up to a wall. The other has to get his ring closer to the wall to win. The experts can be seen with their winnings hanging round their necks . . . one told me he had 500 rings. I took his word.'

Immediately after the Second World War, when there was a shortage of picture cards, boys made 'stiffies' or 'skaters' by flattening the cigarette packets and folding them into themselves by an ingenious method which, as one boy put it, was 'handed down'.

KNOCK 'EM DOWN

'Knock 'em Down' was even more popular than 'Nearest the Wall', for a player has a chance of winning any number of cards at one time.[5] One or more cards are stood up against a step or wall, and the players, usually just two, take turns to flick their cards at them from 3 or 4 feet away, flicking with an overhand motion in the hope of achieving greater accuracy. A fortune in cards may be at stake since, as an 8-year-old put it, 'You might go for along time with out nocking the last one down and then some one would knock it down. That person would take all the sigerret cards that had mist.' However, one boy points out that you can play 'either for keeps or for lends'.

Empty cigarette packets are sometimes set up for the targets, and the thick bubblegum cards are particularly suitable for knocking them down. If the players run out of cards they pick up the card that is nearest to them and use that. Occasionally they make the game more of a gamble by playing with eyes shut, or flicking with the left hand, but the flight of a cigarette card being, to an adult eye, as erratic as a butterfly's, this seems scarcely necessary.

[5] 'Knock 'em Down' is listed in Douglas, *London Street Games* (1916), 124. 'Knock a Door Down' was played at Kirkby-in-Ashfield, *c*.1925.

Names: 'Knocks', 'Knocksie', 'Knocksie Down', 'Knock 'em Down', 'Knock Down Dolly' (Peterborough), 'Flicksies', 'Fall It' (Bristol), 'Wallie', 'Chuck a Card' (Helensburgh).

LONGSIE

' "Longsie" is when you flick a card and the one that goes the furthest is the winner' (Southwark).[6] The game is known by any of the generic names, 'Flick', 'Faggies', 'Ciggies', and so on (see 'Nearest the Wall') or, more specifically, 'Fars' (Birmingham), 'Longsie' or 'Longs' (general in London), 'Throwie' (Aberdeen), and 'Beats' (Grimsby), because they have to beat the other card or cards. There is a slight or imagined advantage in being the last to play, so the game is liable to be prefaced by cries of 'lardy' or 'leggy' as the children claim last go.

TOPSY

'Topsy', also known as 'Flickers', 'Fliers', 'Flivvers', 'Skimmy', 'Shortsies', 'Covers', and 'On Top', was the usual game that two boys played with each other when both were in possession of a good wad of flick cards. The object is to flick a card so that it lands on top of another already on the ground, an occurrence that is usually more providential than the player who flicked the card cares to admit. The game is played with both players standing beside each other and flicking their cards either into an open space or at a wall about 6 feet away. Occasionally the players face each other, about 10 feet apart, and flick into the space between them, when they may, to obtain greater propulsion if not accuracy, balance the card on their left fist, or some suitable object, and flip the edge of the card with their finger so that it flies off spinning.

One card is put on the ground, and the players stand 6 feet away. Each throw or flick their cards in turn. The object of the game is to try and get one card on the top or on the corner of any of the cards on the floor. The longer the game proceeds the more cards will be on the ground, and the eventual winner wins the lot, except one which is left down to be thrown at again. The game goes on until one player has won all the cards.

There are certain complications. If a card merely touches another, or does not overlap more than the white border, it is said to be 'Tipsy', and

<hr>

[6] *N & Q* 146 (1924), 337, 'A game of skill which involves flicking cigarette pictures to a distance is known by London boys as "scouring" or "skerring".'

is liable not to count unless it has been declared that 'Everything counts'. The other player may cry 'No tips!'; or in some places another turn is allowed. If the players do not agree, 'then a fight begins and when you stop fighting you scrabble for the cards'. If a card goes underneath another it is 'slips' and does not count. If a card lands upright against the wall the player can claim 'knocks' (West Ham) or 'flicks' (Brixton) and try to hit the card with his finger onto another card, but if his opponent says 'farthest' he must knock it onto the farthest card from the wall to win. Sometimes the rule is that the card must be 'bridges', that is to say, joining two cards together, before the game can be won, and the game is then likely to be called 'Bridges' (Walworth).

Variants include:

Picture a Go. If the game is being played with valuable cards, the successful player takes only his card and the card he landed on. In Poplar, east London, this was called 'Picture a Go'.

Certain One. The first card flicked or set down is nominated 'certain one' and it is this card which has to be covered to win the lot (West Ham).

Droppie, Droppers, or Dropsies. Instead of being flicked, each card is held against a particular brick or spot in the wall, about 3 feet from the ground, and then allowed to flutter down. If the card lands on another, the player usually takes all, but sometimes is entitled to pick up his own card and the one covered, and to have another turn. The game, which was moderately common in the 1950s and 1960s, was played mostly by girls.

Span. Both players put a card on the ground. The first player flicks his card and if it falls near enough to another card for him to be able to span the distance with his hand, he picks up both cards and has another turn. He continues until he fails to make a span, when it is the other player's turn (Bishop Auckland).

Norman Douglas lists 'Touch-Card', 'Getting On', 'Overlappings', 'Drop Them' (or 'Drops'), and 'Spans' (or 'Spanners') amongst games with 'fag-cards' in *London Street Games* (1916), 123–4. Correspondents confirm that 'Topsy', then known as 'Faggies', 'Flick', 'Skimmering On', and names not recollected, also 'Droppings' or 'Wall Drop', were known before or during the First World War. 'Droppie' appears to be shown in the distant right of Brueghel's picture of children's games, 1560, but played with marbles, where one boy is holding a marble up against a wall and another is measuring the distance between two marbles.

MENKO

Games similar to the American and European cigarette-card games were being played in Japan in the 1890s, with round paper cards called 'Menko'. The pictures on them were various, and of a powerful and heroic nature; during the Japanese–Russian war of 1904–5 they depicted famous generals.

Konosuke Fujimoto, in his *Dictionary of Outdoor Games* (1994), 228–31, describes the 'Menko' games he played as a boy *c*.1940, when the cards were oblong-shaped, or in the shape of the pictured human figure. There were four different ways of playing:

1. Throw your card forcefully onto the ground. If it slides under your opponent's card, you win it. This way is for younger children or beginners.

2. Throw your card down beside one already on the ground. If the wind created by your card turns the other card over, you win it. Sometimes a player puts his foot just beside the target card, which increases the effect of the wind. Or some people strike the edge of the target card to make it turn over.

3. Draw a circle (40 cm. in diameter). Each player puts two or three cards in the circle as a stake—whatever is agreed. Each player aims a card at one of those in the circle, trying to push it out of the circle and win it.

4. Players stake equal numbers of cards, and pile them in a column on a wooden box or chair. Players throw a card at the column, in turn, so that a card or cards falls on the ground. If a falling card lands so that it overlaps one on the ground the thrower wins that card.

The second and fourth ways are, taken together, strongly reminiscent of the game of 'Pogs' which swept the United States in 1992–4 and then invaded Britain (see above). Since 'Pogs' is supposed to have begun in Hawaii around 1970, it seems as if elements of the game may have travelled from Japan eastwards to Hawaii.

8

Ball-Bouncing

The girls speak of ball-bouncing with devotion. It is 'the favourite of the popular games', 'the mostest game I play'. 'At this time of year [May] nearly every girl brings two balls to school with her. You can't walk beside a wall without getting in the way of people bouncing their balls.' 'We're not allowed balls at school. We play for hours and hours at home though.' 'Is it any fun on your own?' we asked. 'Oh yes, I play every night.'

A summer evening's walk round northern alleys confirms what they say. There are young girls, on their own or in groups of two or three, their faces turned to a wall, intently juggling a couple of balls, whamming them against the wall with the full force of their shoulder muscles, either repeating a routine over and over again until they can master it, or competing against each other, gasping an accepted chant as they do so, and revealing a dexterity it seems almost unnatural for an ordinary child to possess. 'Two-balls', the most common form of ball-bouncing, otherwise known as 'Twosie Balls' (Birmingham), 'Two Baller' (County Durham), 'Two Bally' (York), and 'Doublers' (Scotland) is, they say, 'a very interesting game'. This comment is recurrent. Ball-bouncing is, indeed, less a competitive game than a pursuit in which skill is acquired for personal satisfaction. The ball is used here not as an object that can be rolled, or kicked, or fought for, or hit, or run after, or thrown to or at another person, or chucked for a great distance, but as a precision article which almost instantly rebounds from a firm vertical surface (a smooth wall or door) in a predetermined manner. They say 6 years old is none too early to start learning to play; and despite the fact that most ball games—from polo to marbles—are associated with the male sex, and that this game is exclusively feminine, it is possible that more people spend more hours acquiring a skill in ball-bouncing than in training for any other ball game in the kingdom.[1]

[1] The only game like this countenanced by boys is 'Glasgow Headers', played in Dundee, where boys spend long periods heading a medium-large ball against a wall, returning it to the wall again and again without letting it drop to the ground. The élite in this sport are able to accomplish it simply by neck-work and, with seeming nonchalance: they put their hands up in front of them and rest upon the wall while continuing to head the ball.

The amount of two-balls that takes place in a school playground depends not only on the availability of a suitable wall but on the forbearance of the head teacher. In 1952 the headmistress of St Andrews Junior School, Laverstock, Wiltshire, was able to describe the outside wall of her classroom—34 feet long with no windows or doors—as 'a wonderful wall for ball games'; and this despite the fact that 'it isn't soundproof, and the noise heard from inside is shattering'. In consequence a class of 9- and 10-year-olds was found to be in possession of fifteen rhymes for two-balls, and their play had evolved so that they recited them sixteen times over in different positions: Plainsy, Dumbsy, Standstills, Right hand, Left hand, Standing on right leg only, Standing on left leg only, Crosses (right foot over left), Crosses again (left foot over right), Little Dad (crouching down—the children said 'coupy down'), Big Dad (standing on toes), Thin Dad (close up to wall), Middle-sized Dad (two strides from wall), Fat Dad (four big strides from wall), Dancing Dolly (jigging up and down), Jumping Jinny (jumping up and down).

Gradually, since the mid-1970s, ball-bouncing has been disappearing from the south; but in the north, and especially in Scotland, the tradition continues.

BACKGROUND TO BALL-BOUNCING

The ancient Greeks seem to have known every form of ball game. Pollux (ix. 105) describes the game *Aporraxis*, in which a ball had to be thrown hard on the ground and caught on the rebound, the one who scored the highest number of catches being the winner. 'One could also', he said, 'throw the ball against a wall.' Juggling balls in the air probably goes back at least as far, and may well have been practised by children. It was, apparently, a popular children's pastime early in the nineteenth century, for the author of *Youthful Sports* (1801), saw 'a little girl keep up two or three balls at once, and that so cleverly, that it was quite astonishing'; and the illustrator of *The Book of Games* (1805), depicts two girls standing together, each concentrating on juggling her own two balls, perhaps in competition with the other, and one of them has her left hand behind her back. Curiously, this juggling in the air is not mentioned thereafter, and seems to have been overtaken by the more popular juggling against a wall.

In the eighteenth century, girls were already juggling balls against a wall. 'Miss playing with her Ball', in John Marchant's *Puerilia*, 1751, might be any present-day schoolgirl:

> This little Ball
> Against the Wall
> Or up and down I toss;
> It mounts aloft,
> And down as oft,
> It nimbly comes and goes.

And Jane Taylor was surely reporting a common sight when she wrote in *Original Poems for Infant Minds* (1804):

> Then ascends the worsted ball;
>   High it rises in the air,
> Or against the cottage wall,
>   Up and down it bounces there.

'Worsted balls', made by binding worsted wool around cork, were serviceable bouncers, as were the north-country 'keppy balls', made of segments of leather and stuffed with bran, and 'cuck a balls', made of particoloured rags, which were used with fortune-telling rhymes.

Fortune-telling with a bouncing ball lies at the heart of the history of ball-bouncing. About 1946, schoolgirls in Northamptonshire were patting a ball on the ground while chanting, 'Bally, bally, tell me true, How many years shall I go through? One, two, three . . .', and hoping to assure themselves of a ripe old age. Near Montgomery, in 1952, girls played the same game ('. . . How many years shall I live now?'). Likewise, in Ruthin, Welsh-speaking children chanted:

> Pel bach dywed y gwir,
> Pwy ydi cariad *Robin* bach ni?
> A, B, C, CH'D . . .
>
> Little ball, tell me the truth,
> Who is the sweetheart of our little *Robin*?
> [or any other name]
> A, B, C, D . . .

In the nineteenth century—and certainly long before—girls bounced a ball against a tree as many times as they could, asking, 'Keppy ball, keppy ball, Coban tree, Come down the long loanin' and tell to me, The form and the features, the speech and degree, Of the man that is my true love to be . . . How many years old I am to be, One a maiden, two a wife, Three a maiden, four a wife . . .' (G. Tate, *History of Alnwick* (1866), i. 436). Or a ball was thrown against a wall while the player chanted 'Cook a ball, cherry-tree; Good ball, tell me, How many years I shall be Before my true love I do see? One and two, and that makes three; Thank'ee good ball, for telling of me' (J. O. Halliwell, *Popular Rhymes* (1849), 220–1).

About 1840, in Aberdeen, the words were, 'Gowf ba', cherry tree, Catch a bird an' give it me, Let the tree be high or low, Let the weather be frost or snow' (*Folklore*, 25 (1914), 355). A version of this invocation was printed in an American children's book of English origin, *Nancy Cock's Pretty Song Book*, 'Printed by you know who, in Sugar Island', in 1786, and may thus claim to be the earliest ball-bouncing rhyme ever to be printed: 'Cuckoo, cherry tree, Catch a bird, And give it to me; Let the tree be high or low, Let it hail, rain or snow.'

The name 'cuckoo' or 'cuckoo ball' for a child's ball, still in use by schoolchildren in Leeds and some other parts of Yorkshire in 1952, is a corruption of 'cuck a ball', from *cook, cuck, chuck*, to throw or toss.[2] Attempts to connect it with the old divinations made from the first call of the cuckoo in spring may be discounted.

It was not until the long-delayed arrival of the indiarubber ball that balls could be said to spring from the ground 'as if alive'.[3] Indiarubber balls seemed to have remained novelties, and expensive ones, until the 'balloon ball' made of gutta-percha appeared on the market. This was hollow and air-tight (or supposed to be), and made a good lightweight bouncer. *Cassell's Book of Sports and Pastimes* (1882), was enthusiastic about it:

A balloon ball may be kept bouncing a long time by a small child after short practice. The ball should be grasped firmly in the right hand and thrown to the ground with just that amount of force that will make it rebound to a height about on a level with the player's elbow. When at that height the ball should be sharply struck with the open hand and sent again to the ground, and so on again with each rebound until the player misses.

Lady Gomme includes a game played by patting a ball on the ground in *Traditional Games*, ii. 406: 'Game, game, ba', ba', Twenty lasses in a raw [row], Nae a lad amon them a', Bits game, game, ba', ba'', from Fraserburgh and Dalry, Galloway. It was still played in that way by lassies in Scotland until at least the mid-1970s, for example in Maud, Aberdeenshire, 1975, 'Stot, stot, ba', ba', Forty lasses in a ra', Nae a laddie among them a', Stot, stot, ba', ba'.'

---

[2] First recorded in Grose's *Provincial Glossary* (1787): '*cook*, to throw. "Cook me that ball". Gloucestershire'; *cuckoo-ball* is first found in Robert Forby's *Vocabulary of East Anglia* (1830).

[3] The Indians in South America used to amuse themselves with heavy balls of vegetable gum in the time of Columbus, and the early Spanish explorers were startled by the way the material rebounded. Nevertheless caoutchouc was not commercially imported into Europe until the 18th-c., and then not for its elastic properties but—as its popular name implies—for rubbing out pencil marks, for being a 'lead-eater'. An Indian rubber ball is mentioned in [John Ayrton Paris], *Philosophy in Sport* (1833), 425; and when 'Mr Dowler bounced off the bed as abruptly as an India-rubber ball' (*Pickwick Papers* (1837)) he was being likened to a modern phenomenon—something of a novelty.

Bouncing or patting a ball on the ground is now out of fashion, except in the games above and 'Alairy' (p. 136). The sharp rebound of a rubber ball thrown to a wall, the variety of throws and stances possible, and the accretion of verses to regulate the throws, have made two-balls a far more stimulating challenge. Further, the sophisticated play current today was facilitated by the introduction of the Sorbo sponge rubber ball in 1920. In the 1950s and 1960s a 'Sorbo sponge bouncer', 2½ in. diameter, weight 2⅓ oz., price 1s. 9d. (1965) was standard equipment, although a variety of similar sponge rubber balls, painted and unpainted, could be bought for 1s. 3d., or a shilling, or even less. In mid-May the keeper of a village store told us that all the girls from the nearby school had been coming in and asking for his ninepenny rubber balls. Proudly, but slightly puzzled, he added, 'They buy them *two at a time*!' In the 1970s, tennis balls—the cheaper, yellow 'practice balls'—were more used than rubber (or synthetic rubber) balls.

SEVENSY

It is evident from *Cassell's Book of Sports and Pastimes* that a canon of fancy throws was already becoming established in 1882. The game of 'Sevens' is described as 'catching a ball seven times consecutively in some one particular way, and then again seven times in some other way, and so on'. Among the ways 'frequently adopted' were:

First catch the ball seven times with both hands.

Then throw the ball with the right hand and catch it with the same . . .

Next throw with the left hand and catch with the left . . .

Again throw, and clap hands in front during the interval that lapses between the throw and the catch.

Vary this again by clapping the hands at the back.

Last suggestion. Turn round between the throw and the catch.

When this was completed, or if a catch was missed, 'the ball should be passed on for the ingenuity of the next young player'—an injunction maintained to this day in ball-bounce etiquette.

By 1898 Lady Gomme's correspondents were reporting similar movements when throwing a ball against a wall (*Traditional Games*, ii. 64–5, 405–6); for example at Hexham in Northumberland children played a game called 'Pots', a name apparently borrowed from Northumbrian hopscotch:

Throw to wall, let it bounce, and then catch. Repeated three times.

Throw to wall, let it bounce while twisting hands, then catch.

Throw to wall, let it bounce while clapping hands in front, behind, and in front again, then catch.

Throw to wall, let it bounce while turning round, and then catch.

Two more movements followed in which the ball was thrown to the ground and then caught.

In the 1950s–1970s 'Sevens' or 'Sevensy' was—and probably still is—the basic exercise with a single ball, usually carried out with throws that decrease in number as the actions become more difficult to do. Thus in Alton, Hampshire:

Seven throws to wall.

Six throws to wall, letting ball bounce on ground on return before catching.

Five throws to wall, letting ball bounce on ground, then smacking it back to the ground before catching.

Four throws to wall, hitting ball back to wall with palm of hand, and then catching.

Three throws under leg to wall.

Two throws to wall, letting ball bounce on ground, hitting it back to the wall with hand, and then catching.

One throw to wall, letting ball bounce on ground, hitting it up in the air, then hitting it back to the wall and catching on rebound.

In some routines it is the manner of throwing the ball which becomes increasingly difficult, as in Cleethorpes:

Seven plain throws to the wall.

Six throws with legs apart, throwing the ball from behind between the legs to the ground so that it bounces to the wall and returns.

Five throws, bouncing the ball between legs as before but with back to wall.

Four throws, standing sideways to wall, throwing ball from behind back onto ground so that it bounces to the wall and returns.

Three throws, standing sideways, from behind back directly to wall.

Two throws with legs apart, bouncing the ball between legs, turning away from wall, and hitting the ball back to the wall through the legs, and then catching it.

One throw, starting with back to wall and bouncing ball through legs, turning round to face the wall with legs apart, hitting the ball back to the wall from behind through the legs, and then catching.

If a girl manages to accomplish this sequence, she may then attempt it 'dumbsie' (without speaking), 'stillsie' (without moving), with right hand only, with left hand only, 'clapsie' (with a clap after each throw), and 'dancing dolly' (jigging from one foot to the other).

Sequences like this are sometimes played up to 'Tensy', for example at

Binbrook, Rushmere St Andrew, Radcliffe, Oundle, and Welshpool. On the whole, however, these 'exercises', as the children are apt to term them, appear to be less common than they used to be. The proportion of children who described 'Sevensy' and 'Tensy' in our surveys, in the 1950s–1970s, was significantly lower than that of our adult informants; in fact, no reports have come in since 1975—but this may be simply because the major surveys were over.

PLAINSY, CLAPSY

The current trend with a single ball is for quick, rhythmic routines in which no two succeeding throws are the same; thus in Edinburgh the ball is either thrown against a wall or thrown in the air as they repeat:

Plainie (a plain throw)
Clappie (hands clapped before catching)
Roll up in (hands rolled round each other in wheel-like movement)
Tobaccy (hands clapped behind back)
Right hand (ball thrown and caught with right hand)
Left hand (ball thrown and caught with left hand only)
High sae toosh[4] (ball caught in cupped hands held above head)
Low sae toosh (ball caught in cupped hands held low on the tummy)
Telephone (action of telephoning, using both hands)
The answer (clapping over and under leg)
Touch your heel (heel touched)
Touch your toe (toe touched)
Through you go (ball thrown under leg from outside)
Bring it back (ball thrown under leg from inside)
Wee birly-O (hands crossed on chest)[5]
Big birly-O (twirl round before catching)

Such a sequence is not untypical of the northern half of Britain. In Bishop Auckland they chant and enact: 'Clappo, Rollio, To Balcony, Highsetina, Lowsetina, Cup, Basket, Right hand, Left hand, To touch your knee, To touch your toe, To touch your heel, And under you go— Butterfly!' (in which 'the arms are crossed over the stomach'). In Ballingry, Fifeshire, they chant: 'Plainie, Clappie, Lappie, Backie, Right hand, Left hand, Through the woods (under right leg from inside), Back again, Telephone, Answer, Low sottouche, High sottouche, Wee birlio (hands twirled), Big birlio (turn around).' And in Flotta, one of the Orkney Islands, they 'shout': 'Plainie, Clappie, Rollie, Backie, Right hand, Left hand, High schottische, Low schottische, Touch the toe, Touch the heel, Touch the ground, Turn right round.'

[4] Possibly 'highsie toosht' (*toosht*, to toss about, *EDD*).
[5] *To birl*, to revolve rapidly. Current since late 18th c.

Just how little the sequences vary from one part of Britain to another may be seen by comparing recordings from the toe of Cornwall, the East End of London, mid-Wales, and the far north of Scotland:

Plainsie, clapsie,
Roll-a-ball to backsie,
Right hand,
Left hand,
Touch your heel
Touch your toe,
And under you go.

(Girl, 10, Pendeen)

Plain, clap,
Twisty, toboganny,
Right hand,
Left hand,
Touch your heel,
Touch your toe,
Touch the ground,
Twirl around.

(Girl, 12, Welshpool)

Plainy, clappy,
Roll the wheel to backy,
Right hand,
Left hand,
Touch your heels,
Touch your toes,
Touch your knee,
And under you go.

(Girl, 11, Hackney)

Plainie, clappie,
Rollie, backie,
Right hand, left hand,
Highsie, lowsie,
Touch your heel,
Touch your toe,
Touch the ground,
Burl right round.

(Girl, 14, Golspie[6])

In the 1950s–1970s, when we were collecting these formulas, the most noticeable divergences were not in the formulas themselves, but in the children's speech when describing them. A 10-year-old on Flotta (Shetland) said, 'Plainie means heaving the ball up in the ordinary way'. 'Heaving', pronounced *havun*, is the customary word on Flotta for throwing upwards; and 'ball' is pronounced 'ba'' or 'bal', rhyming with *Sall*. In Bristol, and eastwards to Trowbridge and Wootton Bassett, children do not ordinarily say they 'bounce' the ball—they 'dap' it.[7] In Orkney they 'dart' it. In Abergavenny they 'tamp' it, or did so until recently; and in Bishop Auckland, County Durham, they 'stounce' it. In most of the north country, including the City of Durham, Cumberland and Northumberland, and virtually throughout Scotland, children use the word 'stot', a word that has been current since the Middle Ages, and which possibly meant 'to bounce' before 'bounce' ever did. Balls are 'stotted'. A 'stottie-ba'' is a ball for bouncing, and may be declared to be

---

[6] Cf. also, in Dublin: 'Claimy (plain), Clappy, Rolly, Foldy (arms folded), Hippy (hands on hips), Slippy (foot pushed along ground), Highsicky (high on wall), Lowsicky (low on wall), And a basket (turn around).'

[7] *EDD* gives 'to dap a ball' as a term current in Gloucester, apparently at the end of the 18th c. William Clarke speaks of 'dapping' a ball in the first edition of *The Boy's Own Book* (1828), 4. He was not a Londoner born, and is believed to have come from Bristol.

a 'braw stotter', for a game of 'stotters'. An exception is Glasgow and the surrounding industrial area, where the word is pronounced 'stoat' and is written so by the children.

ALAIRY

There is also linguistic interest in the only widely played game in which a ball is continuously patted to the ground. The words always begin 'One, two, three, alairy', no matter how deviant the succeeding lines; and the tune usually resembles that of the singing game 'Dusty Bluebells' (see Opie, *Singing Game*, pp. 366–7). The special feature is that each time the player sings 'alairy' or its rhyme-word (or supposed rhyme-word), she cocks up her leg and swings it over the ball. When the player is skilful it is a dainty performance, almost a dance. Since the 1950s, instead of the ball being bounced on the ground, it is often bounced against the wall, and is allowed to bounce on the ground on the last word of each line, so that a leg can be swung over it and the ball caught.

One, two, three, alairy,
My ball's down the airey,
Don't forget to give it to Mary,
Early in the morning.

(Welwyn and Enfield, 1950s; Isle of Dogs, East London, c.1905; Seavington, Somerset, 1922. Cf. Opie, *Lore*, 'Charlie Chaplin', pp. 108–9. 'Airey', area)

One, two, three, a leerie,
Four, five, six, a leerie,
Seven, eight, nine, a leerie,
Ten a leerie, postman.
Open the gates and let him in,
He is soaking to the skin,
Open the gates and let him in,
Early in the morning.

(Lanarkshire, c.1902; Forfar, 1910 and 1952; Caithness, c.1915.Wellington, New Zealand, 1950. Golspie, 1953, ball patted on ground. First four lines only, well known in Scotland and northern England. Also USA, including New York City)

One, two, three, alaira,
I saw my sister Sarah [or Clara],
Sitting on her bumalaira,
Eating chocolate babies.

(Chingford, Dovenby, Dulwich, Radcliffe, Swansea, and elsewhere in the 1950s)[8]

One, two, three, a-leerie,
I spy Mrs Peerie [or Pirie],
Sitting on a rocker cheerie
[rocking chair],
Eating chocolate babies.

(Aberdeen, c.1915; still current 1954)

One, two, three and a larry,
My husband's name is Harry,
If you think it necessary
Look it up in the dictionary.

New York City, 1938; Aberdeen, 1952; Montreal, Canada, 1962; Montrose, 1974[9]

[8] Cf. Opie, *Lore*, p. 115, 'One, two, three a-leary, I saw Wallace Beery, Sitting on his bumbaleerie, Kissing Shirley Temple', from Edinburgh, c.1940, and Kirkcaldy, 1952; an earlier version of this ended 'Kissing Sophie Tucker'.

[9] Deviants, both 1952: 'One, two, three, alairy, I saw Aunty Mary Sitting on the lavatairy Eating chocolate dainties' (Manchester); 'One two, three, alaira, Four naughty boys with Sarah, Sarah's nice, they like spice, One, two, three, alaira' (North Shields).

It is interesting that the mysterious word 'alaira' (or 'a leerie') was accepted without question for most of the century, and only in the mid-1970s did it begin to be rationalized into 'a lady'. (For instance, in 1975, Dundee and Glasgow, 'One, two, three a lady . . . Ten a lady, postman'; Perth, '. . . postman's knock'. An Edinburgh version, 1990, went 'One, two, three old lady'.) In the USA, in the 1940s, the word was being written down as 'O'Leary', and probably the children thought they were singing about an Irishman (*JAFL* 60 (1947), 48).

'One, two, three alaira' is the pattern for one of the many sequences of fancy throws. In the 1950s our survey produced versions of the basic rhyme from thirty-two places, and versions of the sequence verses from six. This is a typical sequence (from Sale, Manchester):

You have one ball or two balls and toss them against the wall and catch them. On the numbers you just throw the balls on the wall but at 'a-loopy' you throw them overarm. It goes, 'One, two, three, a-loopy . . . Ten a-loopy, catch the ball.' Then you repeat it and you do, 'One, two, three a-baker' [one hand behind back], then 'a-curtsey' [curtsey], 'a-bowy' [bow], 'a-dropsey' [ball bounces on ground before catching], then 'upsy' [ball thrown high against wall].

Sequences were popular at least into the 1970s, when the children of Peabody Buildings, London WC2, were recorded (1974) singing 'One, two, three one-leggy . . . Ten one-leggy, drop the ball' (throw under right leg), followed by '. . . the other' (throw under left leg), 'one handy', 'the other', 'shovsie' (pat ball against wall with right hand), 'the other' (the same with left hand), 'upsie' (ball thrown into air, not onto wall), 'downsie' (bounce ball on ground), 'overs' (overarm onto wall), 'dumbsie' (saying nothing), 'stillsie' (stand rooted to spot while playing—difficult, as the player has to stretch for the ball).

In *London Street Games* (1916), 5, Norman Douglas lists 'One-two-three-and-a-lairy' and remarks, 'I wish I knew what a-lairy meant.' The word is variously pronounced to rhyme with 'Mary', 'Beery', 'Pirie', 'Sarah', and 'Harry', and the children say it means just what they do—crooking their leg over the ball. The only record of the word 'airy' known to the *Oxford English Dictionary* occurs in the alliterative poem *Piers Plowman*. In the A text, of 1362, Piers goes to see how his half-acre of land is being ploughed and finds that some who have undertaken to help him are malingerers: 'Summe leiden þe legges a-liri as suche losels cunne' (vii. 115). In other words, 'some crooked their legs beneath them as such scoundrels can', so that they could moan to Piers that they were unable to work. It seems possible that when children play 'Alairy' they are repeating and enacting a Middle English term which was discarded by adult society centuries ago. The *OED* Supplement says that perhaps *aliry* comes from Old English *lima lyre*, loss of (the use of) limbs, thus giving the 1362 quote a meaning of 'made their legs lame, acted as if paralysed'. It may further be noted that the other texts of *Piers Plowman* give the variant readings *a lery* and *a lyry*, and that an apparently related term appears in *The Tale of Beryn* (*c.*1400), line 309: 'He stappid into the tapstry wondir pryuely And fond hir ligging lirylong.'

TWO-BALLS AND A SINGLE PLAYER

Looking through the ball-bounce phrases and rhymes we collected in the 1950s, 1960s, 1970s, and, spasmodically, up to the present day, they seem to be ragged, random, and ridiculous. This is, of course, the case. The point of most ball-bouncing is not the poetry of the words but the sequence of different throws that must be performed in time to the words without making a mistake. The basic juggling movement is performed so smoothly that an adult observer finds difficulty, at first, in following it. In all two-ball games, the ball in the left hand is passed into the right hand and the rebounding ball caught in the left hand; this is achieved all in one fluid, continuous movement which suggests pulling rather than catching. The subsequent movements follow the local sequence, and each has its own local name. Words are scarcely necessary: they are simply a more entertaining way of timing the throws than, perhaps, counting 'One, two, three' and saying the name of the action. A phrase of four counts is needed, such as 'For-get-me-not' (Liss, 1970) or 'Black cat sat on a mat' (Rochford, Essex, 1957; Dulwich, 1976), and an action is performed on each beat. (A couplet or quatrain that gives four beats to a line is also used, and then the actions are performed on the rhyme-words.)

Nevertheless, the words must have *some* appeal (a child would not bounce a ball to the four beats of 'I walked up the road', for instance). They must have pleasant associations and, nearly always, rhyme ('Dickie Whittington Lord Mayor of London', Feltham, Middlesex, 1957; 'Mother in bed eating bread', Forfar, 1952); or provide the satisfaction of spelling out a word ('Peter Pan, Peter Pan, Peter P-A-N', Glasgow, 1961); or have the authority of advertising slogans ('Brooke Bond Divi-dend Tea', known nationwide; 'Drinka pinta milka day, M-I-L-K', Arbroath, Montrose, and Greenock, 1974; 'Alka-seltzers, speedy Alka-seltzers, Alka-seltzers, take the pains away', Glasgow, 1975). However, the most popular formula of all turned out to be, simply, 'North, South, East, West'.

In Liss playground, in February 1970, Jane Pay, aged 11, explained how to play 'For-get-me-not' ('Though I don't like it. "For-get-me-not", that's all you say.')

You do plainsie first. You get two balls, and you just throw them on the wall twice. Then you go to overs. You throw the balls overarm, against the wall. For upsie you throw the right-hand ball straight up in the air, then when it comes down, the left-hand ball should be in your right hand and you throw that one up. You just keep moving them along, like juggling. Downsie is like upsie, but you bounce the balls down on the ground. Dropsie—you throw the right-hand ball on the wall, overarm, and let it drop on the ground and catch it with both hands, though you are still holding the other ball in your left hand. Then you swap over balls and do it with the other ball. Then crashes, or crash bang. You bang the ball so that it will go into the very bottom edge of the wall, and it bounces back at you. You do that one ball at a time, too. For crash drop, you throw the ball like for crashes and let it drop and catch it. Crash upsie—you just add crash and upsie. Crash downsie, you crash the ball and do downsie. After that you do crash allsie, where you do all the things you've done all together—plainsie, overs, upsie and everything—you do them all in turn without stopping.

'For-get-me-not' is certainly no verbal gem. A couplet can offer more in the way of entertainment:

> Dopey Dinah went to China
> In a packet of semolina.
>
> > (Forfar)
>
> The Boys' Brigade are too afraid
> To drink a bottle of lemonade.
>
> > (Nottingham)
>
> Soldier blaw, soldier blaw,
> Soldier blaw yer kilt awa'.
>
> > (Aberdeen)
>
> Lord Mayor sat on a chair,
> One, two, fell through.
>
> > (Hythe, Hampshire; specially
> > recommended for 'Three-balls')

The significant words are the rhyme-words. For instance, in the most popular of these formulas, 'Shirley Temple is a star, S-T-A-R', they are *star* and *R*. At the first recital, all eight throws are plain—i.e. the basic juggling movement; but thereafter on *star* and *R* the throws will be fancy ones. The girl chants 'Shirley Temple is a star, S-T-A-R' as before, but this time the throws are: Plain, plain, plain, *overs*, Plain, plain, plain, *overs*. This is followed by throws of Plain, plain, plain, *dropsies*, or whatever special throw comes next in the local sequence. Thus at Bishop Auckland, where they commonly chant 'Shirley Temple' or 'Maid Marion thinks she's good, 'Cause she's married to Robin Hood', although

Girl playing 'Two-bally' in Workington, 1962.

the sequence begins in the same way as at Liss with 'plainsie', then 'overs' (at 'good' and 'Hood') for the first repeat, then 'upsie', the subsequent actions are 'drapsie' (let ball bounce on ground before catching), 'over-drapsie', 'upsie-drapsie', 'stauncie' (staunce = bounce; bounce ball on ground, not against wall), 'right leg' (ball under right leg onto wall), 'left leg', 'sidie' (bounce ball on ground, to side), 'right leg stauncer' (bounce ball under right leg and catch), 'left leg stauncer', 'backsie' (clap hands behind back before catching), 'frontsie' (clap hands in front).

Similarly at Bristol, when they do 'Shirley Temple is a star' or 'Tiny Tim swallowed a pin, That was the end of Tiny Tim', they go through plain, then at *pin* and *Tim* they follow with 'over', 'upsy', 'dropsy', 'basket' (catch the ball in interlaced fingers), under the right leg, under the left leg, then 'digsy' (ball thrown into angle between wall and ground), then 'fountain' ('You throw the ball into the air, let it bounce, then catch').

In Birmingham, where they often choose to play 'Shirley Temple' or 'Lord Mayor be *fair*, Beefo, Oxo', the sequence is 'droppies', 'overs', 'ups', 'slams' (throwing ball to ground so that it bounces onto the wall and returns—the opposite of Bristol's 'digsy'), 'twirly rounds' (high onto wall and turn around), 'jig abouts' (keep jigging while throwing), under right leg, under left leg, under right hand, under left hand, stand on right leg and throw with right hand, stand on left leg and throw with left hand, stand on left leg and throw with right hand, and stand on right leg and throw with left hand. 'If the child gets through the whole of this,' said our careful observer at Golden Hillock Road, 'the idea is to repeat with two "droppies", two "overs", two "uppies", and so on.'

It may be useful to set out here a large enough number of the rhymes used with ball-bouncing sequences, chiefly from the 1950s and 1960s, to indicate why the 8–10-year-old ball-bouncers thought them not only worth repeating but even 'very funny' or 'very clever'. Humour is an obvious ingredient. Food is a favourite subject, especially such beloved foods of childhood as ice-cream, candy, and jam. The adventures of characters from books and comics ('Milly Molly Mandy'; 'Pansy Potter') are related; and also brief narratives making fun of grown-ups ('Mrs Brown' and the other ladies). Rhyme and rhythm are considered clever in themselves (and why not?); and the ultimate in cleverness is a rhyme invented by oneself or by one's best friend. It can be a second line added to an overfamiliar formula ('Brooke Bond Dividend Tea, That's what the doctor ordered for me', Radcliffe, Lancashire, 1952; or '. . . Half a pound for one and three', Bishop Auckland, 1961), or it can be an entirely original composition, such as:

Bread and butter makes the tucker,
If you've nothing else for supper.

(Aberdeen, 1952)

Marzipan tea cakes,
That's what the Guides makes.

(New Cumnock, 1961)

Many of the following, besides, have resonances of other, older traditional verses, or had or still have other functions, such as counting out. Children do not know, or care, that a folklorist's task would be made easier if ball-bouncing rhymes were only used for ball-bouncing (and skipping rhymes only for skipping; counting-out rhymes only for dipping). When a child needs a rhyme, any rhyme will do. Nor do they know what problems they cause when they nonchalantly add two lines of one song to two lines of another, or take the first line of an orthodox chant like 'Each, peach, pear, plum' and, on a whim, add a new line of their own— 'I like bubble gum'. It is their right, of course. But should the folklorist unravel the hybrid rhyme and identify the various elements? Not to a minute degree, I think; it would be too tedious. Cross-functional or historical notes have been added to the following rhymes where they seemed interesting; and all the rhymes are normally used for sequences of actions performed on the rhyme-words—'except,' they say, 'if you don't want to.'

Mary Ann, bread and jam,
Marmalade and treacle.
A bit for you, a bit for me,
A bit for all the family.[10]

(Wootton Bassett. Almost the same in
London Street Games (1916), 62)

Milly Molly Mandy
Bought a stick of candy,
When it was gone, she bought a
  scone,
Milly Molly Mandy.[11]

(Sale, Manchester)

Ice-cream cone,
Ice-cream cone,
You only pay a penny
  For an ice-cream cone.

(Aberdeen)

Three wee tatties in a pot,
  Lift the lid an' see if they're hot.
If they're hot, cut their throat,
Three wee tatties in a pot.[12]

(New Cumnock)

[10] c.1890, 'A bit for you, and a bit for me, And a bit for all the familee' (or, '. . . And a bit for Punch and Judy') were the last two lines of a four-line skipping verse beginning 'Up and down the ladder wall, Ha'penny loaf to feed us all' (see *Traditional Games*, ii. 202–3).

[11] Milly Molly Mandy is the heroine of a series of easy-to-read and very popular books by author-illustrator Joyce Lankester Brisley. The first, *Milly-Molly-Mandy Stories*, was published in 1928, and the last in 1967.

[12] A popular counting-out rhyme, see Opie, *Games*, p. 38. The first couplet was recorded by H. C. Bolton, *Counting-Out Rhymes of Children* (1888), no. 716, from Philadelphia.

Oranges, oranges, four a penny,
Mother's ill and can't buy any,
Father's drunk and hasn't a penny,
Oranges, oranges, four a penny.[13]

(Aberdeen, Glasgow, New Cumnock)

Winnie the witch fell in a ditch,
Picked up a penny and thought she
    was rich.[14]

(London)

Pansy Potter the strong man's
    daughter,
Went to school without her jotter.
'Oh,' said teacher, 'that's not
    proper,
Coming to school without your
    jotter.'

(Perth. Similar at Aberdeen, Bressay, and
Helensburgh)

Mrs Brown went to town
With her knickers hanging down;
Mrs Green saw the scene,
And put it in a magazine.[15]

(Versions from all over Britain, including
our own children)

Keyhole Kate she went to skate,
She didn't come home till half-past
    eight.
Half-past eight is far too late,
For a little girl like Keyhole Kate.

(Aberdeen and Bressay. Keyhole Kate
was the heroine of a comic strip in
Dandy)

Tommy Thistle blew a whistle,
On a Sunday morning,
A policeman came and took his
    name,
And Tommy said, 'Good morning'.

(Dunoon and Paisley. Tune: 'Yankee
Doodle')

My mother and father were Irish,
And I was Irish too;
They bought a pianner
    For two and a tanner,
And that was Irish too.[16]

(Gloucester and Manchester)

---

[13] Lines that have been around for many years, first appearing as the beginning of a conglomerate counting-out rhyme in Robert Chambers's *Popular Rhymes* (1847), 258, 'Lemons and oranges, two for a penny, I'm a good scholar that counts so many; The rose is red, the leaves are green, The days are past that I have seen . . .'. In 1961 a teacher in Dacca, Pakistan, wrote to say that she had found that the game of 'Oranges and Lemons' (played as in England) was a great favourite with little Muslim girls, the words being 'Oranges and Lemons sold for a penny, All the schoolgirls are so many; The grass is green, The rose is red, Remember me when I am dead, dead, dead . . .'.

[14] A popular rhyme all over Britain, which gave rise to many variations, such as 'Winnie the witch fell down a ditch, Ha! ha! ha! Serves her right, she shouldn't fight, Ha! ha! ha!', Cheltenham, 1960, and 'Winnie the witch fell in a ditch, And cried for *help, help, help*', Ford Houses, near Wolverhampton, 1969.

[15] An indispensable rhyme, apparently. It is still repeated just for its naughtiness. Similar lines have been used as part of a clapping rigmarole since *c*.1900 (N. W. Thomas (ed.), County Folklore, 4. *Northumberland* (1904), 120). Popular for ball-bouncing, it spawned numerous ephemeral variations: 'Mrs Brown wore a crown, It was yellow, white and brown', Forfar; 'Mr Ross thinks he's boss, Because he's the owner of H. P. Sauce', Aberdeen; 'Mrs Grey she went to stay, With her husband in the USA', Aberdeen; 'Mrs Red went to bed, In the morning she was dead,' Montrose; and so on.

[16] This is the remnant of a song, 'The Ninepenny Fidil', by Joseph Campbell, current 1915.

Upstairs, downstairs, in a caravan,
You only pay a sixpence to see a
  funny man;
The funny man broke, tied to a
  rope,
One, two, three, and a hop, hop,
  hop.

(Aberdeen[17])

I went to Morecambe for the day,
I lost my ticket on the way,
The porter said I had to pay,
For riding on the railway.

(Manchester)

There was a wee melodyman,
A rufty tufty toady man,
I always do the best I can
To follow the wee melodyman.

(Common in the north country. Comes
from the singing game 'Follow my gable
'oary man', see Opie, *Singing Game*,
407–8)

Open the gates for the Red Cross
  nurses,
Open the gates for the Red Cross
  nurses,
Open the gates for the Red Cross
  nurses,
They must do their duty.
D-U-T-Y stands for duty,
D-U-T-Y stands for duty,
D-U-T-Y stands for duty,
They must do their duty.

(Bishop Auckland and Aberdeen)

Please keep off the grass,
And let the ladies pass;
The first remark in Hanley Park
Is 'Please keep off the grass'.

(Hanley, Stoke-on-Trent. Similar in
Liverpool, but 'Sefton Park')

Rabbie Burns was born in Ayr,
Now he stands in George's Square,
If you want to see him there,
Jump on a bus and pay your fare.

(Aberdeen, Dunoon, Perth, Glasgow,
Forfar)

A wee wee woman and a wee man,
A wee wee kettle and a wee wee
  pan;
'You boil the kettle while I boil the
  pan,'
Said the wee wee woman to the
  wee wee man.[18]

(Radcliffe, Shrewsbury, Stoke, Kendal,
Lincoln, Amlwch (Anglesey), St Helens,
Lancashire. Players crouch down as far as
possible while throwing balls)

Walter, Walter,
Lead me to the altar,
I want to be a bride,
Stamp, Gibraltar.

(Glasgow,[19] Dunoon, Aberdeen)

---

[17] 'At "upstairs" you throw it high up the wall. At "downstairs" stot it. At "hop, hop, hop" you hop.' The rhyme has been current in Scottish playgrounds for most of the 20th c., see *Miscellanea of the Rymour Club*, 1 (1911), 192.

[18] This comes in the traditional story of 'The Wee Bunnock', in Robert Chambers's *Popular Rhymes of Scotland* (1869), 86: 'I fore-ran, A wee wee woman and a wee wee man; A wee wee pot and a wee wee pan; And sae will I you an I can.'

[19] Girl, 14, Glasgow: 'You play this by stoating the balls off the ground against the wall. [Bouncing the balls off the ground onto the wall is Glasgow's own particular way of ball-bouncing and is the reverse of the method used everywhere else.] When you come to the line "Stamp, Gibraltar", you stamp your foot and stoat the ball behind your back. This action is called "jibby".' 'Stamp Gibraltar' is not just a rhyme for 'Walter' and 'altar', but a recognized Scottish action, appearing e.g. in a Dunoon version of the skipping song 'Charlie Chaplin went to France', in which 'Heel, toe, stamp Gibraltar' is one of the actions that 'Charlie taught them'. The words 'Walter, Walter, lead me to the altar' are from Gracie Fields's famous song, 'Walter, Walter', 1937, fitted with an action ending and rhyme-word.

Mrs Masaline at the pawn shop
    door,
With a bundle on her back and a
    bundle on the floor.
She asked for ten shillings but she
    only got four,
So she nearly pulled the hinges off
    the pawn shop door.

(Dunoon and Aberdeen)

Where's Mary? Where's Mary?
Has anyone here seen Mary?
She went away six weeks today
And has not yet returned.
So send the bellman round
To every village town
Crying, 'Mary, Mary, bonny Mary,
Mary must be found.'

(Scarborough)

Little Betty Bouncer,
Loves an announcer,
Down at the BBC.

(Bristol and Enfield[20])

Underneath the spreading chestnut
    tree,
Gilbert Harding said to me,
If you want to join the BBC
You must have a voice like me.[22]

(Kintore, Aberdeenshire)

Sweet sixteen, little Angeline,
Always dancing at the village green.
The boys pass by,
You hear them cry,
'Sweet little Angeline, sixteen!'[21]

(Glasgow, 1952 and 1975; Paisley, 1975)

I love a lassie, a bonny bonny lassie,
As sweet as the heather in the dell,
As sweet as the heather, the bonny
    bonny heather,
Mary, my Scotch Bluebell.[23]

(Dunoon)

Explaining a ball-bounce ritual, a little girl in Penrith said, 'You can use any rhyme. There are so many they are countless.' Over the years we were sent 244 rhymes and songs chiefly used for ball-bouncing; some were sent only once, most were backed up by several or many other recordings. Some scraps of verse and invention sent were not even deemed worthy of separate entry: ' "Threepenny ices, three!"—that's all there is of it' (Swansea); ' "John Smith, Potato crisp, Who was the girl I kissed?", then you say it again, except it ends "Up and over, let it drop", and it goes on, a different ending each time' (Leeds); ' "Chik chik alonca, bonca, donca, skena skena", and you do it plainsy, over, dropsy, one-

---

[20] An echo of the song written by B. C. Hilliam ('Flotsam') and sung by Flotsam and Jetsam in 1928.
[21] 'Poor little Angeline', by Will Grosz and Jimmy Kennedy, 1936, accompanied the novelty dance of that year, the 'Palais Glide'.
[22] One of the many parodies of 'The Chestnut Tree,' the 'Novelty Singing Dance Sensation' of 1938. See Opie, Lore, p. 101. Gilbert Harding, broadcaster and television star, died in 1960.
[23] A straight rendering of Harry Lauder's song 'I love a lassie', 1906, which is better known in the parodies used for counting out , 'I love an aipple' and 'I had a sausage'; see Opie, Lore, p. 92.

hand, upsey-over, under, under other leg, hopsy, jumpsy, downsy, backsy and necksy, and you drop a ball at each skena' (Stornoway). Nursery rhymes were also used, as one would imagine, among them: 'One, two, buckle my shoe', 'Little Jack Horner', 'Taffy was a Welshman', 'Peter Piper', 'Miss Muffet', 'My maid Mary', 'Inky pinky ponky, Daddy bought a donkey', and 'Dr Foster'.

TWO-BALLS IN SCOTLAND

Scotland has the strongest ball-bouncing tradition in Britain, stronger even than in the north country, and rhymes which are normally used for skipping are often commandeered for 'doublers'. In parts of Scotland, too, they have a sequence of actions known as 'legs', which is not known south of the border. In Glasgow the first girl 'stoats' the balls against the wall while they all chant to the tune of 'Bobby Shaftoe':

> Tommy rise and light the fi-er,
> Turn the gas a wee bit high-er,
> Send ye for the Black Maria,
> Faither's gone to jail.[24]

If she 'makes out' during the chant (i.e. makes a mistake) she hands the balls to the next player; but if she completes it, she starts on the first of six 'legs', each leg lasting the time it takes to sing the song:

Right leg. Lifting the right leg, the balls are thrown under it to the ground so that they strike the wall and return.

Left leg. The same, lifting the left leg.

Front Bridge or Throughie or Splitsy. Standing slightly sideways with right leg well forward, but both feet on ground, the balls are stoated between the legs.

Back Bridge or Backie. The same with left leg forward.

Jibby. 'You throw the balls behind your back and hope they doesn't go squinty.'

Birly. Throwing the balls straight to wall, turning round and catching them.

---

[24] More usual versions concern a baby; e.g. sung to the tune 'Men of Harlech', in Chesterton, near Newcastle under Lyme, *c.*1910: 'Jack get up and light the fire, Turn the gas a little higher, Go and tell your Aunt Maria, Baby's got a tooth'; and in Glasgow, for balls, 1975: 'Ma get up and put the fire on, Turn the gas a wee bit higher, Phone Doctor MacIntyre, Because baby's got the flu. F-L-U spells flu.'

'The Old Grey Mare' was a Glaswegian favourite for 'legs' from the 1950s until at least the mid-1970s, and had its own special punishment for the loser:

> The old grey mare she ain't what she used to be,
>   Ain't what she used to be, ain't what she used to be,
> The old grey mare she ain't what she used to be,
>   Many long years ago—banjo! [or, 'Ever since the
>   old man died—paralysed']²⁵

When each girl had achieved as many 'legs' as she could, in spite of attempts by the others to distract her, the one with the fewest was dubbed the 'grey mare' and could, said a 13-year-old, 'be bumped on the back any time it pleases anybody till they find another grey mare by starting the game again'.

In Dunoon when they play 'legs' they sing:

> Away up in Holland, the land of the Dutch,
> There lives a young lassie I love very much,
> Her name is Susanna, but where is she now?
> She's up in the Highlands a-milking her cow.²⁶

In Aberdeen, where they perform either 'five' or 'six' legs, they also like a touch of pathos:

> If you listen I'll tell you a story
> Of a boy who was taken from home,
> Just to fight for his king and his country,
> And to fight for his own folks at home.

In Kirkcaldy, however, where 'legs' was known as 'dollars' in 1952, it was described as 'a ball game with a name but no words'.

It is noticeable that Scottish girls have a taste for romance.

> I'm away on the train,
> An' you're not comin' wi' me,
> I've got a lad o' my ain,
> Ye canna tak' him frae me.
> He wears a tartan kilt,

---

²⁵ A veteran song. The melody was originally known as 'Down in Alabam', with words and music credited to J. Warner, copyright 1858. The second stanza began 'Old blind horse come from Jerusalem', which gradually, through many versions, evolved into the well-known 'Old grey mare she ain't what she used to be' in about 1915. See James Fuld, *Book of World-Famous Music* (rev. and enlarged edn., 1971).

²⁶ Alternatively, in Aberdeen, 'Away down in Holland, the land of the Dutch, There lives a fair maiden I love very much. Her name is Susanna, her age is sixteen, A-rocking her baby to sleep on the green'; in Moss Park, near Paisley, 'Away up in Scotland, the land of the Scots, There lives a wee lassie who makes porridge oats; She makes them for breakfast, she makes them for tea, The bonnie wee lassie she makes them for me.'

He wears it in the fashion,
An' every time he birls aroon
Ye see his next week's washin'.

> (Aberdeen. Similar at Forfar and Paisley. From
> the folksong 'I know where I'm going')

My lad's a bonnie lad,
My lad's a dandy,
My lad's a bonnie lad,
And he likes sugar candy.
If yer gaun to gie a dram,
Dinna gie him brandy,
Tak' the bottle frae his moo'
And gie him sugar candy.

> (Braes of Enzie, Buckie. Similar in Aberdeen.
> Current Banffshire in 1910)

A big ship was leaving Bombay today,
All set for the Isle of Man.
Margaret was standing with tears in her eyes,
Along came a sailor and said in surprise,
'Oh darling, it won't be for long, for long,
Oh darling, it won't be for long.'
He turned round and kissed her,
And said he would miss her,
'Oh darling, it won't be for long.'[27]

> (Aberdeen; also Cumnock, Glasgow, Paisley, etc.)

'A big ship was leaving Bombay' proliferated into less romantic versions such as this, from Dundee, 1975:

There's a ship that is leaving Bombay,
Far, far away.
Here stands *Pauline* with tears in her eyes,
Andrew is holding her hand as she cries,
'Oh my darling, I love you I do,
Be faithful, be honest, be true.'
He bent down to kiss her,
The silly thing missed her [or, in Glasgow,
      'She ducked and he missed her'],
'My darling I love you I do.'

The following from Aberdeen in 1952 seemed just a curiosity, and a clumsy one at that:

---

[27] This is from the first stanza of the well-known wartime song 'Bless 'Em All', written in 1916 by Fred Godfrey while serving in the Royal Naval Air Service. The publishable version begins, 'There's many a troopship just leaving Bombay, Bound for old Blighty's shore.' See C. H. Ward-Jackson, *The Airman's Song Book* (1945), 17–18.

In New York City where I did tell,
A butcher's boy I loved so well;
He courted me my heart away,
And then with me he would not stay.
When father dear came one night,
And found his daughter out of sight,
He ran upstairs and broke the door
And found his daughter by a rope.
He took a knife and cut her down
And on her back was this he found:
'Oh father dear, what a fool I've been
To hang myself for Bobby Breen.'

However, in *Ballads and Songs Collected by the Missouri Folk-Lore Society*, ed. H. M. Belden (1940), 201–7, 'The Butcher Boy' is shown to have been a well-known ballad in Britain and America; and as only in America is the defaulting lover a butcher boy (in English analogues he is 'a brisk young lad', 'a sailor lad', and the like), it seems that the Aberdeen 12-year-old had her ball-bouncing song from an American source.

WORDS CHANGING WITH ACTIONS

In another type of ball-rhyme the words change automatically with the name of the action being performed, a device which leads to some juvenile-appreciated stultiloquence. Thus in many places throughout Britain, and throughout the 1950s–1970s, they sang to the tune 'Knees up Mother Brown':

Plainsy, Mrs Brown, plainsy Mrs Brown,
Plainsy, plainsy, plainsy, plainsy,
Plainsy, Mrs Brown.

This is followed by 'Upsy, Mrs Brown' and so on according to the local rules. The longest sequence was collected in 1975 in Montrose: 'Plainy Mother Brown, Plainy Mother Brown, Plainy, plainy, plainy, plainy, Plainy Mother Brown—Hoi!', and then 'Overs', 'Upsy', 'Upsy and over', 'Over and upsy', 'Dropsy', 'Downsy', 'Bang-crash', 'Under-leggy', 'Second under-leggy', 'Front dwelly', and 'One handy'.

'Plainsy Billy Boy', Bishop Auckland, 1961, was another game of the same sort; it was 'Plainsy Billy Baloo' in Salford, 1974, 'Plainsy to Waterloo' in Arbroath, 1974, and 'Plainy ball balloon' in Greenock, 1975. ('After plainy', said the Greenock 11-year-old, 'you do Loopy, Baker, Raindrop, Snowdrop, Bouncy, Slammy, Skinny, Fatty, In-you-go, Out-you-go, Right hand, Left hand, Jumpy and Dancy.' 'Plainsy Mickey Mouse' was popular in Workington, 1960, and Wolverhampton, 1970. 'Plainsy Wall's Ice-Cream' was the formula in Montrose, 1975, and Huddersfield, 1978. A 10-year-old in Market Weighton, East Yorkshire, volunteered a four-line verse:

> Plainsy to America,
> Plainsy to Japan,
> Plainsy to the Isle of Wight
> And plainsy back to land.

Then you go 'Overs to America', and then you go Upseys, Unders, Slamsies, Downsies, Bounces, Kingsies, Queensies, Arches. Well, for Kingsies you stand with your legs apart and your back to the wall and throw the ball between your legs and turn round and catch it. Queensies is the same but you stand sort of crooked and throw the ball crooked [diagonally]. Arches, you stand with your back to the wall and put your left leg up and throw the ball under it so it bounces on the ground and up onto the wall, then you turn round and catch it. I can't do it, but I've seen other people do it.

Another four-liner, from Workington, 1960, and Liss and Poole, 1975, goes:

> Plainsie at the bus stop,
> Plainsie at the sea,
> Plainsie on the merry-go-round,
> One, two, three,

which is only an echo of the counting-rhyme 'One, two, three, four, five, six, seven'.

In Blackburn, 1952, and Accrington, 1960, 'Mary Ann swallowed a pan, Twice as big as the Isle of Man' was followed by 'Mary Upsy swallowed an upsy, Twice as big as the Isle of Upsy', and so on through the sequence. This nonsense, it is said, 'makes the game funny as well as difficult'. Likewise in Liss, Chelsea, Edinburgh, Llanbryde (near Elgin, Moray), Poole, and Montrose, they find that 'Plainsy Jim swallowed a pin, That's the end of plainsy Jim' creates such absurd situations that their task is hilarious rather than Herculean. By the time they have worked through 'Dropsy Jim', 'Bouncy Jim', Crashes Jim', 'Crash drop Jim, 'Drop crash Jim, 'Under left leg Jim', 'Under right leg Jim', 'Under both legs Jim', 'Change hands Jim', 'Dumbsy Jim', 'Stand still Jim', 'Stand still dumb Jim', 'Twirl around Jim', 'Farsy Jim', 'Middlesy Jim', 'Nearsy Jim', 'Little man Jim', 'Hopsy Jim', and 'Dollies Jim', their brains are addled, and their bodies exhausted and shaking with giggles.

TWO-BALLS AND SEVERAL PLAYERS

In 1957 a Spennymoor girl, describing two-balls as the most popular game in the neighbourhood amongst 11- and 12-year-olds, said the usual course was that the owner of two balls came out into the street and began to play with them.

Soon a few other girls come to talk, and ask if they may join in. Then the game really starts. The girls shout out 'foggy' which means first, and when they have decided who shouted first, the remaining girls shout 'secky' and the first girl to shout it plays second. This goes on until the girls are in proper order to play. Then the first girl picks up the balls and throws them, one after the other, against the wall, in chorus with a rhyme such as,

> When I was a laddie I lived with my grannie,
>   And what a good grannie she was to me;
> She tied me up in my liberty bodice,
>   And tied my knickers below my knee.

Those who were watching chanted the words with her, and if she dropped a ball, or if she forgot the words or one of the actions, she was out.

The rhymes are important as regulators. They not only mark the occasions when a fancy throw should be made, they measure the achievement of each player and, in some games, dictate when a player should pass the balls on to the next girl. Indeed the rhyme is the game. When the girls agree to play 'Nebuchadnezzar' or 'Eachy Peachy' or 'Matthew, Mark', they are thinking of games of varying length or style rather than the names of different rhymes. From one end of Britain to the other, during the 1950s, 1960s, and into the 1970s, clusters of little girls might be found chanting the same weird and fascinating words, as in Scotland, and probably in the north country, they still do:

> Nebuchadnezzar, the King of the Jews,
> Sold his wife for a pair of shoes;
> When the shoes began to wear,
> Nebuchadnezzar began to swear;
> When the swear began to stop,
> Nebuchadnezzar bought a shop,
> When the shop began to sell,
> Nebuchadnezzar bought a bell;
> When the bell began to ring,
> Nebuchadnezzar began to sing—

When he begins to sing, the player has to adopt a 'little man' or similarly awkward position:

Doh, ray, me, fa, so, la, te, doh,
I've got cramp in my big toe;
I'll have to stay at home from school,
And rest it on a great big stool.[28]

The likelihood of the player dropping a ball and being 'out' or, as they say in Birmingham, 'down', is now excitingly increased, and the next player confidently awaits her turn. If a girl misses a catch because the balls hit each other in mid-air, she is likely to claim another turn, saying it was bad luck, and if the others insist she is out, 'she may go off in a huff'. But in Bishop Auckland, where the accident was known as 'bumps', she was allowed to continue, even if it happened a second time, provided she said 'Forgets forgives'.

Any rhyme that is of suitable length, and that perhaps is felt to be a bit impudent, is used:

There once lived an Indian maiden,
    And she stood about ten foot
        high,
And the colour of her hair was sky
        blue pink,
    And she only had one eye.
        Rick-a-doo-dum-dae,
        Rick-a-doo-dum-dae,
        High, low, rick-a-doo-dum-dae.

(Aberdeen)

Hey, Tony Curtis,
    How about a date?
I'll meet you round the corner
    At half past eight.
I can do the rhumba,
    I can do the twist,
I can wear a petticoat
    Right up to my hips.[29]

(Dunoon)

Please, miss,
My mother, miss,
Told me, miss,
To tell you, miss,
That I, miss,
Can't, miss,
Come to school tomorrow, miss,
Pains across my chest, miss,
Coughing all the night, miss,
A-tishoo, miss,
A-tishoo, miss,
There I go again, miss.

(Radcliffe, Lancashire, and Vale, Guernsey)

---

[28] 'Schoolboys', remarked *Punch*, 16 May 1857, p. 193, 'until birched for their irreverence, have a habit of chanting a lay setting forth that 'Nebuchadnezzar, the King of the Jews, Had three pairs of stockings, And four pairs of shoes.' 'Nebuchadnezzar', 1950–75, was recorded as 'Abracadabra' (London W2); 'Alla Balla Boosha' (York); 'Pontius Pilot' (Shrewsbury; Manchester); and 'Archibald-bald-bald' (Edinburgh), when it was often introduced with the opening words of George Robey's song 'Archibald, Certainly Not': 'My mother called me Archie, my father called me bald, To settle all the argument they called me Archibald.'

[29] Tony Curtis, film-star 'pin-up boy' of the 1950s and 1960s. Cf. 'Hi, Roy Rogers!', Opie, *Lore*, p. 116. The last four lines are borrowed from the skipping song 'I'm a little Scots girl'.

Each, peach, pear, plum,
I spy Tom Thumb,
Tom Thumb in the cellar,
I spy Cinderella,
Cinderella at the ball,
I spy Stuart Hall,
Stuart Hall in the stable,
I spy Betty Grable,
Betty Grable is a star
S-T-A-R.[30]

(Wythenshawe, Manchester, and thirty-
four other places: also used for skipping)

When the war was over
And Josephine was dead
She wanted to go to heaven
Wiv the crown upon her head;
But the Lord said, 'No,
You've been a naughty gel,
You can't go to heaven
But you can go to hell.[31]

(Stepney, London, and six other places:
also used for skipping)

Perhaps the most exciting, and certainly the most co-operative, of ball-bounce games is 'Matthew, Mark', in which the girls get in a line one behind the other. The girl in front bounces the balls against the wall, while the others chant relentlessly, over and over again at speed, without seeming to take breath:

Matthew, Mark, Luke, and John,
Next-door neighbour carry on.

Each time they come to the word 'on', the next girl in line steps forward, takes over the balls as they come off the wall, and continues playing, a more difficult feat than they make it appear, for the change must be made time and time again without missing a beat. 'You can't do this until you are very very good at two-ball,' commented a typical 7-year-old, 'but I shall be able to do it when I am a bit older.' The game seems to be popular almost everywhere, sometimes with varying words. The following versions were current during the 1950s–1980s:

Matthew, Mark, Luke, and John,
Put your dirty breeches on.

(London NW)

Matthew, Mark, Luke, and John,
Went to bed with their trousers on;
One shoe off and one shoe on,
Let your neighbour carry on.

(Glasgow. Cf. Opie, *Lore*, 21)

---

[30] 'Who's Stuart Hall?' I asked the children at Rackhouse Junior School, May 1975. 'He's a broadcaster and TV personality. He's been here for a meal,' they said. Usually, in the 1970s, Henry Hall, the 1930s dance-band leader, was still 'in the stable'. In this book we do not usually give rhymes which have appeared in *The Lore and Language of Schoolchildren* (q.v., p. 115). This is included because it was probably the most popular ball-rhyme of the 1950s–1960s, although in the mid-1970s it was often reported as a skipping rhyme. In the ball versions Betty Grable sometimes remained the 'S-T-A-R' (e.g. Perth, 1976) and sometimes was replaced by the singer Perry Mason (e.g. Paisley, 1975) or Larry Grayson ('Shut that door!'), the TV comedian (e.g. Montrose, 1979).

[31] A song known since the First World War, when it began 'When the war is over and the Kaiser's dead', see Opie, *Lore*, p. 104. During the Second World War it was 'Hitler will be dead'. Quite how 'Josephine' crept in no one knows, but she appears in all the most recent renderings.

Mafu, Mark, Loke, John,
Bless the balls and carry on.

(9-year-old, Alton)

Matthew, Mark, Luke, and John,
Read the Bible, pass it on.

(Cumnock)

and, mnemonically,
Matthew, Mark, Luke, and John,
Acts and Romans follow on.

(Sheffield)

Matthew, Mark, Luke, and John,
Next door neighbour carry on;
Next door neighbour's got the flu,
So I pass it on to you.

(Bloomsbury, London, and Birkenhead,
Merseyside)

Matthew, Mark, Luke, and John,
Catch these balls while I move on.

(Tipton, Staffordshire)

Matthew, Mark, Luke, and John,
Throw it up, pass it on.

(Cheltenham)

Matthew, Mark, Luke, and John,
Bless the bed that I lie on;
Four angels round my bed,
One to watch and one to pray
And two to keep me safe always.

(Girl, 13, Langholm. A version of the
evening prayer repeated continuously
since the sixteenth century, see ODNR,
no. 347)

In 1965–6 the Beatles temporarily replaced the Evangelists:

Ringo, Paul, George, and John,
Next-door neighbour follow on.

ACTION RHYMES

While many of the rhymes used for two-balls could equally well be used
for skipping, or even dipping, and sometimes are, there are a number
which lend themselves to enactment or have built-in descriptions of the
feats to be performed, and which are traditionally exclusive to ball-
bouncing—the ubiquitous game 'Oliver Twist', for instance, which is
really too complicated to be performed with two balls, although it has
been attempted:

Oliver Twist, you can't do this,
So what's the use of trying?
Touch your knee,
Touch your toe,
Bounce your ball
And under you go.

(Birmingham)

Oliver, Oliver, Oliver Twist,
Bet you a penny you can't do this.
Bend your knees, stand at ease,
Quick march, over the arch,
Oliver, Oliver, Oliver Twist.

(West Ham, London)

Similarly, in the popular playlet 'Mademoiselle' (or 'Madam O'Sel')
which has been enacted for fifty years (it is listed in *London Street Games*,
95) the performer always used to be content with one ball:

Mademoiselle	*Ball thrown against the wall and caught.*
She went to the well,	*One step back and thrown again.*
She did not forget	*Thrown and allowed to bounce on ground.*
Her soap and towel.	*Bounced twice on ground.*
She washed her hands,	*Thrown to wall, and hands rubbed.*
She wiped them dry,	*Thrown to wall, and hands wiped on skirt.*
She said her prayers,	*Thrown to wall, kneel before catching.*
And jumped up high.	*Thrown to wall, jump high before catching.*

A new, shorter version, which could be played with two balls, became
more popular than the old, perhaps because of the attractive ending in
which the girl catches the balls in her skirt (Aberdeen, Fulham, Halifax,
Kirkcaldy, Louth, Market Rasen, Welshpool, and Sydney, Australia):

> Mademoiselle went to the well
> To wash her face and dry it;
> She tossed the soap right up in the air,
> And caught it in her basket.[32]

The well-loved fracas of 'Over the garden wall' provides exactly the
kind of script that children like:

> Over the garden wall,
> I let the baby fall;
> Mother came out and gave me a clout,
> I asked her who she was bossing about,
> She gave me another to match the other,
> Over the garden wall.

'On the word "over" you throw the ball "overs", and when you drop the
baby you let the ball bounce once. When your mother hits you, you hit
yourself.' This is the way the game was played 1950–75, and probably
longer, in Aberdeen, Alton, Amlwch, Bellshill, Birmingham, Blackburn, Cumnock, Dundee, Forfar, Glasgow, London, Market Rasen,
Montrose, Netherton (Worcestershire), Newcastle, Paisley, Penrith,
Shrewsbury, Swansea, and Welshpool, and apparently throughout Australia and the United States (many variants). It has been current at least

---

[32] 'Mademoiselle' is Continental in fact as well as title. French girls bounce their balls to the
words: 'Marie-Madeleine, Va-t-à la fontaine, Se lave les mains, Les essuie bien, Monte à sa
chambre, Joue à la balle; Un peu trop haut, Casse un carreau; Un peu trop bas, Tue son petit chat;
Sa mère lui dit Comme pénitence Tu me feras Trois tours de danse. En voici un, En voici deux, En
voici trois' (Jean Baucomont *et al.*, *Les Comptines* (1961), 301). The words have been adapted from
the Catalanian and French song of Ste Marie-Madeleine. Marie-Madeleine, having killed her little
cat, must do penance by making 'three tours of the dance'. This identifies the song as a medieval
carole; see Opie, *Singing Game*, pp. 10–17, esp. 'Maria Julia', pp. 14–15. The ball-bounce rhyme is
also known in Germany.

since the 1920s, and may stem from Harry Hunter's minstrel song 'Over the Garden Wall' which, Flora Thompson recalled in *Lark Rise*, was so productive of parodies.

The following, though sometimes used for skipping and occasionally for clapping, is overwhelmingly a two-ball rhyme. It concerns a well-known brand of chewing gum and requires some lively actions:

> PK, penny a packet,
> First you lick it,
> Then you crack it,
> Then you wipe it down your jacket,
> PK, penny a packet.

'On the word lick you must put your tongue out, then when crack is said you must stamp your foot and on the word jacket you must quickly put your hand down your dress' (girl, 10, Stratford, east London). It was also collected from Aberdeen, Alloa ('Wrigleys Spearmint, tuppence a packet'), Basildon, Blaenavon, Cleethorpes, Cumnock, Dundee, Leeds, Lower Gornal, near Dudley, Manchester, Perth, Swansea, and Tredegar.

Actions are easier to remember when they are rhymed with the numerals from one to ten, or one to twelve; or sometimes 'you do it up to your age'.

Number one, I ate a scone, A, B, C.	*Player has to touch her mouth.*
I jumped aboard a Chinaman's ship,	*Jump.*
And the Chinaman said to me,	
'Going under,	*Underarm.*
Going over,	*Overarm.*
Like a soldier,	*Stamp feet.*
Straps on his shoulder.'	*One hand on shoulder and play 'one-handers'.*

And so on through:

Number two, I buckled my shoe, etc.	*Touch shoe.*
Number three, I climbed a tree, etc.	*Pretend to.*
Number four, I kicked the door, etc.	*Pretend to.*
Number five, I sat on a hive, etc.	*Squat.*
Number six, I chopped some sticks, etc.	*Pretend to.*
Number seven, I gazed to heaven, etc.	*Look up.*
Number eight, I slammed the gate, etc.	*Pretend to.*
Number nine, I drank some wine, etc.	*Pretend to.*
Number ten, I sat on a hen, etc.	*Same as number five.*

(Girl, 10, Penrith, 1957)

There are numerous variations in this marathon, which is sometimes 'done to all the different actions—overs, upsy and everything' and sometimes 'just plain'. Some versions have the chorus about the Chinaman's ship, and some simply chant numbers and matching actions: 'When I was one I ate a bun, When I was two I touched my shoe', and so on (though young Mancunians in 1975 had the pleasing formula, 'When I was one I ate a scone, Ate a scone, ate a scone, When I was one I ate a scone, And that was the end of chapter one'). In Birmingham, 1977, 'You have to have a partner. It's rather like a dance:

> When I was one it all begun,
> My father said to me:
> Can you go—
> Under, over, Casanova,
> One, two, three—
> All aboard!

It goes on like that and you make up little rhymes.' They did make them up, too. The fun of playing in pairs was partly that they could watch the other person doing the actions, but even more that they could race to think of the rhyme to be used for each number—though the options for rhyming phrases are few and have all been used before.

The song used as the framework for the sea-going versions must share a common ancestry with 'As I was going to school one day to learn my ABC' (see Opie, *Lore*, p. 22). It has a rollicking tune, a piratical chorus ('. . . the captain said to me, "We're going—This way, that way, forward and back, Over the Irish sea", A bottle of rum to fill my tum, And that's the life for me'), and has a parallel life as a community song.

Matching each year of one's life with a rhyming event is an old pastime. Robert Chambers printed these 'lines of no particular application . . . often heard among children' in *Popular Rhymes* (1847), 290:

> When I was ane, I was in my skin;
> When I was twa, I ran awa';
> When I was three, I could climb a tree;
> When I was four, they dang me o'er;
> When I was five, I didna thrive;
> When I was sax, I got my cracks [got the news];
> When I was seven, I could count eleven;
> When I was aught, I was laid straught;
> When I was nine, I could write a copy line;
> When I was ten, I could mend a pen;
> When I was eleven, I was sent to the weaving,
> When I was twall, they ca'd me brosy Wull [clumsy Will].

STOCKING BALL

'Stocking Ball' can be described as a mechanization of ball-bouncing which contributes nothing to its simplification. With the aid of a nylon stocking, a single player controls a ball which can be rapidly bounced on different parts of a wall without the ball itself being touched. The game was popular in the 1940s and had turned into a craze by the 1960s and 1970s. Characteristically it would arrive in a playground as 'the latest craze' and, after monopolizing the spare time and energy of every girl for six months or so, burn itself out. Whether 'Stocking Ball' (or 'Thumper', Montrose, 1974, or 'Sock-a-Ball', Birmingham, 1976) is still manifesting itself in these sudden crazes, I do not know.

In early days it was sometimes played with a tennis ball on the end of a string; then, almost always, the ball was put in the toe of an old nylon stocking, and it was best if it was wrapped in another stocking before it was inserted. This padding made the outer stocking last longer, although it wore a hole quickly enough, even so, and the ball flew out. When this happened, 'you have to tie a knot in the foot of the stocking, but if it gets too short you have to find another one', and, while the craze was on, mothers who laddered their stockings frequently found they were unexpectedly popular.

The player stood with her back flat against the wall and her legs apart, or stood on her right leg (if she was right-handed) and propped her left leg up against the wall 'or somewhere out of the way'. The leg of the stocking was wound round the hand once or twice to make it the desired length, usually 18–20 inches. 'It would be impossible using the whole length of the stocking—it is difficult enough as it is.' Also the bouncing would not be so rapid. The operator started by swinging the ball across her body, with a flick of the wrist, bouncing it against the wall on either side of her. Nearly always she did this to one version or another of 'Have a cigarette, sir'. In Petersfield they first bounced the ball from side to side, about waist high, in time with the words, and between the legs on 'sir'. Thus a right-handed player would go:

Light a cigarette, sir.	Ball bounced to right, left, and right of body, then down between legs on 'sir'.
No, sir.	To right on 'No', between legs on 'sir'.
Why, sir?	Right and then down.
'Cos I've got a cold, sir.	Right, left, right, and down.
Where d'you get your cold, sir?	Right, left, right, down.
At the North Pole, sir.	Right, left, right, down.
What were you doing there, sir?	Right, left, right, down.

Catching polar bears, sir.	Right, left, right, down.
How many did you catch, sir?	Right, left, right, down.
One, sir.	Right and then down.
Two, sir [and so on to 'Nine, sir']	Right and down alternately.
The tenth caught me, sir.	Right, left, right, down.

Variations follow: for instance, swinging the ball from above the right shoulder to outside the right knee, and bringing it between legs again for 'sir'; swinging from above left shoulder to outside left knee, and from right shoulder to under left arm. A skilful player can lie flat on the playground and do it, which from an adult point of view may be preferable, for the noise a stocking-propelled ball makes on a wall can inhibit all intellectual activity within 50 yards.

9

Skipping

EARLY HISTORY

ONE might think there would be an abundance of early references to an activity as universal as skipping. However, there is no evidence that the pastime of skipping over a turning rope was known before the seventeenth century. The mystery is partly explained in *Frederick and George: or the Utility of Playground Sports*, *c*.1810: 'Skipping', it says, 'is performed with the short and long rope, and the hoop.' Once a hoop is allowed to be an instrument for skipping, the early references begin to appear, though not as early as ancient Greece and Rome—for although in classical times hoops were trundled by athletes, as well as by boys and girls, there is no indication that they were used for skipping.[1]

Gargantua, when he played 'au cercle', might well have been skipping with a hoop (Rabelais (1534), i, ch. 22), for in *Les Trente-six figures* (1587), plate 12, 'Sauter dans le cerceau, & autres ieux', a boy is indeed shown jumping in a hoop (while another boy is shown trundling a hoop). This is confirmed by a woodcut in Olaus Magnus' *Historia de gentibus septentrionalibus* (1555), which shows some young people jumping through hoops decorated with bells, as a sort of sideshow to a band of sword-dancers. Again, in Stella's *Jeux de l'enfance* (1657), there is no skipping with a rope, but skipping with a hoop is depicted (plate 39, 'Le Cercle et le Bilboquet'): the accompanying verse runs 'Et j'aime bien mieux les postures De ces Sauteurs dans le Cerceau; quand ilz prenent mieux leurs mesures que le beau meusnier à l'Aneau'.

The earliest description of rope-jumping is in Jacob Cats's verses 'Kinder-spel', included in his book of emblems, *Silenus Alcibiadis*, commonly known as *Emblemata* (1618). Skipping, he says, is comparable to the art of managing time:

---

[1] The only 'evidence' for rope-skipping in antiquity is the bronze statuette of a faun, of Greek origin, found in 1810 on the Côte d'Or (see Amaury Duval, *Monuments des arts du dessin* (1829), i, pl. 30). The faun was holding in his right hand what seemed to be a fragment of rope and, although his left hand held no similar fragment, he was provided with a 'restored' skipping-rope. M. Duval considered that the fragment was more likely to be the end of a thyrsus (a short staff tipped with an ornament like a pine-cone), and that, as in other examples, the faun was a representation of a tightrope dancer.

> Het coorde springhen leert den vont,
> Om wel te vatten tyt en stont,
> Soo ghy cont springhen op de maet,
> Niet al te vroech, niet al te laet,
> Niet al te traech, niet al te snel,
> Soo sydy meester van het spel.

Rope-jumping teaches the art, Of rightly grasping time and hour, If you can skip in time, Not too early, not too late, Not too slowly, not too quickly, Then you are master of the game.

In the second edition, also published in 1618, an engraving (probably by J. Swelinck) after Adriaen van de Venne shows children playing in the courtyard of the Abbey of Middelburg, and among them is a boy skipping neatly (if somewhat woodenly) in a long rope turned by two other boys. When Cats used 'Kinder-spel' in a later book, *Houwelyck* [Marriage] (1625), the verses were illustrated with a street scene in which a boy is seen skipping along in a single skipping-rope.

Presumably boys were also rope-jumping in Britain in the seventeenth century, but no evidence for it seems to exist. A description is not found until 1737, when, in typically flowery eighteenth-century verse, 'youthful striplings' are shown disporting themselves in the courtyard of Bristol Grammar School:

> Part arm'd with scourges vex the flying top;
> Part whirl from head to foot the circling rope.[2]

In a print entitled 'Sport upon Sport or Youth's Delight', engraved about 1750, one of the twelve sports depicted is 'Ropes'—i.e. skipping-ropes—with the juvenile dialogue 'Can you cross it?'—'You saucy Fellow lets see you.'[3] 'Crossing' is one of the regular fancy steps in the single rope; thus the 1737 and *c.*1750 quotations show that single skipping, later looked upon as a young ladies' activity, was, in the first half of the eighteenth century, considered primarily a sport for boys.

The words used for skipping and skippers provide their own historical summary of the game. 'Ropes', in the print mentioned above, is by far the earliest; and in Scotland skipping is still commonly referred to as 'ropes'. 'Jumping Rope' is in *Mrs Lovechild's Book of Three Hundred and Thirty-Six Cuts for Children* (1810); the plates are dated 1779. Scotland and the USA retained 'jump rope' or 'jumping-ropes' (though in the USA 'rope-skipping' is also used). Thus, in G. McIndoe's *Poems and Songs* (1805), 'At three year auld he crys for . . . jumpin'-rapes'; Mrs

---

[2] *The Exercises Performed at a Visitation of the Grammar-School of Bristoll . . . To which are added, Verses on the Grammar-School Spoken at a former Visitation* (published by A. S. Catcott, Master of the said school, 1737).   [3] *N & Q*, 10th ser., 8 (1907), 72.

From Olaus Magnus, *Historia de gentibus Septentrionalibus*, 1555.
In Sweden the Shrovetide rituals included sword dancing and skipping in hoops.

Skipping in hoops, called 'Le Cercle', in Jacques Stella, *Les Jeux et plaisirs de L'enfance*, 1657.

Above: Illustration to the poem 'Kinder-spel', in Jacob Cats's *Silenus Alcibiadis* (known as *Emblemata*), second edition, 1618; engraved, probably by J. Swellinck, after Adriaen van de Venne. A boy is skipping in a long rope on the far left; in the centre foreground other boys are whipping a top and playing marbles.

Right: Girls skipping in a long rope at Greatham, in Hampshire, 1996.

Child's *Girl's Own Book* (1831) refers to 'Jumping Rope'; and in Emerson's *Journal* (1834), 'One is reminded of the children's prayers who in confessing their sins, say, "Yes, I did take the jumprope from Mary."'

One of Ackermann's comic prints, apparently by Rowlandson and dated July 1799, provides the first quotation for 'skip' in the required sense: a grotesquely large-headed maiden is shown skipping with a rope and saying, 'I think I shall skip myself into some of your good Graces.' *The Infant's Library* (1800), gives the earliest quote for 'skipping': 'Skipping. This is a very healthful play in winter; it will make you nice and warm in frosty weather.' 'Rope-skipper' appears in Thomas Surr's novel *Splendid Misery* (1801): 'Her Ladyship is the best rope-skipper we have.' 'Skipping-rope' is first found in Dorothy Kilner's *The Village School* (1783), vol. ii: 'a new skipping-rope', and later in Fanny Burney's *Journal* (1802), where she writes of her little son's interest in some paper cuttings sent by a friend's children—'the Cuttings, especially of the skipping ropes'. In England 'skipping' was, and is, the standard word, except in Ipswich, where it is 'skips', and in some northern places (e.g. Bishop Auckland, Spennymoor, and Workington) where it is 'skippy', a term also current in parts of Australia.

SINGLE SKIPPING

Girls start skipping young, often at 5 years old, and usually begin with a short rope 6 or 7 feet long for single skipping. A short rope with wooden handles (the traditional decoration seems to be spaced red and blue stripes) could, in the 1950s, be bought in Woolworth's for 2s. 9d., while ropes with ball-bearing handles could cost 7s. 6d. In 1958 there was a rope in the toy shops which automatically counted the number of times the rope went round. In 1964 one manufacturer produced an 'Olympic Skipping Rope' for 2s. 11d., with handles shaped like Olympic torches. There was also a sixpenny plastic 'rope' with dragon's-head handles, made in Hong Kong, which was surprisingly serviceable.[4] By the 1980s

---

[4] Skipping-ropes have been manufactured for children ever since the 18th c. One of the things Jacob Steadfast would like to buy, in Dorothy Kilner's *The Village School* (1783), ii. 9, is 'a new skipping-rope'. In *The Book of Games* (1805), 141, Mr Fielding gave his eldest daughter 'a very nice skipping-rope, with turned and varnished handles' as a reward for good behaviour. On 9 November 1875 J. A. Crandall of Brooklyn patented a 'Centennial Jumping Rope' in which the skipper had to jump over a second loop when the first loop was uppermost, but despite its celebratory status this novelty did not catch on. In 1911 Gamages of Holborn advertised 'The Patent Daisy Skipping Rope, Revolving Handles', at 4s. per dozen, and the 'Musical Skipping Rope' (with bells) amongst other gymnastic apparatus such as dumb-bells, Indian clubs, and callisthenic wands; and, in their 1913 *Christmas Bazaar* catalogue, 'The New Velocity Ball Bearing Skipping Rope. Now used by principal Schools Colleges and Gymnasiums through the Country'.

Adriaen van de Venne's famous picture of children playing in a cobbled street, engraved by Experiens Silleman; illustration to the poem 'Kinder-spel' in Jacob Cats's *Houwelyck*, 1625. A boy skipping quickly along in a single skipping-rope can be seen towards the right. Other pastimes shown are (left to right) blowing up a bladder for a football, flying birds on strings, playing 'House', bowling a hoop, whipping a top, running along with a 'windmill', blowing bubbles, playing 'Horses', standing on one's head, flying kites, playing at 'Train-bands', with flag, fife and drum, riding a hobby-horse, walking on stilts, leap-frog, loggats, blind man's buff, and playing with 'lazy tongs'.

plastic was the usual material for 'rope' and handles, and Woolworth's were selling a 'super-fast, fluorescent skipper' called the 'Rippa Skipper' for £2.99. Later, 'real' materials came back into favour, and when the 1993 Woolworth's skipping-ropes came in (as always, in the summer) the best-seller, made by 'Classico', had varnished wood handles and a cotton rope and cost 99p.

In single skipping a girl can run along while in the rope, but this is not much done now—the days when girls skipped along the road to school are long since past. Usually she skips on one spot, and sees how many skips she can do without tripping in the rope, or tries fancy footwork, skipping 'hopsie', or jumping while crossing the feet, or 'walking up stairs' by skipping first on one foot then on the other (known to children in Regency days as 'climbing the ladder'), or twirling the rope to one side and then skipping again, sometimes known as 'side bumps' (about 1810 it was called 'winding the jack', and in 1858 'turning the mangle'[5]) or crossing the arms over while skipping, jumping through the loop and then uncrossing them, sometimes known as 'French loops' (Sheffield) or 'crossie' (in Scotland), and in County Durham (particularly Spennymoor) still known by the nineteenth-century term 'crossing the buckle', which earlier, according to Brockett's *North Country Words* (1825), 50, was 'a peculiar and difficult step in dancing . . . To do it well, is considered a great accomplishment.'[6] 'There is such fun in skipping,' says a 9-year-old. 'In the holidays I was skipping all morning, and I didn't want to go in for dinner I was enjoying it so much. Next time you play skipping you see if you can get up to a hundred. I can tell you it works up your appetite. Your mother will be pleased if you eat a lot.'

In the playground, however, which is an inescapably sociable place,

[5] *Frederick and George*, 51; R. K. Philp (ed.), *The Corner Cupboard* (1858), 57–8.

[6] Dancing 'cross the buckle' is also referred to in Anderson's *Ballads in the Cumberland Dialect* (1805), 64. In corrupt form the term is probably preserved in the skipping chant 'Cross the bible, one, two, three', current, for instance, in Cwmbran, Monmouthshire, in 1954, Canberra, New South Wales, in 1959, and Hoylake, Merseyside, in 1991, in which two players skip towards each other in a long rope, crossing over at the centre. It may be noted here that *c.*1800 dancing in a skipping-rope became fashionable as an elegant pastime. Charles Dibdin made fun of it in 'Skipping Ropes' (*Song Smith* (1801), 65–6): 'Your ladies of fashion who freely subscribe To ev'ry whim folly may chance to imbibe, With skipping-ropes pleasantly pass time away, And skip up and down just like kittens at play'. Adam Buck depicted Sophia Western with a skipping-rope in an aquatint of 1800; and a fashion plate in *The Ladies' Own Memorandum Book* for 1799 shows a young lady wearing a high-waisted Empire dress and skipping with crossed arms in a single rope. When George Keats, in 1818, was concerned about his sister Fanny's chilblains, he sent her a skipping-rope with the suggestion 'perhaps Mr Bourke will teach you the skipping-rope hornpipe' (*More Letters of the Keats Circle*, ed. H. E. Rollins (1955), 11). By the end of the century the healthiness of skipping was not always accepted: 'Skipping ropes should be rigorously excluded from the girls' games. It possesses not one single merit, and is always attended with injury. A little later in life, women's modern ailments are attributed to the climbing of stairs, but truly spring from skipping rope games' (*Farm and Home*, 15 August 1891).

the owners of short ropes seldom skip on their own. They accept a challenge to an endurance test, or teach someone a trick, or play 'Calling in'. While turning a rope over herself, a girl calls in a friend; 'I call in my very best friend, her name is *Mary*.' Mary runs in without stopping the rope and skips in it, facing the girl who called her in.[7] Occasionally a second girl is called in, who jumps in behind the first skipper, so that three people are jumping in the short rope at once—not an easy accomplishment. Sometimes two people skip together in a short rope side by side, with linked arms or holding onto each other's waists, and each turns one end of the rope. This is generally called 'gipsy skipping', but in some places (e.g. Hackney) it is known as 'Andy Pandy', although the rhyme 'Andy Pandy sugary candy' is not used. Frequently they employ the old nursery rhyme 'Two little dicky birds sitting on a wall' (for instance in Swansea and Ruthin, and consistently in Liss). When the words 'Fly away Peter! fly away Paul!' are said, the skippers step nimbly out of the rope, and return when summoned by 'Come back Peter! come back Paul!'[8] While the skipping craze is on, the girls practise endlessly, skipping at breaks, at dinner time, and 'in the cool of the evening'. 'We go on till we're exhausted' is a recurrent phrase.

THE LONG ROPE

Once a girl has mastered the rhythm and speeds of a short rope, she prefers the long rope. It is more sociable and offers more possibilities. The girls appropriate their mother's old clothes line, as Flora Thompson's schoolmates did a century ago. 'Everyone', remarked a 10-year-old in Bishop Auckland, 'seems to get out their skipping-ropes just after the bad weather has finished. The girls are always bringing their mother's old washing line to school to see who has the longest rope, and some of them stretch half-way across the playground.'

Girl after girl declares that 'skipping is the game we most often play', yet it is a game that has seldom been carefully observed by adults. A spectator of 10- or 11-year-old girls performing in the long rope is likely to see something very different from the usual conception of children skipping. Here is no leisurely gambolling in the rope, no airy-fairyness. The expert skipper reminds one not of a fluttering butterfly but a

[7] There was as close a link between the games of American children and British children in the 19th c. as there is now, possibly closer. The New York contributor to *The Girl's Own Paper*, 12 March 1892, p. 373, describes this as 'Going a-Visiting'. She describes 'Climbing the ladder' under the name 'Skipping the ladder'.

[8] Ibid. This 'pretty form of Skipping the Rope' was known as 'Going to School' in New York *c.*1892.

machine gun. The two 'enders' turning the rope stand surprisingly close to each other. The rope screams as it is turned, and cracks as it hits the ground. The skipper seems to be suspended in the air; she rises barely an inch from the ground as the rope whips under her feet. She seems to rebound from the ground, rather than jump from it. She may seem unconcerned (perhaps she has her hands in the pockets of her overcoat), but her feet are working like pistons.

Before the girls run into the rope they get into its rhythm by ducking their heads at it as it comes towards them, rocking their bodies at the same time. Usually they duck their heads twice, and then run in. If they are scared they may continue ducking and rocking, missing turn after turn, until the others are fed up and make them hold an end.

The normal movement of the rope is down towards the incoming skipper; but as a Cumnock girl explained, 'You can go in a "looper". This is when you jump in when the ropes are turning away from you.' Ritchie, in *Golden City*, p. 117, says, 'There are two sides to a falling rope. It depends which side you happen to be standing on when you are about to jump in. The rope may be falling towards you (that's the "plainie side") or it may be rising away from you (that's the "dykie [wall] side".' In Spennymoor, likewise, 'loupy-dyke' means that 'the rope is turned away from you'. (To 'loup-the-dyke' is to leap a wall or ditch, hence the transferred meaning is 'undisciplined, wayward'.)

The people who turn the ropes are regionally known as cawers, coilers, turner-uppers, twiners, twirlers, twisters, or winders, according to the local word for the action of turning. In Cumberland, Northumberland, Durham, and North Yorkshire they 'twine' the rope, and in Darlington, Windermere, and South Yorkshire (Leeds, Selby, Barnsley) they 'twind'. In Lancashire they usually 'turn up', although in Accrington the 'turner-up' is known as 'hold', as a 10-year-old explains:

If a person treads on the rope she or he is out and that person is hold. If two people are out they are both hold. One person who is hold of the rope has to be first end. What I mean is, if anyone is out they've to take hold of the rope to turn up and whoever is first end has a skip.

In East Anglia and south-east England the word 'twist' is favoured, so the rope-turner is a 'twister'. In Cornwall she is a 'twizzer'. In Scotland, where they caw the rope, the person is a 'cawer' or sometimes a 'coiler'. But the majority who turn the rope are 'enders'.[9]

[9] In the Petersfield district of Hampshire they 'purl' the rope, and have done so for eighty years; this is apparently the same word as in 'purl one, knit two'. However, Alice E. Gillington's young informants used the Chaucerian 'trill' (*Old Hampshire Singing Games* (1909)) and she refers to 'trilling the rope'. In New Zealand, according to Sutton-Smith in *The Games of New Zealand Children* (1959), the turners are 'chorers' (i.e. 'cawers'), which once again shows that the dominant immigrants were Scottish.

TAKING ENDS

Turning the rope is not a popular job, and determining who shall take ends, at the start of a game, can itself be something of a ritual. Often, when the game is proposed, the players quickly shout 'No end' or 'No endie' or, in the north country, 'Nobby ender'; the two who shout last have to take ends, and compete with each other who shall be 'first end', 'firsy ender', or 'foggy ender', thus being first to be relieved of the duty of turning the rope when one of the skippers fails, and leaving the other to be 'seccy' or 'seggy' or 'laggy' ender. Sometimes they dip for enders, or the rope is given to the last two people to arrive. But a more satisfactory method, some think, is 'taking loops'. 'You roll up the rope and everyone takes up a loop, then we pull, and the two nearest to the ends holds them.' Then, likewise, the girls who find themselves with the ends cry 'firsy ender' or, in Norwich, the girl who soonest says 'first end' and turns the rope over her head and touches the ground, is the first to be relieved from 'twisting'. 'You can't just stand there and argue about it,' our informant expostulated. 'This decides it.'

STUMBLING IN THE ROPE

When a skipper makes a mistake—when, for instance, she does not take her cue for coming into the rope, or fails to get out of it on time, or trips and stops the rope altogether—she is 'out' and may have to 'take an end'. In Scarborough, and Wakefield, and Cleethorpes, a person who catches her feet in the rope is said to have 'clicked' (a Scarborough 10-year-old said, 'I usually click when I've got my eyes closed'). In Glasgow she is 'made out'; thus in an alphabet game, 'When they have made out they have to remember the letter they were at when they made out.' Often a skipper is disqualified if she so much as touches the rope. There is a chorus of 'Out!' from all the other players; while in Cumnock, 'If the rope catches in your skirt the other girls shout "Tails!"'

SPEED OF THE ROPE

The speed of turning the rope is generally graded 'salt—mustard—vinegar—pepper'. 'Salt' or 'slowsies' is skipping with a rebound. 'Mustard' is faster. 'Vinegar' is faster still, and 'peppers', 'peps', 'pepping', or 'peppering' is very fast single-jump skipping. However, in North Shields and elsewhere 'peppers' are sometimes called 'fasties', in Glasgow 'Belgiums', in Forfar 'firies', in Swansea 'rashes', in Aberdeen and Luthermuir 'spicies', and in Cwmbran 'wackers'. In

Weston-super-Mare they call out 'bacon' for slow and 'eggs' for fast.[10] In
Dublin and Perth the formula is often 'Salt, mustard, ginger, pepper'; in
Shrewsbury and Wellington 'Salt, mustard, cayenne pepper'. In Lancashire it is often, perhaps usually, 'Pee, po, pie, pepper' (or 'Pee, po, pie,
pep'). In County Durham it is 'Pitch, patch, pepper', in Lincolnshire
and sometimes elsewhere 'Pitch, patch, pine, pepper'. Sometimes these
formulas lead into a skipping competition 'to see which person can do the
most "peps" ', a game which can be called 'Beat the Number' or 'Beat the
Rope'. Sometimes the formulas conclude a skipping chant, and the rope,
which was turned at a moderate speed during the chant, is speeded up at
the end:

> Oliver Cromwell lost his shoe
> In the battle of Waterloo
> By a hitch, pitch, patch, pepper.

This is notably so in one of the most popular rhymes of earlier generations of schoolchildren:

> Knife and fork, lay the cloth,
> Bring me up a leg of pork,
> Don't forget the salt, mustard, vinegar, pepper.
>
> (Marylebone; Gomme, ii, (1898), 204)

> Lay the cloth, knife and fork,
> Don't forget the salt, mustard, vinegar, pepper.
>
> (A. E. Gillington, *Hampshire Singing Games* (1909), p. vi)

> Lay the cloth, knife and fork,
> Bring me up a leg of pork.
> If it's lean, bring it in,
> If it's fat, take it back,
> Tell the old woman I don't want that.
>
> (*London Street Games* (1916), 57)

> Lay the table, knife and fork,
> Bring me up a leg of pork.
> If it's fat, take it back,
> Tell the old woman I don't want that!
> If it's lean, bring it in
> And don't forget the Salt, mustard, vinegar, pepper.
>
> (London. Mark Benney in *Home and Away* (1948), 22)

---

[10] Cf. F. C. Husenbeth, *The History of Sedgley Park School* (1856), 106, referring to the period
1803–10, ' "Long-rope" was when two boys held and turned a long rope, and a boy had to step in,
and skip, and step out of this rope, without touching it or disturbing its turns. He called to the
turners, to signify at what pace they were to turn. The slowest was "cabbage", then "faster", then
"bacon", and finally "double ones".'

Mabel, Mabel, set the table,
Don't forget the salt, vinegar, mustard, pepper.

(Poughkeepsie, New York, 1926. *JAFL* 39 (1926), 84)

One of the most popular rhymes for skipping before the First World War seems to have been the tailpiece to the old nursery song 'Boys and girls come out to play', two lines of which,

Up the ladder and down the wall
A halfpenny loaf will serve us all,

had been printed in *The Famous Tommy Thumb's Little Story-Book* about 1760. The whole had appeared in *Songs for the Nursery* (1805):

Up the ladder and down the wall,
A half penny roll will serve us all.
You find milk, and I'll find flour,
And we'll have a pudding in half an hour.

The words children skipped to in London, and thereabouts, three generations later, were little different:

Up and down the ladder wall,
Penny loaf to feed us all;
A bit for you, and a bit for me,
And a bit for all the familee.

(Marylebone. Gomme, ii (1898), 203)

Up and down the ladder wall,
Ha'penny loaf to feed us all;
A bit for you, and a bit for me,
And a bit for Punch and Judy.

(Paddington Green. Gomme, ii (1898), 202)

But gradually, or perhaps swiftly, the word *pepper* seems to have intruded:

Up and down the city wall,
Ha'penny loaf to feed us all;
I buy milk, you buy flour,
You shall have *pepper* in half an hour.

(Deptford. Gomme, ii (1898), 203)

Up the ladder, down the wall,
Halfpenny loaf to feed us all.
You buy milk and I'll buy flour:
I'll make a pudding in half-an-hour.
Salt, mustard, vinegar, *pepper*.

('Games of Soho', *St Anne's Soho Monthly*, June 1907, p. 152)

Up the ladder, down the wall,
Ha'penny loaf to feed us all,
I'll buy milk and you buy flour,
There'll be *pepper* in half an hour.

(*London Street Games*, 1916, p. 57)

Up the ladder and down the wall,
Penny an hour will serve us all.
You buy butter and I'll buy flour,
And we'll have a pudding in half an hour.
With—salt, mustard, vinegar, *pepper*.

(London, *c*.1910)

The point was, as Alice Gomme noted before the turn of the century, 'At pepper turn swiftly.'

'Pepper' is also the term for fast skipping in Canada, Australia, and New Zealand. In the United States it is frequently 'hot peppers' or 'red hot peppers'. In 1892 a writer in *The Girl's Own Paper*, p. 373, stated that 'Pepper, salt, mustard, cider, vinegar' was 'the favourite game' of American girls. 'As soon as they have pronounced "vinegar" they begin to turn the rope as quickly as possible.' In France the term for fast skipping is 'vinaigre', and has been so for at least a hundred years. In Germany they say 'Öl, Essig, Wasser, Sprit' (Oil, vinegar, water, alcohol). In England 'pepper' seems to have been current throughout the twentieth century.

BUMPS

A 'bump', said a girl in Edinburgh, 'is when you jump high and the people cawing have to caw double the speed they usually caw at. They bring the rope round twice while you jump over it once.' The skill of the turners thus has to match the skill of the skipper. 'Bumps' are performed at the end of certain rhymes, when a word is spelt, or counting takes place; for instance, 'At David Greigs I bought some eggs, And they were B-A-D . . .', and 'Big Ben strikes ONE, Big Ben strikes TWO . . .'. To match the bumps, special emphasis is given to each letter or word that counts as a bump. A 'bump', being difficult, is a way of getting out an expert skipper so that someone else can have a turn.

'Bumps' is the usual present-day term, but we had 'Doubles', or 'Doublers', from Birmingham; Bristol; Berry Hill, Gloucestershire; Kirkby in Ashfield, near Nottingham; Spennymoor, County Durham; Leamington Spa; Luncarty, near Perth; Montrose; Oldbury, Worcestershire (and the eccentric 'Hubbles' from Wolstanton, Staffordshire; Stoke-on-Trent; and Newcastle under Lyme); as well as 'Double Loop' from Ipswich, 'Double Jump' from Cleethorpes, 'Double Coy' from Dunoon, and 'Double Unders' from Radcliffe, Lancashire (also from the Durham area, 1925 and 1943, and Sussex, pre-1914). 'Doubles' is the older term, for F. C. Husenbeth recalled that the final test of a skipper's ability in the long rope, *c.* 1805, was 'double ones' (*History of Sedgley Park School*). 'Bumps' arrived as an alternative name for 'Doubles' early in the twentieth century. Dot Starn, for example, played 'Bumps' and 'Double Bumps' in Stoke Newington *c.* 1905 (*When I Was a Child*, p. 9); and Violet Ellis, in *London Lore*, i, pt. 4, who was skipping at about the same time, said 'There was "Doubles" or "Bumps" . . . also done with an individual rope, but if you could turn doubles on a big rope you were in great demand.'[11]

[11] In addition some local names exist: 'Bucks' in Oxford, 1953; 'Beats' at Ford House, near Wolverhampton, 1969; and, possibly acknowledging an importation from the Netherlands,

Certainly it is easier to do a 'bump' in a single rope, which is under your own control, and this is the way it was done when first noted. In *Youthful Sports* (1801), skipping is described as 'A pretty play for active boys! How nimbly he turns the rope, now twice, now three times at a jump!'

Expert skippers sometimes decide to use bumps to emphasize the rhyme-words of a skipping song, or to do bumps instead of peppers when the counting begins in prognostication chants. The rhymes given here, however, owe their existence entirely to the fact that they are recognized bumping songs:

> What ho! she bumps,
> She's got the mumps,
> She's got the M-U-M-P-S.

There are bumps on 'bumps', 'mumps', and on each letter of 'mumps'. The catchphrase at the turn of the century, 'What ho! she bumps', was popularized by the song of that name about a tug boat, written by Harry Castling, 1899. Norman Douglas gives the verse as a new skipping game:

> What O she bumps,
> She skips and jumps,
> If she don't jump
> I'll make her bump.

The temptation is to derive the term 'bump' from this song, but evidence is lacking.

London Street Games (1916). Uckfield, *c.*1945, as text. Derivations: Wootton Bassett, 1950, 'What-ho she One, om-pom, Two om-pom', etc.; Swansea, 1952, 'Ah doo, doo ah, One bom-bom, Two bom-bom', etc. Both with bumps.

> Big Ben strikes one, tick, tock,
> Big Ben strikes two, tick, tock,
> Big Ben strikes three, tick, tock,
> Big Ben strikes four, tick, tock . . .

The words vary slightly in different places, and sometimes even in the same place. They may be 'Big Ben strikes one, bom, bom', or 'Big Ben strikes one coming on for two', or simply 'Big Ben strikes one, two, three.' But whatever the formula, bumps are almost invariably performed. Since the game is played throughout Britain, and is known

'Hollands' in Grimsby, *c.*1920; 'Belgian skipping' in Glen Gairn, *c.*1900 (A. S. Fraser, *Dae Ye Min' Langsyne*, p. 128), 'Belgiums' or 'Bell Jumps', Cleethorpes, 1952; 'Bell Jumps' (for ordinary fast skipping), Glasgow, 1975; and 'Belly', Caistor, Lincolnshire, 1952.

anyway to go back to the First World War, and since even so trivial a formula needs a starting-point, it may well have begun with a music-hall song, 'Big Ben Struck One', by A. J. Mills and Bennett Scott, published 1897. The significance of Big Ben's repeated striking is revealed in the chorus:

> Big Ben struck one! two!! three!!! four!!!!
> Jones was waiting in the rain
> For his darling Martha Jane;
> Waiting, waiting till half past ten,
> Underneath the shadow of great Big Ben.

London Street Games (1916). 'Big Ben strikes one' listed amongst skipping games. Uckfield, *c.*1945. Since 1950: versions from forty-two places. (Less often used, since around 1914, for counting the throws at ball-bouncing, but simply 'Big Ben strikes one . . . without the 'tick, tock'.) Variants: Newcastle, Staffordshire, *c.*1935, 'The Grandfather clock says, one, two, three'. Glasgow, 1952, 'My Grandfather clock goes one, tick tock'. Swansea, 1953, 'Tick-tock, tick-tock, goes my father's big clock [slow skipping], But my grannie's little clock goes tick-tock, tick-tock, tick-tock [faster]'.

> In Liverpool
> There is a school
> And in that school
> There is a room
> And in that room
> There is a desk
> And in that desk
> There is a book
> And in that book
> There is A B C D . . .

'You do bumps on the fourth beat in every line, and bumps through the ABC.' The rhyme owes its popularity to its linked lines, an age-old song formula found in the sixteenth-century Corpus Christi carol, and in 'This is the key of the kingdom'. The rigmarole about television pro-grammes, 'Bronco Layne had a pain, So they sent for Wagon Train', suddenly and hugely popular in 1960, and the spooky story 'In a dark, dark wood' (Opie, *Lore*, p. 36) are constructed in the same way; indeed the latter, which goes back at least to 1900, has influenced 'In Liverpool', for some versions end, like the 'dark, dark wood', with a box in which there is a G–H–O–S–T (or, at Berry Hill, Gloucestershire, a G–O–A–T).

Versions from twenty-nine places since 1947. Cwmbran, Monmouthshire, 1954, 'In Pontypool there is a school'. Bristol, 1960, 'In this world, there is a country . . . town . . . street . . . school . . . room . . . desk . . . some books . . . the alphabet'. A frequent variant (e.g. in Stromness, Orkney, 1991), is 'In Leicester Square there is a school'. Sometimes children substitute the location of their own school; and in the Birmingham area they chant 'In Birmingham, There is a pram, And in that pram, There is a B-A-B-Y.'

I fell in a box of eggs,
All the yellow went up my legs,
All the white went up my shirt,
I fell in a box of E-G-G-S.

The skipper does bumps when spelling eggs.

London Street Games (1916). Wootton Bassett, Wiltshire, 1950, as text. Cleethorpes, 1952, 'I was born in an Easter egg, And all the yolk ran down my leg.'

Mrs D, Mrs I, Mrs FFI,
Mrs C, Mrs U, Mrs LTY.

Played in various ways, but usually with bumps for the letters.

Forfar, *c.*1910. *London Street Games* (1916), 'Difficulty' listed as a skipping game. Brighton, *c.*1920. Greenburn, Aberdeenshire, *c.*1920. Kent, 1934. Versions from eleven places since 1950.

Red Riding Hood,
Went to the wood,
And in the wood,
She met a W-O-L-F.

Bumps on end words of lines, with a bump for each letter of 'W-O-L-F'.

Versions from Welwyn, Hertfordshire, London, and Swansea in 1952 (also Swansea, 1960). Radcliffe, Lancashire, 1952, ends: '. . . She saw a wolf, And this is how to spell it, W-O-L-F.'

Hokey-pokey, penny a lump,
The more you eat the more you jump.

In the 1950s the skippers did bumps on *lump* and *jump*; and in London and Birmingham they chanted exactly the same words as at the beginning of the century. In Aberdeen, a delightful verse was current, repeated while the girls skipped fast:

Mammy, gi'es ma bankie doon,
Here's a mannie coming roon
Wi' a basket on his croon
Sellin' hokey pokey.
Hokey pokey, a penny a lump,
That's the stuff tae make ye jump,
If ye jump yer sure tae fa',
Hokey pokey that's it a'.

However, *c.*1910, said a correspondent to the *Aberdeen Press and Journal*, December 1959, the hokey-pokey verse was introduced by the words,

> Hally-bally-bally, hally-bally-bee,
> Sitting on his mammy's knee,
> Greetin' for a wee bawbee
> To buy a Hokey-pokey,

and those words usually belong to the well-known dandling song which runs '. . . Greetin' for another bawbee, Tae buy mair Coulter's Candy'.

The beauty of hokey-pokey, which seems to have been introduced to Britain in the 1880s, was that it was a hard ice-cream, and very cold. A lump of it could be taken away to be enjoyed, as opposed to the soft ice-cream formerly on sale, which was served in a glass, and had to be eaten there and then with the vendor's spoon.

St Anne's Soho Monthly, June, 1907; *London Street Games* (1916). Versions from ten places in 1950s, sometimes 'Ice-cream a penny a lump'.

> Eggs a penny each, eggs a penny each.
> Look, Mum! Buy some,
> Eggs a penny each.
>
> Eggs at tuppence each, eggs at tuppence each.
> Look, Mum! Buy some.
> Eggs at tuppence each.
>
> Eggs at thruppence each, eggs at thruppence each.
> Look, Mum! Buy some.
> Eggs at thruppence each.

In this example of rapid inflation, the skipper has to do a bump each time the price of an egg is mentioned. The verse also sometimes becomes part of 'Jelly on a plate' (see p. 222).

Northampton, 1941. Wootton Bassett, 1949. Hackney, 1952. Essex, 1955. Edinburgh, 1975, 'A penny Easter egg, a penny Easter egg. You buy one, you try one, A penny Easter egg. A tuppenny Easter egg, a tuppenny Easter egg . . .' sung to the tune 'Knees up Mother Brown'.

> My pink pinafore, my pink sash,
> Fell into the water with a splash, splash, splash.
> Sailing on the water like a cup and saucer,
> My pink pinafore, my pink sash.

This game, in which there were bumps on 'splash, splash, splash' and the final 'my pink sash', seems now to be entirely overlaid by the ubiquitous dipping or counting-out rhyme 'Dip, dip, dip, my blue ship' (or 'Dip, dip, dash, my blue sash') which also sails on the water like a cup and saucer.

Versions from twelve places in the 1950s, none since. Some versions much contracted, e.g. 'My pink petticoat, my pink sash, Sailing on the water with a splash, splash, splash' (Welwyn, 1952). Cf. Opie, *Games*, p. 36.

> You naughty boy,
> You stole my toy,
> You named it Roy.
> You naughty B-O-Y,
> You stole my T-O-Y,
> You named it R-O-Y,
> You naughty B-O-Y.

Much played in Scotland. 'You do medium ropes most of the time, and belgiums [bumps] at boy, toy, Roy, B-O-Y, T-O-Y, R-O-Y, and B-O-Y', explained a 7-year-old Glaswegian.

London Street Games (1916), 90, 'You naughty flea, You bit my knee' given as the beginning of a chant. Grimsby, 1920 or earlier, 'You naughty girl, You stole my curl, You made me C-R-Y'. Versions from seventeen places since 1950, including 'You naughty lady, You stole my baby, You named it Sadie, You naughty L-A-D-Y' (Swansea and Edinburgh), and 'You naughty Yank, You stole my tank, You gave it Frank, You naughty Y-A-N-K' (Staffordshire, 1953).

> At David Greigs I bought some eggs,
> And they were B-A-D.
> I took them back and got some more,
> And they were G-O-O-D.

Spelling games are numerous and, to an adult, spell-bindingly unoriginal: 'A bottle of pop, A bottle of pop, a bottle of P-O-P' (Birmingham); 'I like jam, I like jam, I like J-A-M' (Spennymoor); 'I want some jam for my young man, I want some J-A-M' (Kirk Ella, Yorkshire); 'I want some more, I want some more, I want some M-O-R-E' (Alton, Hampshire); 'R. White's ginger beer goes off pop, goes off pop, goes off P-O-P (Eastcote, Middlesex, 1954; and Norman Douglas listed the game 'R. White's ginger beer goes off pop' in 1916). The skippers do bumps as the words are spelt; and the attraction of these games must lie in that first magic, as it seems, of realizing that the separate letters of the alphabet will make a word when arranged in a particular way. David Greig (pronounced Greg) is featured purely because his name rhymes with 'egg'; but for the record he was a family grocer who opened his first shop in

Brixton in 1887, and had a chain of more than 200 shops in the south of England when he died in 1952.

Camberwell, c.1937, 'I took them back, He gave me a smack. Which made me C-R-Y'; also Wootton Bassett, 1950. Laverstock, Wiltshire, 1951, as text.

SWAYING GAMES

These are games in which the rope is swayed from side to side while the player jumps back and forth over it.

Boys already had a name for this form of rope-jumping in George III's time. At Sedgley Park School in Staffordshire it was called 'Bells'. F. C. Husenbeth, who arrived at the school in 1803, recalled that, as well as ordinary skipping in the long rope, 'there was also "bells", where, instead of turning the rope, they swung it backwards and forwards on the ground, and a boy kept stepping over it and back again in measured time' (*History of Sedgley Park School* (1856), 106).

Rope-swaying is usually associated with particular chants, and the terms mostly seem to come from the chants. In addition to 'Bells' the terms for rope-swaying, and the chants from which they come, include the following:

Bingo. Cleethorpes.

Bluebells. Various places in England, Scotland, and Wales, perhaps particularly in Scotland. Thus an 8-year-old girl in Edinburgh, explaining a game: 'You can either play it in bluebells or cawing. In bluebells the rope goes under, in cawing it goes over.' The term comes from the well-known game and chant:

> Bluebells, cockle shells,
> Eevy, ivy, over.

'You sway the rope backwards and forwards, and then when it goes over they turn the rope fast till you're out' (10-year-old girls, Manchester, 1975). Sixty years earlier the game and chant were recorded by Mrs Florence Kirk in *Rhythmic Games and Dances* (1914).

Cockle shells. Maryon Park, London SE7. From 'Bluebells, cockle shells', see above.

Cradle or *Cradle Way*. Golspie, Sutherland, and Ruthin, north Wales. Not common despite its descriptiveness.

Pavie. Glasgow. Recollected as being 'paisley' in the 1920s. Derived from *pas vif*, a lively step. In 1598 Robert Birrel described in his *Diary* a slack-rope dancer who performed on a rope fastened to the top of St Giles's Cathedral: 'the lyk was never sene in yis countrie, as he raid doune the tow and playit sa maney paviess on it.'

Rockie. Parts of Scotland and the Isles. Thus a girl on the Isle of Lewis: 'The rope is going rockie and you jump in.' Also comes in the formulas: 'High, low, medium, dolly, rockie, pepper' (Forfar), and 'High, low, slow, medium, dolly, rockie, hoppie, pepper' (Edinburgh). Norman Douglas, *London Street Games* (1916) has 'Rock the cradle', a term that has had currency in the USA. It was recorded in Washington, DC, in 1888, and described amongst the outdoor games of girls in New York in *The Girl's Own Paper*, 5 March 1892: 'In "Rock the Cradle" the rope is not turned completely over, but is given a motion like to the pendulum of a clock. The long sweeps it takes makes it difficult for the skipper to avoid tripping.'

Swing. 'We also play another kind of skipping called "Swing". In this the rope is not turned but swung from side to side' (girl, 12, Spennymoor). The term also features in 'High, low, swing, dolly, pepper' (York). Norman Douglas, *London Street Games* (1916), listed 'Swing-swong'. Compare *Swish swosh*.

Swish swosh. Also rationalized as 'Swiss watch'. Cleethorpes, Lincolnshire; Titchmarsh, Northamptonshire. In the south of England, *c.*1914, children used to chant:

> Swish swosh, barley wash,
> Turn the bucket right *over*.
> Swish swosh, barley wash,
> Turn the bucket right *under*.

They swayed the rope until they reached *over*, then they turned the rope for ordinary skipping until they came to *under*, when they began swaying the rope again. In 1975, a group of 10-year-olds in Langton Matravers, Dorset, were chanting a hybrid verse from which it appeared that through the association of actions, but without regard to sense, the old swish–swosh game had become attached to the once separate skipping game 'Daddy is a butcher', now usually a part of 'Bluebells, cockle shells':

> Swish swash, golly wash,
> Turn the mangle over.
> Daddy is a butcher,
> Mummy cuts the meat,
> Baby's in the cradle fast asleep.
> How many hours does baby sleep?

Tissy-tossy. Swansea. In the 1930s and 1940s, when swaying the rope, the girls chanted:

> Tissy tossy, nore and nossy,
> Eevery, ivory, over;

or, in the Penlan district:

> Tissy tassy, O my lassie,
> Eevery, ivory, over.

About 1914 the term was 'Tisty-tosty'.

Waves, wavie, or wavies. Almost exclusively a Scottish term: Cumnock, Dunoon, Forfar, Glasgow, Johnstone in Renfrewshire, Kirkcaldy, St Andrews, Spennymoor (County Durham), and the Isle of Arran in the 1920s. It probably goes back to early in the century. At Dirleton, East Lothian, in 1913, children were skipping to lines now generally associated with rope-swaying:

> Wavy, wavy, turn the rope over,
> Mother's at the butcher's buyin' fresh meat;
> Baby's in the cradle, playin' wi' a radle,
> One, two, three, and a porridgee.
>
> (*Rymour Club*, 2:3 (1914), 146)

This idea, that the swaying rope represents waves, accounts for the employment of two sea-going rhymes: see 'Christopher Columbus' and 'Glasgow Waves', below.

Weaver. Welshpool. 'Weaving', Coventry. 'Weaving is swinging the rope to about waist-level' (girl, 13). Just possibly the term comes from the rhyme 'Eaver Weaver, chimney sweeper' (see p. 185). An informant has recalled that when she was young in Southport, about 1905, children used to chant a version of the rhyme:

> Eeper Weeper, chimney sweeper,
> Had a wife and could not keep her;
> Had another, did not love her,
> Up the chimney he did shove her—
> Please turn over my head.

Until the children came to the word *please* they used to sway the rope while the skipper jumped back and forth.

BLUEBELLS, COCKLE SHELLS

> Bluebells, cockle shells,
> Eevy, ivy, over.

Couplet repeated while the rope is swayed from side to side, as a prelude to the ordinary turning of the rope at the word *over*. By tradition, only certain verses follow this 'bluebells' opening. The most usual (recordings from twenty-seven places) apparently dates back to the early years of the century (correspondent, Inverness, *c*.1910). At the end of the verse, when the counting begins, the rope is turned fast:

> Bluebells, cockle shells,
> Eevy, ivy, over.
> My father is a butcher,
> My mother cuts the meat,
> The baby's in the cradle
> Sound asleep.
> How many hours did the baby sleep?
> One—two—three—four . . .

Variants include:

Bluebells, cockle shells,
Evy, ivy, over.
Mother does the washing,
Father does the work,
How many hours does the baby
  sleep?

(Pendeen, Cornwall)

Eevy, ivy, turn the rope over,
Mummy's at the market,
Buying a penny basket;
Baby's in the cradle
Playing with the ladle.
The ladle broke,
The baby choked,
One, two, three.

(Kirkcaldy)

Eva, weva, ivor, over,
Father's in the haystack
Cutting up the hay.
Mother's in the kitchen
Cutting up the meat.
Baby's in the cradle.
How many hours does she sleep?

(Ruthin)

Eevie, ivy, turn the rope over,
Mummy's at the butcher's
Buying some beef;
Baby's in the cradle
Playing with a rattle,
A rickety stick, a walking stick,
One, two, three . . .

(Forfar)

> Bluebells, cockle shells,
> Eevy, ivy, over.
> Mother's in the kitchen
> Doing the cooking,
> Father's in the workhouse
> Mending all the toys.
> Baby's in the cradle,
> How many hours can she sleep?
> One, two, three . . .

(Wool, Dorset)

In *Folklore* (1914), 364, the following was reported as being a common lullaby in Newmarket, and it would be appropriate if the chant had stemmed from a lullaby:

Sleep like a lady!
You shall have milk
When the cows come home.
Father is the butcher,
Mother cooks the meat,
Johnnie rocks the cradle
While baby goes to sleep.

Alternatively, interest concentrates on the mother's needlework:

Bluebells, cockle shells,
Eevie, ivy, over.
Mother's in the kitchen
Doin' a bit of stitchin',
How many stitches did she do?
Five, ten, fifteen, twenty . . .

And in Birmingham she may have to escape from an intruder:

Bluebells, cockle shells,
Evee, ivy, over.
Mother's in the kitchen
Doing a bit of knitting,
In comes a robber
To send her out.
Where does she run to?
Nobody knows—
   England, Ireland, Scotland, Wales,
   England, Ireland, Scotland, Wales . . .

More often, 'Mother's in the kitchen' is part of an ousting game (p. 275), and a briefer form of the 'Bluebells' chant, as follows, is one that has probably had a longer life:

Bluebells, cockle shells,
Eva, iva, ova.
My mother said that I was born
In January, February, March . . .

   (Devon, c.1915, and girls, 11, London,
   SE7, 1953)

Eeny, einy, over,
Kitty's in the clover.
Are you coming out tonight?
Yes, no, yes, no . . .

   (Boy, 11, Eyton, Shropshire)

Bluebells, cockle shells,
The pot is boiling over.[12]
My mother said that you must have
   Vinegar, salt, mustard, pepper.

   (Girl, 13, Welwyn)

Sea shells, cockle shells,
Evy, ivy, over.
How many boys did you kiss last
   night?

   (Four Marks, Hampshire; Elgin, Moray;
   Liss, Hampshire; Wareham, Dorset)

[12] Cf. Douglas, *London Street Games* (1916), 92, 'Evie, Ivy, over, The kettle is boiling over—'.

In Liss the skippers of 'Sea shells, cockle shells' were not more than 8 years old, and were gasping with excitement over their amorous excesses.

CUPS AND SAUCERS

Another chant that, for no discernible reason, is associated with swaying the rope, is,

> Cups and saucers, plates and dishes,
> Little black man in calico britches.
>
> (Bideford, Bedford, and Beadnell,
> Northumberland)

In some places, such as Caistor and Market Rasen, where the words are:

> Cups and saucers, plates and dishes,
> Here comes Sally in her calico breeches.
> How many stitches in her breeches?
> One, two, three, four . . .

the rope is swayed from side to side, the skipper jumping back and forth until the counting begins, when the rope 'is turned right round and peppered'. In other places, as at Beadnell, where the instruction forms (or formed) part of the 1950s chant, the player in the middle at first crouches while the rope is swayed over her head:

> Cups and saucers, plates and dishes,
> Little wee man in calico breeches,
> My mother says 'This rope must go over my head'.

In Scarborough, about 1915, the instruction was similar:

> Pots, pans, plates and dishes,
> Little men dressed in corduroy breeches.
> Please put this rope over my head.

The practice dates back to Edwardian days and, moreover, the observations of a London parish priest of that time make sense of, or at any rate account for, two little snippets picked up from children in New York State in the 1930s (Howard MSS):

> Bessy's little cups and saucers,
> Please go over my head.

> My cup and saucer shall go over my head,
> My cup and saucer shall go under my feet.

In June 1907, in *St Anne's Soho Monthly*, p. 154, the Revd T. Allen Moxon recorded that amongst the young in the streets of Soho:

A very favourite rhyme is as follows:

> My cup and saucer goes over my head.

At this the rope is swung slowly to and fro over the head of the skipper; and she starts skipping in an ordinary way to the words:

> O-V-E-R, all the pots are boiling over;
> My cup and saucer goes under my feet.

Traditional Games, ii (1898), 204. Sheffield, 1949, 'Our old man wears calico britches, Shirt hanging out for want of stitches'. Spennymoor, 1952, 'Wavy, wavy, cups and saucers'. Knighton, 1952, 'Here comes the man with comical britches'. Glasgow, 1961, and Edinburgh, 1975, 'See the wee man in the tartan breeches'.
  South Africa: Pretoria, 1971, 'Here comes the man with the carry carry dishes.'

CHRISTOPHER COLUMBUS

Another classic swaying-the-rope verse is,

> Christopher Columbus was a very great man,
> He sailed to America in an old tin can.
> The can was greasy,
> And it wasn't very easy,
> And the waves grew higher, and higher, and *over*.

'The first four lines are "cradle" skipping,' explained a Golspie girl. 'At the word "higher" the rope gets higher, and at *over* it goes over.' The game is particularly well known in Scotland (twenty-seven recordings, as opposed to seven in England), and it is almost always played with a swaying rope. Sometimes it starts 'Robinson Crusoe is a very brave man', and at Stornoway, on the Isle of Lewis, it goes:

> Robinson Crusoe was a poor old soul,
> He went to sea in a sugar bowl,
> The waves grew higher and higher
> And *over* he went.
> This was the end of Robinson Crusoe.

This may be compared with the earliest version known to us, said to have been current in Chesterton, near Newcastle under Lyme, about 1910:

> Sam, Sam, the bogie man,
> Sailed down the river in a corn beef can;
> The can was greasy,
> It went down easy,
> And so did the bogie man.

A chant with similar words was picked up in Oban in 1974. In Australia, appropriately, the very brave man who sailed the ocean was 'Captain Cook'.

GLASGOW WAVES

Another song that connects the swaying rope with the waves of the sea is:

> Glasgow waves go rolling over,
> Rolling over, rolling over,
> Glasgow waves go rolling over,
> Early in the morning.

This, reported only from the Cumnock area, 1961, is played with alternate swaying of the rope and plain cawing, and is followed, cheerfully and inconsequentially, by:

> The milkman's bell goes ting-a-ling-a-ling,
> Ting-a-ling-a-ling, ting-a-ling-a-ling.
> The milkman's bell goes ting-a-ling-a-ling,
> Early in the morning.

'Glasgow Waves' is the last remnant of a singing game that staggered down the years in various forms from at least 1847 and is now extinct (see Opie, *Singing Game*, pp. 156–8).

EAVER WEAVER

The old rhyme 'Eaver weaver, chimney sweeper' was commandeered for sway-skipping at the beginning of the twentieth century:

> Eaver weaver, chimney sweeper,
> Had a wife and couldn't keep 'er;
> Had another, didn't love 'er,
> Up the chimney he did shove 'er.

The rope is swayed from side to side in 'weaver' style. The girl in the rope jumps back and forth until she misses a jump, or until the enders ask: 'How many miles did he shove her? One, two, three, four . . .', or 'What do you think they had for supper? Salt, mustard, vinegar, pepper', or demand on her behalf: 'Please let the rope go over my head', and start turning the rope fast. In the nineteenth century the lines were used for counting out; and clearly they are related to the nursery rhyme 'Peter, Peter, pumpkin eater', which was current in Boston, USA, about 1825.

For counting out: *Mill Hill School Magazine*, 5, (1877). G. F. Northall, *English Folk-Rhymes* (1892), 345. Sawrey, Lancashire, 1937.

For skipping: Southport, *c.*1905. Clifton, Lancashire, *c.*1910, 'What do you think they had for supper? Rotten eggs and stinking butter'. *London Street Games* (1916), 53. Liverpool, *c.*1920. Versions from eighteen places since 1950, including 'Eaper weaper', 'Heaper sweeper', 'Weaver, weaver', and 'Sweeper, sweeper, chimney sweeper'.

CONTINUOUS SKIPPING

The skippers run into the long rope from the side, skip once, run round an 'ender', wait until it is their turn, and jump into the rope again. The path they take is a figure of eight, and skippers enter the rope alternately, first someone standing by one 'ender', then someone standing by the other 'ender'. They bound into the rope with a long leap, and shoot out of it again like a bullet from a gun. The number of times they skip in the rope before leaving it varies according to the game being played; and sometimes they rush through the rope not skipping at all, hoping the rope does not touch any part of their clothing, or they will be out. The rope turns continuously, and the important thing is to 'keep the kettle boiling', never to leave the rope empty, and never to miss a beat. If anyone misses a beat they have to take an end, or are 'out' and have to stand aside until one person is the winner.

The basic phrase to accompany this game is 'Keep the pot [or kettle] boiling'. It is interesting to see how the expression developed from 'to boil the pot' or 'keep the pot boiling', meaning to earn one's livelihood, to the use first described by J. T. Brockett, *North Country Words* (1825): 'KEEP-THE-POT-BOILING, a common metaphorical expression among young people, when they are anxious to carry on their gambols with more than ordinary spirit.' Just when the formula began to be used for continuous skipping, and when the kettle took the place of the pot, is not certain. A pot and a kettle were more or less synonymous in days gone by; but now that the old-type kettles survive only as 'fish kettles', and 'tea-kettle' has therefore been shortened to 'kettle', the word 'kettle', to a modern schoolchild's ear, has a more familiar sound than 'pot'. Norman Douglas's street children were probably thinking of a tea-kettle when they chanted their swaying-the-rope rhyme 'Eevy Ivy over, The kettle's boiling over' (*London Street Games* (1916), 92).

Frederick Johnson's 'Keep the kettle boiling, One, two, three', from Everett, Massachusetts (*JAFL* 42 (1929), 306) was probably used for ordinary jumping; although continuous skipping is part of the American skipping tradition, with slogans such as 'Keep the kettle boiling, Empty rope's a miss' (P. Evans, *Jump Rope Rhymes* (1954)). A game called 'Keep the pottie boiling' was being played around 1900 in Glen Gairn, Aberdeenshire (A. S. Fraser, *Dae Ye Min' Langsyne*, 128), the skippers going through the long rope in rapid succession to a chant of 'One no miss, two no miss'. However, not until *c*.1930 is there evidence of figure-of-eight skipping to the call of 'Keep the kettle boiling, One, two, three' (Abergavenny, *Farmer's Weekly* letter, November 1952).

During our surveys in the 1950s, 1960s, and 1970s, the kettle had almost supplanted the pot, and the usual exhortation was 'Keep the kettle boiling, Miss a loop you're out' (or, 'Mustn't miss a loop'). In local variations the second line was 'Miss a bow you're out' (Harrogate); 'Till mother comes home' (Cwmbran, Monmouthshire, and Berry Hill, Gloucestershire); 'Till I come back, sir' (London, Leicester, Birmingham); 'Waste no precious time' (Plymouth); and 'Don't waste steam' (Bristol). Sometimes the instruction is simply 'No missing the loop' (Abersychan, Monmouthshire, 1954), or 'No missing a link' (Cleethorpes, 1952); and sometimes it adopts an almost moralistic tone and is expanded to:

> Missy one takes an end,
> It does not matter who it is,
> Or what it is,
> Or where it is,
> Missy one takes an end.
>
> (Castle Eden, Durham, 1943;
> Spennymoor, 1952)

Occasionally they 'Miss one, skip one'. As a Spennymoor 14-year-old explained, 'This is a simple procedure and incurs only missing one rope and skipping every second one.' In Bacup, 1960, the words were 'One and miss one'; in Wordsley, Staffordshire, 1970, 'No nil, one nil, two nil . . .'; Perth, 1975, 'One and a miss, Two and a miss' up to 'Ten and a miss'.

Another symbol of continuous action is the mill. In Forfar, c.1910, skippers had to 'keep the mannie's mill gaein''. In Hanley, Staffordshire, c.1915, a long line of children ran through the rope without jumping, then jumped once, twice, and so on to a count of 'Mill one', 'Mill two', and so on; and this game was still being played in Wolstanton, Staffordshire, in 1961. In Cheshire, in the mid-1930s, the rhyme for the game was, 'If you miss your miller-gate, you're out', which may account for the mysterious 'If you miss your malagate [or 'Marley gate'], Your malagate, you're out', which was found in several schools in Guernsey, 1961–2, for many Guernsey children were evacuated to Cheshire during the war.

Perhaps, chain-link fashion, 'Miss' has been connected with 'Mississippi'. 'Up [or 'Down'] the Mississippi, If you miss a beat [or 'link' or 'loop'] you're out' has long been a popular chant for continuous skipping, 'Mississippi' sometimes being rendered 'Mrs Sippy' (twenty-one places, 1952–91). From four other places the chant was reported as a vehicle for a series of actions; for example in Scarborough, 1975, 'Down

Mississippi if you miss a link you're out' was followed by 'Up Missis-
sippi . . .', 'Turn around Mississippi . . .', 'Hop Mississippi . . .', 'Splits
Mississippi . . .', and 'then you do "Down, Up, Turn around, Hop,
Kick, Splits Mississippi if you miss a link you're out".' There is also
'Down the Mississippi where the boats go PUSH' ('You push the person
out and then somebody pushes you out') which is better known in
America (Abrahams, no. 120; all references are American, 1940s on-
wards).

Here are some of the slogans that have been shouted to urge the
players through the rope, never missing a beat and pausing only to skip
once or twice according to local custom:

> Thread the needle, thread the needle,
> Miss a loop you're out.
>
> (West Ham, 1960[13])

> Teddy on the telephone,
> Miss a loop you're out.
>
> (Barwick-in-Elmet, Yorkshire, 1975)

> Nae lave the tattie pot,
>   the tattie pot, the tattie pot,
> Nae lave the tattie pot,
>   the tattie pot ava.

Or, 'Nell and the tattie pot . . . The tattie pot's no empty' (Forfar, 1950).

> Quick soldiers never miss,
> They only fade away like this.

Or, 'Quick soldiers never miss, Only when they stop to kiss' (Liverpool,
1956 and 1960).
Or, sung to the tune of 'Old soldiers never die . . . They only fade away',

> Brave soldiers never die,
>   Never die, never die,
> Brave soldiers never die,
>   They all go to heaven.
>
> (Leeds, 1966)

> One, two, salute to the officer.

---

[13] One of the illustrations to 'The Song of the Skipping Rope' in the American *St Nicholas*
magazine, 22 (1896), 549, shows a girl running through a long rope. The caption is 'Thread the
Needle'.

Or, 'Passing sergeant must salute' (Stirling and Edinburgh, 1961: the
first skipper stays in the rope and salutes those who skip twice and leave,
or, in Edinburgh, each child salutes as she goes through).

> Follow, follow,
> Brush your boots and follow,
> Ee, aye, oh.
>
> (Forfar, 1954)

> Ash Wednesday,
> Never leave the rope empty.
>
> (Norfolk, *c.*1947; Blackburn,
> Knighton, and Welshpool, 1952;
> Caerleon, Monmouthshire, 1961,
> 'Stop! Take! Never leave it
> empty. Go to church on Ash
> Wednesday. When you are
> there, Say your prayers. Stop!
> Take! Never leave it empty.' *St
> Anne's Soho Monthly*, June 1907,
> p. 155, 'Go to church on Ash
> Wednesday, And never leave the
> rope empty')

It would be interesting to speculate that this rhyme was connected with
the strong tradition of communal skipping on Shrove Tuesday, still
carried out enthusiastically at Scarborough.

> Matthew, Mark, Luke and John,
> Next-door neighbour follow on.
>
> (Very popular 1950–70, and
> equally popular for two-balls.
> 'Ringo, George, Paul and John'
> sometimes took the disciples'
> place during the years of 'Beatle-
> mania')

The following is, surely, an American import (Abrahams, no. 54, has
three American references, but all in the 1960s):

> California Oranges fifty cents a pack,
> California Oranges, tap me on the back.

It was reported from north London in 1960; and in 1975, in Thurso,
Caithness, the explanation was given:

A person jumps in and says 'California Oranges fifty cents a pack', then a
second person jumps in and says 'California Oranges tap me on the back' and
they tap the first one on the back, and the first one jumps out and it carries on
the same.

Longer rhymes allow a longer stay in the rope:

> Drip, drop, drop it in the sea,
> Please take the rope from me,
> Come, come, come to the fair,
> No, no, there's no fair there.
> I must not miss a loop.

> (Popular (twenty places 1951–71). Variations include
> '. . . Up came a Chinaman and said to me, "How many
> apples on the tree?" I don't know so fancy asking
> me . . .' (Wimblington, Cambridgeshire); '. . . Do you
> know your A, B, C . . .' (seven places, e.g. London,
> Leicester, and Birmingham)

> If you're out, stand aside,
> We'll go sailing doon the Clyde.
> Up your nose and doon your belly,
> That's the way to Portobelly.

> (Edinburgh, 1975)

> Datsie-dotsie, miss the rope you're outie-o,
> If you'd've been where I'd have been
> You wouldn't have been put outie-o.
> All the money's scarce, people out of workie-o,
> Datsie-dotsie, miss the rope you're outie-o.

> (Belfast, c.1950, Colette O'Hare, *What do You Feed
> Your Donkey On?* (1978), p. 10, 'Datesy, datesy, miss
> the rope you're out . . .'. Welshpool, 1952)

For games in which one more jump is made each time a player's turn comes round, some kind of time-keeping chant is felt to be necessary, even if it is only 'One skip and away' (*The Corner Cupboard* (1858), 58), or 'One and no missing a bow, Two . . .' (Windermere, c.1900), or 'One and Out' (*London Street Games* (1916); North Chingford, 1952), or 'One and no miss, Two and no miss . . .' (Edinburgh, 1954; Accrington, 1960; Montrose, 1974). In Cumnock 'you must "keep the cord", that is each person must jump into the ropes immediately the person before her jumps out. Sometimes it's "One in Keepy", and that's quite difficult. When playing o to 10 when you are out you have to wait till everybody is out before you get back in again.' In Kilmarnock, also in Ayrshire, the chant is 'One two you must keepy, If you don't you're out.' Judging by the small number of times it was mentioned, this accumulative skipping is not very popular. However, 'Building up Bricks' has been endemic in Liss, Hampshire, playground since at least 1960, and also in Burnt Oak, Middlesex, 1960, Ilford, 1969, and Montrose, 1974. ('Pile of Bricks' was the name for 'Higher and Higher' in Washington, DC, in the 1880s— *Lippincott's Magazine*, 38 (1886), 332.)

Perhaps it is more fun to treat the turning rope as a lottery, for instance chanting the alphabet, and 'the person who gets Z goes to the end of the queue' (Aberdeen, 1960); or, 'If a girl happens to skip on a letter which is the first letter of her name she takes an end' (Spennymoor, 1960). Or 'You can call out numbers, and if someone is in the rope on the number of her birthday, she is out' (Forfar, 1950). Or 'You can skip through the days of the week, and whoever gets Sunday skips "rashes" through all the days of the week from Monday to Sunday and then the game begins again' (Swansea, 1953). Or 'You can skip the months and the person who is December, she has to 'pepper' until she gets past the month which is her birthday or she is out' (Bacup, 1960).

Another kind of lottery decides what kind of skipping the skipper must do, according to the word she trips on. The most popular formula (fifty-six places, 1945–75) is 'High, low, dolly, pepper', which is chanted by the enders. 'If she trips at "High", the enders hold the rope off the ground to turn it, and she skips and begins counting and sees how many she gets. "Low" is when the enders bend down to turn the rope. "Dolly", she has to turn round and round.[14] "Pepper" is skipping fast.' Extended formulas are usually on the lines of 'High, low, medium, slow, rocky, dolly, pepper' (Edinburgh, 1975. 'For rocky the rope swings from side to side'). Another format is 'Good skip, what you like, Dolly or a pepper' (Nottingham, c.1915), or 'Slow skip, what you like . . .' (*London Street Games* (1916), 91), or 'Slow coach, what you like . . .' (Radcliffe, Lancashire, 1952), or 'Hop, skip, what do you want?' (Birkenhead, 1975). A variant of this is: 'Anything you like, anything you like, Touch the ground, turn around, Anything you like' (Frodsham, Cheshire, 1960; Swansea, 1962; Montrose, 1974). Sometimes the 'dolly' is threatened: 'Dolly, dolly, wash my clothes, If you don't I'll break your nose, Slow, fast, dolly or pepper' (north Staffordshire, 1953; Wakefield and Pontefract, Yorkshire, 1954).[15]

CONTINUOUS SKIPPING: SPECIAL GAMES

UNDER THE MOON AND OVER THE STARS

In this, a line of players run one at a time straight through a long rope which is turning towards them, not pausing to skip. This is called 'Under

---

[14] The 'Dolly' imitated is not a doll but the home-laundry 'dolly', an appliance with two arms, and legs or feet, once used to stir and twirl clothes in a 'dolly-tub'.

[15] Cf. the lines addressed to an empty Black Cat cigarette packet, when found and trodden upon: 'Black cat, black cat, bring me luck, If you don't I'll tear you up' (Opie, *Lore*, p. 222).

the moon'. When all are on the far side, and lined up again, they make
their return journey by jumping over the rope, which is rising towards
them. This is called 'Over the stars'. Occasionally the game is known as
'Under the moon and over the sun' (Uckfield, 1945; Bristol, 1960), or
'Under the sea and over the moon (Harrow, 1949); or 'Under the moony
and over the girdle' (Aberdeen, 1952; Bucksburn, Aberdeenshire, and St
Cyrus, Kincardineshire, 1975; a 'girdle' is a circular iron baking-plate,
with a bow handle); and in some places it has acquired a rhymed couplet:
'Under the moon and over the stars, That's the way to get to Mars' or
'. . . How many miles is there to Mars? Five, ten . . .' (e.g. Workington,
1960, and Stornoway, Isle of Lewis, 1975).

   In *Frederick and George: or, The Utility of Play-Ground Sports, as
Conducive to Health, Hilarity, and Hardihood* (*c.*1810), one of the two
ways of skipping in the long rope is to 'run under the rope at each
rotation, without touching it; to get back again, we leap over as it rises,
which requires much accuracy.' W. H. Babcock would have agreed about
the 'accuracy'. He described the game, under its American name of 'Fox
and Geese', as 'one of the most intricate of the skipping-rope games'
(*Lippincott's Magazine*, 38 (1886), 332). A line of children skipped
through the rope, and then skipped 'back-door' over the rope from the
other side. They continued like this, skipping one jump, then two jumps,
and so on 'till they are tired'. Dr Dorothy Howard's Maryland students
told her (1948) that they had played 'Fox and Geese' in the same way.
The game was known as 'Chase the Fox' in New York (*Girl's Own Paper*,
13 (1892), 373) and in New Zealand in 1900 (Sutton-Smith, p. 74).

   In the earliest mention of the moon the phrase is, perversely, '*Over* the
Moon' ('where one jumped into the middle of the rope as it was being
turned away from you', Acton Town, *c.*1914, *London Lore*, i, pt. 4;
London Street Games (1916), 49).

TWO IN, TWO OUT

If two skippers go into or come out of the rope at the same time, both
have to 'take end'. The *modus operandi* of Norman Douglas's rhyme
(*London Street Games* (1916), 58),

> Two in the rope, and two take end,
> Both are sisters, both are friend,
> One named (Maudie), one named (Kate),
> Two in the rope and two take end,

was difficult to work out until glossed by later schoolchildren. A 10-year-
old from Castle Eden, County Durham, sent the following rhyme:

Two in, two out,
To leave the rope is out,
Never tell your neighbour
When you're going out.

She explained, 'If two skippers go in or out together they have to take ends'. Another version from Newcastle, 1952, further clarified the situation: 'Leave the rope and clicking tails [catching skirt in rope] are out, Never tell your neighbour when you're going out, For if you do, She's sure to follow you'; and a Spennymoor version warned '. . . Never tell your neighbour, Whether you're going in or out'. The game was being played in various places in England until at least 1960, and also in Australia.

BEWARE

'If you go into the rope you shout "Beware!" and if you go out of the rope you shout "Beware". The rhyme is "Beware! beware! You can't go in, You can't go out, Unless you shout, Beware!"' (girl, 12, Spennymoor, 1952; also Swansea, 1952).

JUMPING IN AND OUT ON SPECIFIC WORDS

The enders chant 'In, Out'; the skippers must jump in on 'In' and, when they want to leave, jump out on 'Out'. Norman Douglas listed 'Inners and Outers', probably the same game, in *London Street Games* (1916), 49. Other, more amusing, words have been used, of which 'Lip Stick' and 'Match Box' have been the favourites (run in on 'Lip' and out on 'Stick'; run in on 'Match', out on 'Box'). In a junior school in Staines, 1969, a selection was on offer: 'Lip! Stick!' or 'Knife! Fork!' or 'Dinner! Pudding!' or 'In! Out!' or 'Match! Box!' or 'Water! Cress!' or 'Somebody! Nobody!' ('Somebody, nobody' was heard in Pokesdown, Bournemouth, in 1927: *Word Lore*, ii. 128). In Bristol, 1966, they chanted the names of the Beatles, 'Paul, John, George, Ringo'; they went in on either 'Paul' 'John' or 'George', and out on 'Ringo'.

COLOURS

'Two people take end and between themselves pick a colour in secret. They sway the rope slowly and the rest of the people who are playing have turns each of jumping over the rope and guessing a colour. The one who calls the right colour takes an end' (girl, 10, Wolstanton, Staffordshire, 1961). The rope may be wriggled near the ground, instead of swayed. The colour-choosing can be quite sophisticated ('tartan', or 'blue with pink spots') according to the enders' determination to stay

in office or the skippers' determination to have a lengthy innings in
the rope. 'Colours' was mentioned amongst other skipping games in
London Street Games. Alice E. Gillington included a similar skipping
game in *Old Surrey Singing Games* (1909), in which a flower was chosen,
not a colour.

SKIPPING ALONG THE ROPE

Two skippers travel along the rope from one end to the other, bouncing
past each other on the way—in fact, changing places. The game may
have a humdrum name like 'Criss Cross' (Market Rasen, 1952), but is
usually known as 'Up and Down the Ladder' (or 'Up and Down Jacob's
Ladder'). The skippers continue as long as they can, and the enders
count the successful changes; for instance in Birmingham, 1977, the
counting went 'Up and down the ladders one by one, two by two, three
by three' up to 'ten by ten', and 'when you've reached ten you're allowed
to go out'. In some places the enders sing an adaptation of the singing
game 'In and out the windows': 'Up and down the ladder . . . As you
have done before. Stand and face your lover . . . As you have done before' (Birmingham, 1952; Workington, 1961; Jedburgh, 1972). More
elaborate rhymes have been used:

> Up and down the house
> Looking for Mickey Mouse.
> If you catch him by the tail
> Hang him up on a rusty nail;
> Send him to the cook
> To make a bowl of soup.
> Hurrah boys! Hurrah boys!
> How do you like the soup?
> We like it very well
> Excepting for the smell.
> Hurrah boys! Hurrah boys!
> We like it very well.
>
> (Cumberland, *c*.1945; Darlington, *c*.1945;
> Kirkcaldy, 1952; Spennymoor, 1960; Glasgow, 1961
> and 1975; Dunoon, 1962)

> Up and down the ladder of the old caravan,
> You only pay a penny to see the funny man.
> The funny man broke,
> Oh what a joke!
> Up and down the ladder of the old caravan.
>
> (Forfar, 1954)

> Kings and queens and partners too,
> All dressed up in royal blue,
> One, two, three.

In this, two players may jump into the rope together, perhaps on the preliminary count of 'One, two, three'; and often they have to change places and vacate the rope when 'One, two, three' is repeated. In Edinburgh in the 1970s lines from 'I'm a little Girl Guide dressed in blue' became attached, and the couplet was followed by:

> Salute to the king and bow to the queen,
> And turn your back on the Union Jack,

and in the same decade there was a fashion for beginning,

> Kings and queens and jelly beans,
> Please jump in.

> (Glasgow; Stornoway, Isle of Lewis)

Versions from Liverpool, *c*.1945, 'Kings and queens are partners too, so are we: In together, out together, Out go we'. Manchester, *c*.1945. Eight places since 1950. J. T. R. Ritchie, *Golden City* (1965), Edinburgh, with description.

Another formula is 'Pass the doctor one by one' (e.g. Cumnock, 1961); and it may be that the American formula 'Changing bedrooms one by one' has become naturalized in Britain, for it was reported from Edinburgh in 1975. In Maybole, Ayrshire, *c*.1960, there was 'another way of playing the same game'. The two skippers spoke to each other as they passed, saying:

> Pass the doctor, how do you do?
> Very well, thank you, how do you do?
> Fine, thank you.

And in the Dudley, Worcestershire, region, *c*.1970, another 'doctor' dialogue was prevalent:

> May day, doctor's day,
> Happy anniversary, May Day.
> Yes, sir, no, sir,
> Pass along the bus, sir,
> Any more tickets today?
> Ding-a-ling!

A dialogue was usual in older versions of the game. For instance, in Wrecclesham, Surrey, *c*.1895:

> Please Father, give me a penny.
> Go to your mother.
> Please Mother, give me a penny.
> Go to your father.

Gomme, *Traditional Games*, ii (1898), 204, has a girl 'skipping back and forth the long way of the rope and asking "Father, give me the key", "Go to your mother", and so on'. An American recollection from Opdyke, Illinois, 1949, gives a more complete story:

> 'Mother, Mother, where's the key?'
> 'Go ask Father.'
> 'Have you washed the dishes.' 'Yes.'
> 'Have you swept the floor?' 'Yes.'
> 'Turn the key in the lock, and run out to play.'
>
> (D. S. McIntosh, *Folk Songs of the Illinois Ozarks*
> (1974), 101)

In the American 'Going a-begging' (*Girl's Own Paper* (1892), 373) one girl says, as they pass each other, 'A piece of bread and butter', and the other replies 'Try my next-door neighbour.' This exchange turned up in Chudleigh village school, on Dartmoor, in 1955, where 6-year-olds, back to back in a long rope, were chanting, 'Any bread and cheese, Ma'am? No, Ma'am, go next door, Ma'am', and 'at the end of the rhyme the skippers turn, and cross over, changing places'.

ENDERS WALK ROUND IN CIRCLE

The particularly agile can skip while the two enders walk slowly round in a circle. This feature sometimes accompanies the chanting of the months at the end of 'My schoolmaster is a very nice man' and the chanting of the 'Tinker, tailor' prognostication at the end of 'Mrs Moore she lives on the shore'. It is also associated with swaying games; for instance Eilís Brady, *All In! All In!* (1975), 88, heard small Dubliners performing 'Eever, eyever, chimney sweeper' followed by 'Mammy's in the butcher's buying a pound of beef', and when the rope was brought over, the enders walked round in a circle intoning 'With a rick-a-rock, a rick-a-rock, A rick-a-rock-a-rover'.

TWO ROPES

Two ropes (or one long one, doubled up) are swung inwards, following each other like the loops of a rotary egg whisk. The skipper has to jump from side to side so as not to get mixed up between the two ropes. 'Most people count in twos to see how long they can skip,' said a South Shields girl, 'but if anybody wanted to they could sing their usual songs.' To succeed, explained a girl in Bishop Auckland in 1961, 'you must get quite thick ropes and make sure they touch the ground, because if they don't you have to jump high and you will soon be out.'

The most usual name for this activity is 'French Skipping', or 'French Ropes' in Scotland (nineteen places throughout Britain, 1952–75; also Dublin, 1953). A less popular name during that period was 'Double Dutch', 'Dutch Skipping', or 'Dutch Ropes' (eight places). During the previous fifty years the names 'French Skipping' and 'Double Dutch' (or 'Dutch') were used about equally; and in some places (e.g. Blacko, Lancashire, and West Ham, c.1910) the game was simply 'Double Rope'. In Aberdeen the name 'Londies' or 'French Londies' (1960) indicates that the game may have come, or was once believed to have come, from London. The *Concise Scots Dictionary*, under *Lonon*, has twentieth-century recordings for this name for the game from Shetland, Aberdeen, and Angus; and, as 'Lonon-ropes', from Aberdeen and Edinburgh. In Cumnock, Ayrshire, in 1961, it was known as 'Bananas'. 'German Ropes', in which the ropes were turned outwards instead of inwards, was sometimes mentioned as a more advanced alternative to 'French Ropes' (for instance in Glen Gairn, Deeside, c.1900, A. S. Fraser, *Dae Ye Min' Langsyne*, 128–9; and in Argyllshire at the same date, *Folklore* (1906), 216).

Amongst the earlier recordings there is enough diversity of names to show that the game was well established. In Portland, Oregon, c.1910, the name was 'Double French' (*JAFL* 61 (1948), 65). Gomme, *Traditional Games*, ii (1898), describes 'Double Dutch', with 'French Dutch' for turning the ropes outwards. In St Louis, Missouri, 1895, it was 'Double Dutch' inwards and 'Double Irish' outwards. And—marginally the earliest recording—in New York the name was 'Chicago' (*Girl's Own Paper*, 13, (1892), 373).

OTHER WAYS

There are other ways of jumping over a rope besides skipping. 'Higher and Higher' is simply high-jumping. 'You have a rope and two people holding it and a line of people. The rope starts on the ground and when everyone has had a turn you put the rope higher up' (girl, 10, Wereham, Norfolk, 1975). 'They display an amazing agility, whisking their bodies into the air by a revolving action and clearing almost their own height' (Edwin Pugh, *Living London* (1903), vol. iii). However, in Argyllshire at this time the highering and lowering was done whilst skipping normally; 'Low Water' was when the rope was sweeping the ground, and 'High

Water' 'when it is as far from the ground as a performer can clear' (*Folklore* (1906), 216). This was still the case in Maryland, 1948 ('Still Water'), and North Queensland, *c.*1953 (both Howard MSS); and in New Zealand (Sutton-Smith, p. 83). 'Snakes and Ladders', or 'Snakes', explained an 11-year-old in Hackney, 1952, 'is similar but instead of highering it or keeping it still you wiggle it along the ground.' It is widespread and popular, though sometimes dubbed 'a baby game'. Norman Douglas was the first to notice 'Snakes', in *London Street Games* (1916), 49.

'Fairies and Witches' is another non-skipping rope game. According to a girl in Lilliput, Dorset, 1976, 'If you're in the middle you loop the skipping rope and swing it round and they jump over it and you go "Fairies, Witches, Fairies, Witches" and if it catches them on "Witch" they have to go in the middle.' At Oxford, 1978, they chanted 'Tinker, tailor, soldier, sailor' and 'if it lands on them at "Rich man", say, that's who they're going to marry'. At Wilmslow, 1962, the game was called 'Satellites'; at Newcastle, 'Electric Wire' (Pandrich, p. 68). Many people remember this as a gymnasium game, but it has a playground life of its own; it has, for instance, been a regular feature of Liss playground for at least thirty years.

ELASTIC SKIPPING

This is not skipping in the usual sense, for there is no turning rope to jump. Instead, the two enders stand with feet apart inside a loop of elastic, which passes round their ankles and is thus stretched into a long oblong frame between them. The role of the enders is completely static, and their place can be taken by dustbins or chairs. The performer stands sideways to the stretched elastic, usually with the elastic to her right, and goes through a series of actions. She lifts the farther strand of the elastic over the nearer strand with the pointed toe of her right foot, whilst hopping on her left foot, and passes it across the other strand and back again so many times. Then she jumps into the frame, facing the long way, jumps her feet apart and lands with feet either side of the frame, jumps feet inside together again, jumps so that her feet land *on* the two strands of elastic, brings them inside together again, and so on, sometimes with the strands crossed, according to the local sequence. When she has performed all the movements without a mistake the elastic is raised to calf height; then it is raised to knees, thighs, and finally waist. (Claims that the sequence can be performed with the elastic round the enders' necks can probably be discounted; indeed, one girl said 'When it's *that* height you don't do jumps you just do cartwheels over it.') The common impression is that the game looks 'like a giant cat's cradle'.

In the summer of 1960 elastic skipping arrived in England as 'an entirely new game', and was for eighteen months, apparently, the exclusive possession of London children. 'This year's craze', said a 10-year-old girl in Fulham, 'is American Skipping. Karen Clark brought American Skipping over from America.' In playgrounds all over London little girls could be seen with their heads bent over the fiddling task of joining a packet of elastic bands into a long loop, or going through the dance-like steps of the game, which is as dainty, and in some ways as skilful, as the Scottish sword-dance, and has a similar look.[16]

However, when a powerful craze comes over from the United States there is not one point of entry but many. American families coming to London undoubtedly brought the game with them; but so did American Air Force families coming to bases in England and Scotland. For instance, when 'Chinese Ropes' was the rage in Dunoon Grammar School in 1962, about fifty of the girls in the school were from the nearby American Air Force base. 'Chinese Ropes' (or 'Rope', or 'Ropies', or 'Skipping'), reflecting the American name 'Chinese Jump Rope', continued to be the term in Scotland (e.g. Jedburgh, 1972; Glasgow and Paisley, 1975).

Elastic skipping spread rapidly in 1963–4. There could scarcely have been a junior school playground in Britain where it was not known. 'French Skipping' was now the most usual name in England and Wales, though Londoners remained faithful to 'American Skipping'. (Any foreign name was felt to be appropriate, however: e.g. 'Dutch Skipping' in Liss, 1964; and 'German Skipping' in Bedford, 1966.) By the mid-1970s the predominant name was simply 'Elastics', and the game is still, in the 1990s, known by that name. Correspondents followed the game's progress with excitement: a teacher in St Helier, Jersey, said: 'Linda, who sent you "American Skipping" in November [1963] tells me she learnt the game in Hampstead "a few years back"'; a parish priest in Workington wrote 'Chinese, or French, skipping went round Workington like wild fire this Easter [1964], and I know that it had hit Liverpool and Preston before last Christmas.' The actions began to vary. The original starting-sequence of lifting one strand a number of times across the other faded away (though remaining in the London version) and the sequence in which the player jumps directly onto the elastic strands became more important and was carried out in different modes,

---

[16] The steps recorded by M. O. Jones in Wichita, Kansas, 1963, follow a different sequence ('Chinese Jumprope', *Southern Folklore Quarterly*, 30: 3 (Sept. 1966), 256–63).

Left: 'Sophia Western', the heroine of Henry Fielding's novel *Tom Jones*, here depicted by Adam Buck, 1800, with a skipping-rope, as the epitome of grace and elegance.

Right: Elegance no longer being a prime requirement, girls are now free to exert themselves unselfconsciously. Elastic skipping, known as 'French Skipping', at Greatham Primary School, Hampshire, 1995.

such as 'Bouncy' (with a rebound after each jump), and 'Hopsy' (landing
on one foot inside the frame instead of both). A further development was
known as 'Diamonds', 'which is really complicated'. The performer
crosses the elastic band with her feet and proceeds to jump round inside
a 'diamond'. Then she jumps out of the diamond so that both feet finish
up outside the rectangle as the elastic is released. Finally, she jumps back
into the rectangle. The complete sequence for 'Norwegian Skipping' at
Grove, near Wantage, in 1963, was: 1. Jump in between the elastic
strands. 2. Jump *on* the elastic. 3. Jump in again. 4. Jump astride, outside
the elastic. 5. Turn to face the other way, taking the elastic between the
ankles. 6. Jump up, doing 'scissors', freeing the elastic and landing in
the middle again. 7. Jump out to one side. 8. Facing the strands, take the
nearest elastic, resting on the feet, to overlap the far elastic. 9. Turn
round, take one foot out, place it on the spot where the two elastics cross.
Do the same with the other foot. 10. Jump in the middle. 11. Jump right
out. This is done 'Oneses' (round ankles), 'Twoses' (round mid-calf),
and 'Threeses' (round knees).

In the early 1960s the enders might chant 'In, out, in, out, In, in, in
out' as the jumps were made; but as the game developed, or perhaps as
the children grew bored with the game as it was, they began to adorn it
with a miscellany of borrowed rhymes: the all-purpose 'Roses are red,
Violets are blue' was used (Worsley, Staffordshire, 1969); and some old
counting-out rhymes, such as, 'Mary at the cottage gate, Eating cherries
off a plate, Two, four, six, eight,' and 'Inky pinky ponky, Daddy bought
a donkey, Donkey died, Daddy cried, Inky pinky ponky.' 'Queen, Queen
Caroline' was revised: 'Kathy, Kathy, Kathaleen, Washed her hair in
Windowlene,[17] Windowlene keeps it clean, Kathy, Kathy Kathaleen'
(Montrose, 1974). Other words are: 'Jingle, jangle, centre, spangle,
Jingle, jangle, out' (Notting Hill Gate, London, 1976); 'England, Ireland, Scotland, Wales, Inside, outside, donkeys' tails' (Birmingham,
1977, and other places); and 'When you do "Double Diamond"', said a
9-year-old in Dulwich, 'which is when there's one bar up here and the
other down there, and it's twisted over and you've got to jump over the
twist, there's a special rhyme that advertises beer':

> Double Diamond works wonders,
> Works wonders, works wonders,
> Double Diamond works wonders,
> So drink one today.

[17] 'Windowlene' is the window-cleaning fluid made by Reckitt & Colman, Hull.

The strangest rigmarole came from North Hinksey, on the outskirts of Oxford, in 1985: 'Itchy me, star shee, Logo hutsy yutsy—kill it', and the game was called 'Itchy me'.

The game reached other countries too. It arrived in Israel in 1960 ('Gummi', Eifermann (1968), 218–20). In Australia it had certainly arrived by 1962, when Ian Turner saw it in Canberra; 'It was called "American Hoppy",' he said, 'then I saw it no more until 1967 in Melbourne, when it was called "Elastics".' Subsequently it was reported in Afghanistan, Austria, the Argentine, Germany, Greece, India, Italy ('Elastici'), Kenya, the Netherlands (1962, when it was called 'the English Twist' or 'the Russian Twist'), Norway ('Hoppe strikk', i.e. 'Jump Elastics'), Turkey, and Yugoslavia—so it would be safe to say it had become worldwide.

When Patricia Carpenter interviewed schoolgirls in Reno, Nevada, in 1964, the 14–17-year-olds had never heard of the game, but the 9–13-year-olds said, 'Oh, you mean Chinese Jump Rope, we play it all the time.' A Chinese-American 10-year-old, born in Nevada, said 'I learned it from my mother who was born in China'; and another 10-year-old, new in Nevada from California, said 'I learned it from a very old Chinese lady who said she used to play it in the alleys in China.' It seemed that 'Chinese Jump Rope' really did come from the Orient.

A correspondent of ours played 'Chinese Skipping' out in Hong Kong in 1956, at an army school and also after school with Chinese children. Another correspondent, who was in Beijing in 1963–4, said the game was so popular there that 'you couldn't walk along the street without seeing it'. She was told that it had come to north China during the time of the Japanese occupation, 1938–45. Another correspondent, who had lived in Tokyo for many years, tried to find out whether elastic skipping was known in Japan before 1938. She said (August 1976):

I have yet to find anyone over 60 who knows the game. People in their forties and younger seem to know it as a matter of course, and it appears to have been played to the rhythm of many different popular songs over the past few decades. One contemporary of mine, born in Nagasaki in 1940, recalls being a 'nuisance' when her elder sisters were playing it, before the end of the war. The song they sang was a rather jingoistic battle air, left over from the Russo-Japanese conflict. Another song current in the northern Kanto district in the 1950s was more fanciful, something about a golden carriage with silver bells.

This still does not prove whether the game began in Japan or China, and although our earliest evidence is from China, from someone who played the game in Shanghai in 1935, it was tempered by the remark that 'the Japanese had already begun to infiltrate then'. It almost seems as if the game sprang up simultaneously in both countries in the mid-1930s.

In their traditional games, Oriental children need agile feet as well as dextrous hands (in their game of kicking a shuttlecock, for instance). A game called 'Awakening Giant', similar in appearance to elastic skipping, was glimpsed briefly in a television film on China ('made recently', March 1975). Two children squatted about 6 feet apart, with a rope in each hand. They crossed hands while a third child jumped in and out of the ropes. Once the rope was trodden on, the giant awoke, the child fled, and it was someone else's turn. The game was said to have been played by old Chinese ladies in their youth. Another game is played with bamboo sticks, and is known to have been brought to the USA and Great Britain by Philippino immigrants, though descriptions are tantalizingly few. Two long bamboo sticks are laid parallel on the ground, and two similar sticks are laid across them, also parallel. The ends of the topmost sticks are held by two people, crouching or kneeling, who tap them alternately onto the lower sticks and against each other, in time to music. Two dancers move in and out of the moving sticks, and if they are caught between the sticks as they close together, they are out. These two games probably represent the tradition that engendered elastic skipping.

ANKLE SKIPPING

The 1960s saw the arrival of another craze—ankle skipping. The apparatus required is a length of string about 2 feet long, with a loop at one end to encircle the skipper's right ankle, and a tennis ball or cotton reel fastened to the other end. The ball swings round in a circle and the skipper's left foot jumps over the string as it comes round. The movement can be started by throwing the ball, and the right foot keeps up the motion by hopping and describing a small circle in the air between hops. The performer is thus turner and jumper in one. She dances lightly and rather quickly from one foot to the other, with a rocking movement, not far from the ground, and the ball moves so fast it becomes a blur.

The game apparently came to England from Belgium. Early in May 1966 a girl who had learned 'Ippy Hoep' on holiday in Belgium wrote to the BBC children's programme 'Blue Peter' and said she would like to show viewers how it was played. She said she thought it would become popular over here, and she was right. By the end of the month it had swept the playgrounds of Britain, usually with the name 'Belgian Skipping'. The craze grew so strong that some schools had to forbid it because the children were getting bruises on their ankles.

Woolworth's got onto the game with commendable promptitude and before the end of the summer were offering rubber balls, pierced, with a yard of plastic-covered cord, for 2 shillings; and they displayed instructions for playing 'Ippyop—the New Skip-with-a-Ball game'. Omac Toys also marketed a ball-on-string under the name 'Ankle Skip, the With It Game for 1966', price 2s. 6d. 'Roto-skip' was also available (seen in Selfridges, summer 1967, dullish yellow, price 3s. 11d.), and 'Roto-skipping' became the predominant name. In 1969 a free Twixaskipper was offered with Twix chocolate snack bars. Late in 1970 an improved 'Roto-skip' appeared, with a flattened spool instead of a ball; and a Hong Kong version with a miniature football, called 'Footsie'.

The craze probably spread world-wide, though our sightings are somewhat random. In Paris it was known as 'Olé Swing' (*Daily Telegraph*, 3 July 1967), and in the same year arrived in Italy as 'Jeu de balle'—although, after the teenage film and TV star Rita Pavone performed it while singing, all the children called it 'palla di Rita Pavone'. In April 1969 Hellenic Cruise passengers saw a little Arab girl playing the game in the old town of Acre, with a red ring round her ankle, a string, and a squashed tin can (squashed so as to swing round better).

BOYS AND GIRLS

Skipping is a girls' game, but boys occasionally join in, especially younger brothers on summer evenings. Some girls are quite kind about the boys' ability to skip. 'My opinion on boys skipping is that if boys would have a little more patience they would be able to learn lots of games from the girls. I think that a lot of boys could be better skippers than a lot of girls.' But more often the girls say: 'If the boys do play with us they usually fool around and act silly.' The boys decry skipping as 'sissy', and especially say that the rhymes are 'girlish' or 'weird'. Public opinion being against them, if they do join in they do so defensively, making sure that bystanders know they are not serious: they gatecrash a girls' game, skip wildly for a few turns, and are chased away. Boys were equally embarrassed a hundred years ago, it seems. Sean O'Casey, in his autobiography *I Knock at the Door*, p. 119, describes girls skipping in a Dublin street, while the boys crept nearer and nearer, then, 'With a defiant shout, weakened with the tone of a shy shame in it, a boy, bolder than the rest, would jump in merrily; the rest would follow him', and they continued the game together.

In the past, was skipping a boys' or girls' game? Early recordings suggest that it was boys who played, although in J. H. Swildens'

Left: Girls skipping in single ropes. *The Book of Games* (William Darton Junior), 1818.

Right: Regency schoolboys enjoy a sedate game of skipping in a long rope. *School Boys' Diversions* (A. K. Newman & Co.), 1820, plate by R. Stennett.

Vaderlandsch A-B boek voor de Nederlandsche Jeugd, 1774 [1781] the woodcut for 'J for Jeugd' shows a little girl skipping in a long rope turned by two other little girls; and in *Mrs Lovechild's Book of Three Hundred and Thirty-Six Cuts for Children* the cut for 'Jumping Rope', dated 1779, shows a boy and girl turning a rope with a girl skipping in the middle. When collecting evidence it is important to distinguish between skipping in a 'skipping-rope' and skipping in a 'long rope'. Single skipping, in a 'skipping-rope', was, as we have seen, thought of as an accomplishment for young ladies from at least the end of the eighteenth century, and perhaps Strutt was thinking of his own youthful experience (he was born in 1749) when he said, in *Sports and Pastimes* (1801), 286–7, 'Skipping . . . is performed by a rope . . . thrown forwards or backwards over the head and under the feet alternately [i.e. single skipping] . . . Boys often contend for superiority of skill . . . and he who passes the rope about most times without interruption is the conqueror.' In *Blackwood's Magazine* for August 1821, pt. ii, 'the Skipping-rope' was described as 'a common game' for girls in Edinburgh. On the whole, it seems that the prevailing opinion was, from the late eighteenth century onwards, that single skipping was for girls, and community skipping in a long rope was for boys. *The Boy's Own Book* (2nd edn., 1828), makes a useful distinction. Under 'The Jumping Rope' (p. 38) it says:

A long rope is swung round by a player at each end of it; when it moves tolerably regular, one, two, or even more boys, step in between those who hold the rope, suffering it to pass over their heads as it rises, and leaping up so that it goes under their feet when it touches the ground, precisely as in the case of a common skipping-rope.

Plenty of evidence exists for girls skipping in single ropes ('skipping-ropes') and for boys skipping in long ropes. However, there is also evidence for what one might call mixed-sex skipping with both sorts of rope, if we can believe that the following remarks mirror real life. *School Boys' Diversions* (1820; 2nd edn., *c*.1823) says that skipping is 'a most charming exercise for cold weather, and proper for either sex'. *The Girl's Week-Day Book*, (*c*.1850), recalling the writer's own schooldays *c*.1820, says 'Skipping was a favourite exercise.' The whole (girls') school skipped together in a long rope; 'I . . . counted fifty-seven regular jumps.' *The Youth's Own Book of Healthful Amusements* (1845), has a story entitled 'Skipping-Rope', written around a woodcut of two girls and a boy skipping in single ropes; the author cannot forbear remarking that 'The skipping-rope was no more made for boys, you know, than the peg-top was for girls, and I do not love to see toys so misappropriated.'

Little Games (*c*.1865), giving an illustration of girls with single ropes, agrees that skipping is 'a very healthy game for all little girls, and the longer you can keep it up, the cleverer you are'.

Cassell's Book of Sports and Pastimes (1888), gives 'Jumping Rope' as 'nothing more than skipping on a scale somewhat larger than before', and I would guess that the game was included simply because it had appeared in boys' books of games throughout the century. *Sports and Games for Girls and Boys* (1889), remarked severely of 'Skipping-Rope', 'This is a game scarcely suitable for lads—cricket will suit them better.' Whether boys were still skipping in single or long ropes at that date is uncertain. A cartoon in *Punch*, (4 June 1859), 'The Advantages of Taking a Short Cut through a Court', shows an old gentleman entangled in a skipping-rope held by two girls. Phil May's *Gutter-Snipes* (1896), no. 39, shows a girl skipping in a long rope, with a boy and girl turning. It may be that the boys' attention was turning to organized games such as cricket and football, for through the mid-nineteenth century the rules of both were being unified: county cricket clubs were emerging in the 1860s; the Football Association was formed in 1863; players became public heroes. Or it may be that the girls more or less appropriated the game when they began skipping to songs, for what boy could involve himself in divinations of his sweetheart's name, or declare that he was 'a little Girl Guide dressed in blue'? At all events, traditional skipping lost favour with boys and now remains only as a part of athletic training.

THE CHANTS

The chants add spice to skipping. Functionally, they are work-songs. They enable the enders to keep time; they regulate the time each person has in the rope; they test the proficiency of individual skippers, and they add the excitement of contest between the players. Most of all they bring fantasy and humour to the long rope, and sometimes a kind of romance. It is remarkable how the girls enjoy re-enacting the daffiness of some doggerel tale, although they have chanted it a dozen times already. Without realizing it, the skippers in the long rope are achieving one of the most satisfying of human rituals, the synchronization of song and action. Onlookers and enders sing the chant or, if it has no tune, say or shout it, and the skipper joins in as she is able.

One of the mysteries in the history of children's games has been the small place given to skipping in the nineteenth century, when today it is one of the most popular playground activities. For instance, the great folklorist William Wells Newell does not mention skipping in his *Games*

and Songs of American Children (1883), and offers a single 'Rhyme for
Jumping Rope' in his enlarged edition twenty years later, and that was
borrowed from W. H. Babcock. When Babcock, inspired by Newell, set
about recording children's games, he found several skipping games, but
only one rhyme to chant while skipping. In his article 'Song-Games and
Myth-Dramas at Washington', *Lippincott's Magazine*, 38 (1886), 332, he
writes, 'While the rope is turning, two girls run in and jump together, all
singing vigorously,

> By the holy and religerally law
> I marry this Indian to this squaw.
> By the point of my jack-knife
> I pronounce you man and wife.'

This strange verse, once a marriage formula used at the end of courtship
games (see Babcock, *Folk-Lore Journal*, 5 (1887), 134–5), has the honour
of being the first published skipping rhyme.

Lady Gomme, in 1898, attached no importance to skipping, devoting
only four and a half of her 964 pages to the game, of which nearly a full
page is taken, without acknowledgement, from the article on New York
girls' games in *The Girl's Own Paper*, 12 March 1892.[18] She does, however, give seventeen English rhymes and skipping formulas, all collected
by the redoubtable Miss Chase, mostly in London. R. C. Maclagan, the
painstaking collector of *The Games and Diversions of Argyleshire* (1901),
commemorates skipping verses with a single couplet: 'Gooseberry, raspberry, strawberry jam, Tell me the name of your young man' (p. 227).
The Revd T. Allen Moxon seemed to have had no difficulty in collecting
twenty-five skipping verses for a 'Games of Soho' series in his parish
magazine (*St Anne's Soho Monthly*, June–July 1907), and Norman Douglas collected at least fifty-one for *London Street Games* (1916); but in
1910, when a list was made of the 137 favourite games of elementary
schoolchildren in London, skipping is described as 'rare' (*N & Q*, 11th
ser., 1 (1910), 483).

Some of the factors contributing to the present-day popularity of
skipping are obvious. The shortening of girls' dresses and hair have made
skipping easier, although, as we have seen, Regency misses often seem to

---

[18]   These are all actions to be performed in a long rope, for instance 'Winding the Clock', in which
the skipper must count to twelve and turn round for each number. The anonymous author observes:
'Curiously enough there are no "skipping-rope" songs, though the nature of the game would seem
conducive to some sort of rhythmical accompaniment. The following is the only verse I have heard
sung by American girls while they are skipping, "Skip, skip, to the barber's shop . . .".' Leah Yoffie,
writing of St Louis, Missouri, past and present, comments: 'In my generation [1895–1900] we did
not recite rhymes to accompany jumping rope, but the children today have many elaborate verses
and many more intricate methods of rope-jumping than we did' (*JAFL* 60 (1947), 31).

have been adepts in the rope. It is possible, too, that the decline in singing games, due partly to the unsuitability of the courtship ring-games for prepubertal players, and partly to their discovery by folksong collectors and subsequent teaching as a school subject, left a vacuum to be filled. Movement to the accompaniment of song is an age-old pleasure, and it may be that during the nineteenth century many little girls sang scraps of song to themselves as they skipped, like Cousin Mattie in a story in *Good Words for the Young* (1868), who 'was making such a noise . . . skipping and singing: "There is nae luck about the house, When the gudeman's awa'."' Surely, though, if a tradition of skipping songs existed in 1874, when Robert Louis Stevenson witnessed 'some quite common children' skipping in the lane beneath his window in Hampstead, there would have been words to go with the movements he described in such loving and lengthy detail (essay, 'Notes on the Movements of Young Children'). Whatever the interpretation of this sometimes contradictory evidence, it is clear that more and more rhymes of various kinds were being used as an accompaniment to skipping, and that there were plenty of established 'skipping rhymes' by *c.*1910.

What are the characteristics of the successful chant (successful in the sense of being long-surviving and widespread)? It is likely to be a four-line verse with four trochaic feet in each line, the first, stressed, syllable coinciding with the slap of the rope on the ground and the jump over it.[19] Very many chants have this pattern, which is, for instance, typical of classics such as 'Mother, mother, I am ill', 'Cinderella dressed in yellow', and 'Charlie Chaplin went to France'.

The words share the characteristics of other folk-songs, in that they are orally transmitted, anonymous, and infinitely variable. Either by accident or design they incorporate directions for getting in and out of the rope ('Granny in the kitchen, Doin' a bit of stitchin', In came a burglar And pushed her out'); or for doing various actions ('I'm a little Girl Guide dressed in blue, These are the actions I must do . . .'); or for finding the answers to interesting questions such as 'Does he love you?' and 'How many babies will you have?' The more banal of these regulatory rhymes were probably custom made; others have been adapted from old rhymes and songs which once fulfilled other needs and were too attractive to relinquish.

Many have no technical qualifications at all to be skipping songs. They are simply part of the common fund of verse and song. They are comic, or romantic, or they make fun of the still unknown and rather frightening

---

[19] This is the same rhythm as the Trobriand Islanders use, as they stamp and chant during cricket matches.

state of adulthood, and some of them have a long ancestry. Children are borrowers and adaptors; they are among the earliest exponents of the recycling principle. The fact is, girls will skip to *anything*: nursery rhymes such as 'Little Miss Muffet' and 'Polly put the kettle on' (both with actions); all-purpose rhymes such as 'Roses are red, Violets are blue'; dipping rhymes like 'Old Mother Ink' and 'Piggy (or Paddy) on the railway' (Opie, *Games*, pp. 57 and 37); 'fun' rhymes like the creepy 'Dark, dark wood' (Opie, *Lore*, p. 36); taunting rhymes like 'Tiddlywinks old man, Suck a lemon if you can, If you can't suck a lemon Suck an old tin can' (Opie, *Lore*, p. 176) and 'There she goes, there she goes, Peery heels and pointed toes' (Opie, *Lore*, p. 182); old street jeers, like 'Aunt Sally sells fish' and 'German boys are so funny'; parodies of popular songs, like the parody of 'Daisy, Daisy' which goes, 'Amy Johnson flew in an aeroplane, Flew to China, never came back again . . .'; football chants like 'Two, four, six, eight, Who do we appreciate?' (Dunoon, 1962, for continuous skipping); and even this memory of an old-style chanted geography lesson: 'The Capital is Edinburgh, Seaport Leith, Portobello, Musselburgh and Dalkeith, Penicuikie' (Forfar, 1954, and *c*.1910). The rhyme shouted at the seeker in the game of 'Kick the Can' (see Opie, *Games*, p. 165) is used to tease a skipper about her boyfriend: 'Hard up, kick the can, *Debbie Houston's* got a man, If you want to know his name, His name is *Gary Wallace*.' The girls use community songs, like 'My father had a barrow and the wheel went round' and 'The old grey mare she ain't what she used to be';[20] also, inevitably because of the words, the old singing game 'Skip to m' lou'' (see Opie, *Singing Game*, pp. 319–20). Square-dance songs, learned in school, have found their way into the skipping repertoire (see 'Green coffee grows on the white oak tops'). In Spennymoor, 1960, they invented a special skipping game to go with a sea chanty taught in singing lessons:

> Were you ever in Quebec
> Stowing timber on the *deck*,
> Where there's a king with a golden *crown*
> Riding on a *donkey*?
> Heigh-ho, away we go,
> *Donkey* riding, *donkey* riding;
> Heigh-ho, away we go,
> Riding on a *donkey*.

[20] This has acquired new verses, e.g. 'The old grey mare she lives in a motor boat, Lifted up her petticoat, In came a billy goat [the next skipper comes in] And out goes she' (Glasgow, 1975). Also see p. 147.

'It is called the Knocking Out game. Everyone skips through the rope, one person per word, *but* if a person skips through on one of the underlined words that person is out. If you want you can start "Were you ever in Baghdad, Kissing someone else's lad?"'

Some popular songs have survived as skipping songs: sentimental Victorian songs ('I am a little orphan girl'); minstrel songs ('Away down East, away down West'); music-hall songs ('Early in the morning at eight o'clock', 'In China there lived a very funny man'); songs from between the wars, for instance the 1930s' foxtrot, 'Horsie, horsie, don't you stop', and Saxie Dowell's 'Three little fishes', beginning 'Down in the meadow in a little bitty pool', a song copyrighted in 1939 which clearly owed a lot to Olive A. Wadsworth's poetical résumé, in the previous century, of life 'Over the meadow': 'Over the meadow, Where the stream runs blue, Lived an old mother-fish, And her little fishes two . . .' (see Whittier, *Child-Life*, p. 51). Some post-war 'pop' songs temporarily appeared in the repertoire, such as 'Poppa Piccolina' ('All over Italy he plays his concertina . . .'), the hit song of 1957, itself based on an Italian children's song from the previous year; and the catchy late-1960s song beginning 'There's a tiny house by a tiny stream', in which each verse ended 'In Gilly Gilly Osan Pepper, Casa Nella Boaga by the Sea', was still well established as a skipping song in the mid-1970s. However, considering how many commercial songs have been published each year and promoted through the media, it is surprising how few have entered the skipping canon, even briefly, though perhaps more were tried out than we ever knew.

It was not until Joyce Terrett monitored for us the skipping games played by the girls of Glanmor Secondary School, Swansea, from 1952 to 1964, that we realized how many potential skipping chants may normally appear in a school playground, be tested, found wanting, and abandoned; and we came to understand that a skipping rhyme that was recorded once, and was of obvious feebleness, was likely to have been a child's own invention in imitation of an established skipping chant. The rhythms of the following, all recorded in the 1950s, are distinctly derivative:

> Pina-one, pina-two,
> Pina-three, pinafore.
>
> (Swansea)
>
> Betty Grey ran away,
> On her mother's washing day.
>
> (Cleethorpes, Lincolnshire)

> Apples are green,
> Apples are red,
> Onions are grown
> In a rhubarb bed.
>
> (Welwyn,
> Hertfordshire)

The village cat sat on a mat,
The village cat sat on a M–A–T.

(Radcliffe, Lancashire)

Swiss milk chocolate,
Penny for a cake,
When I call in
  My very best mate.
Is she in? Yes, no, yes, no . . .

(Helensburgh)

My Auntie Rose,
She pulled my nose,
She pulled it one yard, two
  yards . . .

(Cleethorpes, Lincolnshire)

Johnny Jones went for a holiday,
  Lucky boy! lucky boy!
He went to Scotland for a holiday,
  Lucky boy! lucky boy!
How long did you go?
One week, two weeks . . .

(Swansea)

I'm telling the teacher
You stole a bun,
Halved it and sliced it
And gave me none.

(Edinburgh)

Peter Pan, bread and jam,
Marmalade and treacle,
A bit for you,
A bit for me,
A bit for all the people.

(Middleton Cheney, Banbury)

There were so many of these apparently one-off rhymes, that we decided to include in our main list of skipping rhymes only those which had been reported from two or more widely separated places.

ADVERTISING JINGLES

Girls sometimes skip to advertising jingles heard on television, but it is not common and the rhymes are not felt to be 'proper' skipping rhymes. When a girl in Bristol, 1960, volunteered: 'What's the happiest way of eating fish, Birdseye fish fingers!' as a skipping rhyme, another girl remarked, 'I don't like that, it's very fishy—and in any case you got it from an ITV advertisement.' Girls in Adderbury, Oxfordshire, 1970, reeled off the words of a dozen skipping rhymes, but had difficulty in recalling the jingle for Rowntree's Jelly-tots, which 'we sometimes use': 'Bags I Jelly-tots, All for me. Pink ones, yellow ones, Raspberry too. Bags I Jelly-tots, All for you'. 'Drinka pinta milka day' was briefly tried out for skipping in the 1960s, as was 'Sunlight Soap is the best in the world' around 1910 (although the authentic slogan was 'There is no better household soap'). The Robertson's Golden Shred jingle, used for skipping or two-balls in the late 1960s, seems to have sunk without trace: sung to the tune of 'Frère Jacques', it went: 'See it sparkle, see it sparkle, On the spoon, on the spoon, Hurry down to breakfast, hurry down to breakfast, Golden Shred, Golden Shred.' Unfortunately, advertising jingles, though often catchy, rarely have the right appeal or structure to enter the permanent skipping repertoire. An exception is the Murray Mints television jingle from the late 1950s, which was found as a skipping song in Edinburgh in 1975:

> Murray Mints, Murray Mints,
> The too-good-to-hurry-mints,
> You buy the taste,
> There is no waste
> For minty minty Murray Mints.

To have one's products constantly extolled by little girls in playgrounds all over the country must surely be an ad-man's dream, and, even if the orthodox sales-line is not followed, it is gratifying to hear brand-loyalty occasionally confirmed in a simple statement like 'HP Sauce, HP Sauce, Mummy likes, Daddy likes, HP Sauce' (Kent, 1934), later found as 'HP Sauce, HP Sauce, My mother uses HP Sauce' (Swansea, 1952).

One of the most enduring promotional skipping chants must surely be 'Manchester Guardian, Evening News, I sell edition one', referred to in Opie, *Lore*, p. 5, as going back to the nineteenth century, which continues to come to children's lips in the north-east when they have a rope in their hands; and perhaps even more remarkable, the child in the rope continues to crouch down while they chant 'I sell edition one' and jump up when they start 'Manchester Guardian, Evening News' again.

Equally durable, and more widespread, is the simple formula of 'North, south, east, west, Something-or-other is the best', both in terms of real advertising and for skipping (or sometimes ball-bouncing). One of the advertisements on the back of New Zealand stamps of 1892 was 'Search North, South, East or West, Sunlight Soap is still the best'; and a tin sign of *c.*1900, in the Museum of Advertising and Packaging, at Gloucester, reads 'North, South, East or West, Cadbury's Cocoa is the best'. In the late 1920s children were skipping to:

> North, south, east, west,
> Crawford's [or Baker's] biscuits are the best.

In the 1950s Cadbury's Chocolate (rather than cocoa, which was less popular) was being extolled in the same form by skippers in Birmingham, Radcliffe (Lancashire), Welwyn, and Cleethorpes; and in the 1960s in Lower Gornal (Worcestershire), and Aldershot, and doubtless many other places. Skippers of the 1950s also proclaimed (in Manchester and

Halifax) that Crawford's biscuits and Jacob's biscuits are the best; or, in an extended verse, that,

> North, South, East, West,
> BB [later rationalized to 'Babies'']
>   biscuits are the best,
> When you eat them they digest,
> North, South, East, West.

The BB Biscuit Co. was trading in Leeds in the 1930s. (Compare the ball-bouncing formulas 'Brooke Bond Dividend Tea', 'Drinka pinta milka day', and 'Alka-seltzers' (p. 138).

ACTION CHANTS

An Ipswich 9-year-old enthusiastically lists the different ways she can skip:

> I like playing skips.
> I can skip with one eye clowsd.
> I can skip with two eyes clowsd.
> I can skip with one leg.
> I can skip with one arm.
> I can skip with now arms.
> I can skip with now arms and one leg.
> I can skip with my legs crost.
> I can skip and wrols round.
> I can tuch the ground with my hand
>   when I am skipping.
> I can skip with my hands on my head.

Some chants exist solely because they are the vehicles for a succession of clever feats. 'This chant is a very short one, but you go through it with different actions,' said a 10-year-old, apologetically (Blackburn, 1952). The words were:

> Fishes in the sea, fishes in the sea,
> By a one, by a two, by a three.
>
> Fishes in the hops, fishes in the hops,
> By a one, by a two, by a three.
>
> Fishes in the crosels, fishes in the crosels,
> By a one, by a two, by a three.

The rope is swayed from side to side and the player jumps sideways over it, back and forth. On the word 'three' she straddles the rope. Then she does the same thing hopping; and then 'crosels', with crossed legs. Thereafter she may do it 'twirling' (turning at each jump), and 'blind' (with eyes shut). A girl in Accrington, 1960, explained another action chant: ' "Three little fishes caught in a net, By one by two by splits", and the actions go on in this order, jumps, hops, crosses, twirls, touchgrounds, kicks, fairies, highlanders, knees-up, one eye, blindies.'

The action chants have built-in instructions for the actions required ('I'm a little Girl Guide dressed in blue, These are the actions I must do . . .'; 'Jelly on a plate, jelly on a plate, Wibble, wobble, Wibble, wobble . . .'); or, less specifically, they provide a role which a girl can act, improvising her own actions ('There came a girl from Italy . . .'; 'See those girls in Russian boots . . .'). However, a few special actions need to be described separately, either because they illustrate an older tradition, or because they have their own set of short formulas.

PICKING UP SOMETHING FROM THE GROUND

This is commonly called 'postman' or 'teddy bear' after two long-established games in which versions of this action takes place:

> Early in the morning at eight o'clock,
> You can hear the postman knock.
> He drops a letter: I pick it up.
> Early in the morning at eight o'clock.

And,

> Teddy bear, teddy bear, drop your hankie,
> Teddy bear, teddy bear, pick it up.
> Teddy bear, teddy bear, show a shoe,
> Teddy bear, teddy bear, that will do.

Another name is 'pick the handkerchief' (Ruthin) from the chant 'Lady, lady, drop your handkerchief', a variant of 'Lady, lady, turn right round'. However, modern skippers tend to touch the ground instead of picking up a stone or handkerchief. The feat was known as 'Baker, Baker' in England at the turn of the century, and as 'Bake the Bread' in New Zealand (see William Canton, *In Memory of W.V.* (1901), and Sutton-Smith, p. 74). It was well established in America in the late nineteenth century. 'Mama, Mama, your bread is burning, 1, 2, 3', illustrated by a picture of a girl bending down and picking a stone off the ground while skipping, is one of the vignettes accompanying 'The Song of the Skipping Rope' in the American children's magazine *St Nicholas*, 23 (1896), and the verse speaks of the 'many tricks to do, That our mothers also knew!—"In the Front Door", "Baking Bread", "Chase Fox" and "Needle Thread".' (See also W. H. Babcock, *American Anthropologist*, 1 (1888), 266; *Girl's Own Paper*, 13 (1892), 373.)

CROUCHING DOWN, ROPE TURNED ABOVE SKIPPER'S HEAD

This needs expert turners. In St John's Wood, London, in 1975, girls used a count of '1, 2, 3, up to loo-loo' [an echo of 'Skip to my lou'?] . . . '10, up to loo-loo', and 'Every time the words "loo-loo" come up, the person crouches down while the rope turns above her head and she doesn't skip at all.' Girls in Chiswick, London, in the same year, performed the same feat to the words '1, 2, 3, chichilala . . . 10 chichilala, 10 chichilala 10'. The formula in north Staffordshire in 1953 was 'North Staffordshire railway loopy la, I say number one' and so on, a survival from the 1930s when the words were 'North Staffordshire railway loopline'. 'During "I say number" the child had to crouch under the rope until the number was called, when she jumped over it again.' The action is part of 'Manchester Guardian, Evening News', and 'Andy pandy, sugar and candy' (see pp. 213, 239).

SPLITS

'"Splits" is when you land with your feet each side of the rope.' The action commonly appears in games that end with a 'pop' or a 'bang'; while in some versions of 'Charlie Chaplin went to France' the protagonist is actually said to be doing 'the splits', and in the most recent, and very popular, leg-separating game, 'I'm a little bubble car', each turn finishes with the skipper 'putting on the brakes'. The most popular of the older 'splits' games is:

> Creamy [or Christmas] crackers,
> Penny a packet,
> When you pull them,
> They go bang.

This was well known in the 1930s and 1940s; we received versions from thirty-one places in England and Wales, 1950–80, including the variation 'Crawford's [or Jacob's] biscuits, Penny a packet . . .'.

Around Birmingham, where lore often differs from the rest of the country, the favoured rhyme was: 'I like bom-boms, two a penny bom-boms, When I eat them they go crack' (Dudley, 1969). Seventeen years before, in the City of Birmingham, it had been 'We sell bomb-bombs, two-a-penny bomb-bombs, When we pull them They go bang.' 'Bom-boms' must represent 'bon-bons', the original name for Christmas crackers.

In Scotland the rhyme is, 'I know a lady and her name is Sis, I took her to a dance and she landed like this' (or 'I had a little doll and its name was Sis, And all of a sudden she stopped like this', or 'I had a little dog . . .', or 'There was an old woman . . .', or 'There was an old witch who lived in a ditch, And every time she turned she did the splits'). Possibly the ancestry is American, for although no rope-straddling is involved, a similar skipping rhyme appears in *JAFL* 44 (1929), 305, from Everett, Massachusetts:

> I know a little lady, but her name is miss
>   [miss the jump by stepping on the rope],
> She went round the corner to buy some fish,
> She met a little fellow and she gave him a kiss.
> I know a little lady but her name is miss.

(Probably at 'went round the corner' the skipper ran out of the rope, round an ender and into the rope again, as in 'I'm a little bubble car' (see below).) Dr Dorothy Howard collected this as a taunt, accompanied by grotesque walking motions, *c*.1938:

> I know a lady and her name is Miss,
> Every time she walks, she walks like this.

Other words felt suitable are 'Banana split, banana split, Banana, banana, banana split.' 'At "split" the girl in the rope opens her legs and jumps them together before the rope comes round again.' Perhaps, as seems appropriate for the land which invented the ice-cream-and-banana split in the 1920s, this rhyme too comes from America, though its first appearance is not until 1956, in Margaret Taylor's book of Chicago rhymes, *Did You Feed my Cow?*: 'Went to the Drugstore to get a banana split, One banana, two banana, three banana split!' Around 1970 girls began to 'do it with other fruit'; in Alton, for instance, the next verse was 'Orange squash', with ducking down at 'squash', and then 'Pineapple twirl'.

It was probably in the early 1960s that 'I'm a little bubble car' arrived in Britain. The most common text is:

> I'm a little bubble car, number 48,
> I went round the cor-r-r-ner,
> And slammed on [or pulled on] the brakes.

The 'bubble car' was a three-wheeled, low-priced vehicle introduced in 1957 and later found to be dangerously unstable. The subject of the rhyme can also be a 'little motor car' or, increasingly in the 1970s, 1980s, and 1990s, a 'little bumper car'. Minor variations abound: the car can 'go', 'nip', 'whizz' (frequently), 'dash', 'race', or be taken or driven round the corner. It is a matter of expedience, rather than careful transmission.

The game's great popularity may be due to its ingenious two-part *modus operandi*, as well as the fun of impersonating a motor car. A Salford child explains, 'You're supposed to go out of the skipping rope when they say "round the corner", and you run round one of the enders, and come back in, and then when it's "And put on the brakes" you do the splits so that you catch the rope between your legs.' In the 1929 American recording of 'I know a little lady, but her name is miss' (see above) the words 'She went round the corner to buy some fish' may have been accompanied by the same action.

Everything points to a transatlantic origin. The Canadian folklorist Edith Fowke was the first to note the game (*Sally Go Round the Sun*, no. 102); the words were already, in 1959–60, well diversified:

> Had a little Austin in nineteen forty-eight.
> Turned around the corner and slammed on the brake.
> Policeman caught me, put me in jail.
> All I had was ginger-ale.
> How many gallons did I drink? Two, four, six, eight . . .
>
> > (Vancouver, 1959[21])

> > I had a little car
> > And it went 'Peep! Peep!
> > I took it round the cor-ner
> > And across the street.
> >
> > > (Ottawa, 1960)

> > I had a little Dutch car
> > In nineteen forty-eight.
> > I took it down to Main Street
> > And put on the brake.
> >
> > > (Toronto, 1960)

The probability is that the game was widely known in America as well at that time, though it was not noted before 1968 (Tri-University Project Report, University of Nebraska, August 1968, p. 19).

ACTION CHANTS WITH BUILT-IN INSTRUCTIONS

> I'm a Girl Guide dressed in blue,
> These are the actions I must do:
> Salute to the captain, curtsy to the queen,
> Show me fancy knickers to the football team.

'On the last line', wrote a surprised playground supervisor in *The Bradfield Telegraph*, 1973, 'it's skirts up above their heads.' But she

---

[21] This version turned up in Lhanbryde, a small village near Elgin, Morayshire, in 1975.

added that they were only infants. This skipping game has delighted innumerable children for most of the twentieth century, and has spawned a variety of 'clever' or 'funny' endings. The player in the rope first salutes, and then does a bob for a curtsy. Before 1968 the words were more staid, the actions more numerous:

> I am a Girl Guide dressed in blue,
> These are the actions I can do:
> Stand at ease, bend my knees,
> Salute to the king, bow to the queen,
> And turn my back on the boy in green.

Alternatively the skipper might turn her back on 'the girl in green', 'the gipsy queen', 'the sewing machine', 'the washing machine', or, in 1944, on 'the German submarine'. In 1952 some skippers were still turning their backs on Hitler. More recently, many were turning their back on 'the Union Jack' (because of the internal rhyme). They also turned it on 'the saucy old cat' (because of assonance), and, in Scotland, on 'the sailor boy, as if he were a twopenny toy'. In some places the instruction is 'Turn right round like a ball of string' or 'Turn right round to the LCC' (Hackney), or 'Turn right round to Wall's Ice Cream' (Weymouth). By the mid-1970s the word 'little' had crept into the first line and more often than not the verse began 'I'm a little Girl Guide'. Occasionally the skipper is not a Girl Guide but 'a Boy Scout smart and new', or 'a doll dressed in blue', or, as commonly in America, 'a little Dutch girl'.

After the girls have done 'I am a Girl Guide dressed in blue', the shout may be 'Let's do red Girl Guide', or 'yellow Girl Guide'. The verses about these polychrome Girl Guides have no structural relationship to the 'Girl Guide dressed in blue' but are nevertheless traditional, each having been found in a number of places:

> I am a Girl Guide dressed in red,
> These are the times I go to bed:
> One, two, three, four . . .

> I am a Girl Guide dressed in brown,
> These are the days I go to town:
> Monday, Tuesday, Wednesday . . .

> I am a Girl Guide dressed in green,
> I close my eyes and count sixteen.
> One, two, three, four . . .

> I am a Girl Guide dressed in yellow,
> I go about with a nice old fellow.
> How many times did I kiss him?
> One, two, three, four . . .

> I am a Girl Guide dressed in black,
> I went to the factory and got the sack,
> I went to the station and missed the train,
> I went to school and got the cane.
>   How many whacks did you get?
>   One, two, three . . .
>
>> (Known in Kent, 1934, 'I went to work and got the
>> sack, When I got home I got the strap . . .')

The tradition also included Brownies dressed in brown, presumably because a Girl Guide dressed in brown was felt to be inaccurate:

> I'm a little Brownie dressed in brown,
> See my knickers hanging down [or, if adults are
>   about, 'sockies hanging down'],
> Pull them up, pull them down,
> I'm a little Brownie dressed in brown.
>
>> (Scotland, 1948 onwards, seventeen recordings. Also
>> Wakefield, 1954, '. . . brown, Hanging on a lamp-post
>> upside down')

Another 'Brownie dressed in brown', almost too negligible to mention but in existence for at least forty years, runs '. . . I went to buy some things in town, I got in a bus and sat right down . . .' (Kent, 1934); or '. . . I go shopping in the town, I buy sugar, I buy tea, I buy sweets for baby and me. How many sweets did I buy? One, two . . .' (Welwyn, 1952); or '. . . Every Saturday morning I go to town; I jump off the bus, round the corner, under the bridge and back into the shop, To buy ten bottles of candy-pop, One, two . . .' (Salford, 1970).[22] However, this serves as a good illustration of rhymes genuinely 'made up' by children: banal, and full of resonances of other rhymes and borrowed ideas and actions.

All the records show that before and during the First World War this chant was used for ball-bouncing. The earliest recollection, from Leicestershire, c.1910, is:

> Ball against the wall, I heard the king say:
> Quick march under the arch,
> Salute to your king, bow to your queen,
> Sit down, kneel down,
> Touch the ground, and a birly round.

---

[22] Cf. the rhythm of the calling-in game 'I like coffee, I like tea'; and the action of 'I'm a little bubble car' and its running 'round the corner'.

(The expression 'a birly round' suggests that the chant came from Scotland.) By the 1920s the Girl Guide had arrived, and a typical version, still for ball-bouncing, was:

> I am a Girl Guide dressed in blue,
> All my duties I must do,
> Bow to the king,
> And curtsy to the queen,
> And turn my back on the Kaiser.
>
> (Stoneleigh, Staffordshire)

The rhyme went on being used for ball-bouncing until at least the mid-1970s, but only in the south-west of Scotland, for example in Cumnock, Helensburgh, and Kilmarnock where, in 1975, the words were:

> I'm a Girl Guide all dressed in blue,
> Here are some actions we ought to do.
> Stand at ease, bendies knees,
> Quickly march, through the old arch,
> Curtsy to the queen, and bow to the king,
> And turn your back on the Union Jack.

Skipping game, Britain: M. and R. King, *Street Games of North Shields Children*, 2nd ser. (1930), 23. James Kirkup, *The Only Child*, 149, *re*. South Shields, *c*.1930. Kent, 1934. Liss, 1935. Swansea, 1937–9. Castle Eden, 1943. Rhos, near Wrexham, *c*.1945. Farnham, Surrey, 1947. Britain 1950 onwards, 136 recordings.
  Australia: Turner (1969), 28–9, nine versions.
  New Zealand: Sutton-Smith (1959), 81.
  South Africa: Pietermaritzburg, *c*.1940, 'Turn my back on the Nazi submarine'. Pretoria, 1971. Johannesburg, 1972.
  USA: Abrahams, nos. 205, 206.

> I'm a little Scots girl, I can do the kicks,
> I can do the twirl around, I can do the splits.
> The Queen does the curtsy, the King does the bow,
> The girls show their knickers, and the boys say 'Wow!'

This shares some of the characteristics of 'I'm a little Girl Guide' but is not as widespread. Sometimes the dancer is 'a little Dutch girl', 'a little Chinese girl (Forfar), or 'a Brownie' (Haverfordwest). Sometimes the chant begins, 'I can do the cancan', 'I can do the Highland fling', or, oddly, 'I can do the baseball'. In some places, Glasgow for instance, it begins interrogatively:

> Can you do the sword dance?
> Can you do the kicks?
> Can you do the birlie round?
> Can you do the splits?

The earlier versions ended more romantically: 'and the boys threw a kiss'.

Britain: Castle Eden, 1943, 'Can you do the Scotchie dance?' Thereafter versions from forty places including West Country and Shetland Islands.
   Australia: Turner (1969), 20, 'Can you do the saw dance?', Sydney, 1958; 'I'm a little Dutch girl', Canberra, 1959. Cf. 'Charlie Chaplin went to France'.

❖

Jelly on a plate,
Jelly on a plate,
Wibble wobble, wibble wobble,
Jelly on a plate.

Sausage in a pan,
Sausage in a pan,
Turn it over, turn it over,
   [or 'Frizzle frazzle, frizzle frazzle']
Sausage in a pan.

Baby on the floor,
Baby on the floor,
Pick him up, pick him up,
Baby on the floor.

Burglar in the house,
Burglar in the house,
Kick him out, kick him out,
Burglar in the house.

The most raucous, most infectious, and one of the best-known of the action skipping games. Each player in turn goes into the rope, begins by skipping ordinarily and then, at 'Wibble wobble', 'wobbles her hips like a jelly'. Thereafter, in 'Sausage in a pan' she turns round twice; in 'Baby on the floor' she touches the ground twice; in 'Burglar in the house' she kicks twice. Then she runs out of the rope, the chanting starts again, and the next player runs in. 'It is very puffing for the skipper,' remarked a 10-year-old. It is also a game in which a young skipper may 'get the giggles'.
   The framework of the chant, with its dual repetition of the first line, is such that instant acclaim is guaranteed to anyone inventing a new verse, such as 'Milk in the jug—Pour it out' or 'Clothes on the line—Peg them up'. In practice, though, playground poets rarely have their words immortalized. The four verses above have been in service since the 1930s, and are used over and over again. Although scores of verses have been collected, only a few, such as those that follow, have obtained more than local currency.

Apples on the tree . . . Pull them down.

Bogey under the bed . . . Shoo him out.

Girls in the school . . . Please teacher, please
   teacher, May I leave the room?

Penny [or 'Paper'] on the floor . . . Pick it up.

Pickles in the jar . . . Shake 'em out.

Baby in the pram . . . Cry baby.

Monkeys at the zoo . . . Look Mum! There's one.

Ants on the floor . . . Stamp 'em out.

Marilyn [Monroe] at the show . . . Show a leg.

Monkey up a tree . . . Pull him down.

Sailors on the sea . . . Ship ahoy!

Soldiers on the march . . . Left, right.

Spider on the wall . . . Flick it off.

Ice cream in the fridge . . . Lick it up.

In their actions the skippers either identify with the subject, marching
like soldiers, and showing a leg like Marilyn Monroe; or they do as bid,
stamping on the ants, shooing away the bogey.

Britain: James Kirkup, *The Only Child* (1957), 150–1, South Shields *c.*1930, 'Jelly on the plate',
'Burglars in the house', 'Baby on the floor'. Kent, 1934, as South Shields plus 'Sausage on the roll'
and 'Pickle in the jar . . . Oohja, oohja'. Liss, 1935. Castle Eden, 1943. Since 1950 versions from
every place where enquiries were made, including Cornwall, Channel Islands, and Shetland.
   Australia: Turner (1969), 31, Sydney, 1957, etc., numerous versions as in Britain.
   Canada: Toronto, 1940, 'Jelly on a plate', 'Sausage in the pan', 'Peas in the pod . . . Split them
open'.
   USA: Howard MSS (1938), 'Jelly on a plate' verse. Abrahams, no. 284, 'Jelly in the dish' (or
'bowl', or 'plate'), earliest 1942; apparently not as popular as in Britain.

Lady, lady, turn right round,
Lady, lady, touch the ground,
Lady, lady, do the kicks,
Lady, lady, do the splits.

This game has been a familiar sight in Britain since the nineteenth
century. The skipper has to perform the actions suggested by the words.
(When doing the splits the skipper has usually to land with a leg either
side of the rope.) In 1907 the Revd Moxon instanced the game as showing the remarkable feats girls in Soho were able to perform. The chant
was:

Lady, lady, touch the ground,
Lady, lady, turn right round,
Lady, lady, show your foot,
Lady, lady, sling your hook.

At 'sling your hook', 'she rushes out and the next lady comes in'.

The formula 'Lady, lady, drop your purse, Lady, lady, pick it up', or 'Lady, lady, drop your handkerchief . . .', or, more humorously, 'Lady, lady, drop your baby', was widespread at the turn of the century; skippers had to lay down a small object (usually a stone) and to pick it up while still skipping. This tradition continued in Scotland; for instance, in Langholm, on the Scottish borders, in 1960, girls were swaying the rope and chanting, 'Lady, lady, drop your hanky, Lady, lady, pick it up, Lady, lady, spell your charming name', and the trick was to try to drop the hankie on the rope while it was swinging backwards and forwards instead of on the ground, which made it easier to pick up.

The history of the game, as well as its variety, may be seen in the following recordings:

Lady, lady, turn around,
Lady, lady, touch the ground,
Lady, lady, show your shoe,
Lady, lady, twenty-four, skidoo!

> (Carlisle *Evening Sentinel*, Pennsylvania, 1929 (JAFL 40 (1929), 386))

Dolly, dolly, touch the ground,
Dolly, dolly, turn right round,
Dolly, dolly, reach up high,
Dolly, dolly, touch the sky.

> (London, 1947)

Lady, lady, touch the ground,
Lady, lady, turn around,
Lady, lady, show yer boney knees,
Lady, lady, go out please.

> (Salford, 1970)

Lady, lady, close your eyes,
Lady, lady, reach the skies,
Lady, lady, touch your toes,
Lady, lady, blow your nose.
Lady, lady, touch your chin,
Lady, lady, walk right in,
Lady, lady, touch your heel,
Lady, lady, dance a reel,
Lady, lady, touch the ground,
Lady, lady, turn around,
Lady, lady, bend your knees,
Lady, lady, stand at ease.

> (Weston-super-Mare, 1951)

Alabalabuster King of the Jews
Bought his wife a pair of shoes.
When the shoes began to wear
Alabalabuster began to swear,
And these are the words he said:
Lady, lady, turn around,
Lady, lady, touch the ground,
Lady, lady, show your dirty knees,
Lady, lady, do the strip tease.

> (Ilkley, 1975. 'Alabalabuster' is usually a ball-bouncing rhyme)

Teddy bear, teddy bear, touch the ground,
Teddy bear, teddy bear, turn right round,
Teddy bear, teddy bear, go upstairs,
Teddy bear, teddy bear, say your prayers.
Teddy bear, teddy bear, switch off the light,
Teddy bear, teddy bear, say Goodnight.

One of the most played skipping games in Britain and around the English-speaking world; as also in Germany, Switzerland, Scandinavia, and undoubtedly other places in Europe as well. The chant is so well known (few children were met who were not familiar with it), the wonder is that any similar game such as 'Lady, lady, turn right round', from which it probably emanates, continues to retain a separate identity. The teddy bear toy seems to have acquired its name and popularity following an incident in November 1902, when President Theodore 'Teddy' Roosevelt refused to shoot a bear cub while on a hunting trip, a sportsmanlike act enlarged upon by Clifford K. Berryman in a series of political cartoons in the *Washington Post*.

The earliest definite date for the chant in Britain is Axminster, 1922, when it was being repeated for ball-bouncing:

> Teddy Bear, Teddy Bear, touch the ground,
> Teddy Bear, Teddy Bear, twist right round,
> Teddy Bear, Teddy Bear, throw me a kiss.

The following lines were, however, being used for skipping in the United States in 1926:

> Teddy bear, teddy bear, turn around,
> Teddy bear, teddy bear, touch the ground,
> Teddy bear, teddy bear, show your dirty shoe,
> Teddy bear, teddy bear, twenty-four skidoo,

and in Norway children were apparently skipping to 'Bamse, bamse' in about 1915.

Since then the game has become one of the first children mention when asked about skipping. The lines are repeated with some uniformity throughout Britain, the skipper carrying out the actions suggested by the words, snapping her fingers—or attempting to snap her fingers—at 'switch off the light', and running out of the rope at 'Goodnight'. Occasionally teddy bear is told to 'Go to school, sit on a stool, read a book, sling your hook' (Enfield) or to 'Go to school, pick up your rule, say your ABC, home to tea' (Cheltenham). In Scotland the game is sometimes 'a swaying game not a cawing game' and the teddy bear is told to 'drop your handkerchief' and 'pick it up' as in the old 'Lady, lady' game. Sometimes he has to spell 'Goodnight' rather than say it.

Britain: Reading, c.1917, 'Dolly, dolly, turn right round'. Macmillan Collection, 1922. *JAFL* 39 (1926), 84, with variant 'Mama doll, mama doll, go upstairs'; 47 (1934), 384, variants 'Buster, Buster, hands on head', and 'Butterfly, butterfly, turn around', both from 1929. James Kirkup, *The Only Child* (1957), 149, *re* South Shields, c.1930. Swansea, 1960, 'Golliwog, Golliwog'.
  Australia: Turner (1969), 38–9, six versions.
  Denmark: 'Bamse, bamse, hop nu ind . . .' (Erik Kaas Nielsen, *Leg i Brondby* (1992)).

Norway: Randaberg, *c.*1915, 'Bamse, bamse, hopp, hopp, hopp' (Svein Magne Olsen, *Lek og Sang*, 29). Trondheim, 1976, 'Bamse, bamse, ta i bakken . . .'.

Germany: 'Teddibär, Teddibär, spring ins Seil . . . heb ein Bein . . . mach dich krumm . . . dreh dich um . . . wie alt bist du?' (R. Peesch, *Das Berliner Kinderspiel der Gegenwart* (1957), 30).

> Shirley Temple walks like this,
> Shirley Temple throws a kiss,
> Shirley Temple says her prayers,
> Shirley Temple falls down stairs.

A derivative of 'Teddy bear, teddy bear'. Found only in Swansea, in 1952. The earliest recording is from Connecticut (B. A. Botkin, *Treasury of New England Folklore* (1947), 907).

> Not last night but the night before,
> Twenty-four robbers came knocking at the door,
> I ran out to let them in,
> This is what they said to me:
> Spanish lady turn right round,
> Spanish lady touch the ground,
> Spanish lady do the high kicks,
> Spanish lady do the splits.

Another rhyme on the lines of 'Lady, lady' and 'Teddy bear, teddy bear', in which the role of instructor has, for some odd reason, been given to the twenty-four robbers of the old nonsense rhyme 'Not last night' (see Opie, *Lore*, p. 23). This nonsense rhyme was used for skipping in the United States in 1929, and the combination of 'Not last night' and 'Spanish lady' was first noted there in 1943 (see Abrahams, nos. 309 and 526). The two together make a very effective skipping routine. At 'I ran out', the skipper runs out of the rope and round an ender, coming back in time to perform the actions. The game was apparently not known in Britain before the 1960s, and then achieved a wide and enduring popularity which has lasted to the present day. Variations are few. 'As I went out to let them in' is sometimes followed by 'They hit me on the head with a rolling pin'. In some places the Spanish lady is replaced by a 'ballerina'; a girl in Aberdeen said, 'I think "ballerina" is better. It's better for the dancing.' And in Hackney, London, in 1993, the words were:

> Not last night but the night before,
> Twenty-four robbers came knocking at my door;
> So I went downstairs to let them in,
> And this is what they said to me:

> Christabel, Christabel, turn around,
> Christabel, Christabel, touch the ground,
> Christabel, Christabel, do the splits,
> Christabel, Christabel, have you got the nits?
> With a 'Yes, No, Yes, No, Yes!'

Mutations will go on occurring. In the summer of 1995, in 'the child-filled streets' of Kensal Green, north-west London, Jamaican girls were dancing in pairs to a Caribbeanized version of the song. 'The dance resembled a stylized clapping game, each girl stepping elegantly forward and back, palms outward but never touching those of her partner.' The leader sang a line, and was answered by a chorus: 'Step back, baby, step back!'

> *Solo*: Not last night, but the night before!
> *Chorus*: Step back, baby, step back!
> *Solo*: Twenty four robbers at my door!
> *Chorus*: Step back, baby, step back!

All right, face your partners and here we go!

> Open the door and let them in!
> Step back, baby, step back!
> Hit them on the head with a rolling pin!
> Step back, baby, step back!
> Shoulda seen the way those robbers ran!
> Step back, baby, step back!
> Some ran east and some ran west!
> Step back, baby, step back!
> Some ran over the cuckoo's nest!
> Step back, baby, step back![23]

The last lines come from the old American counting-out rhyme, 'William a Trimbletoe' (see H. C. Bolton, *Counting-Out Rhymes of Children* (1888), nos. 807, 808, 810).

> Cinderella bumberella turn right round,
> Cinderella bumberella touch the ground,
> Cinderella bumberella show your pretty foot,
> Cinderella bumberella take your hook.

Another 'turn right round' rhyme. The skipper turns around, touches the ground, puts forward one foot, and then runs out of the rope. Rhyming or comic reduplications of 'Cinderella' seem to have fascinated children since the beginning of the twentieth century.

[23] Roy Kerridge, 'The Singing Harp', *English Dance & Song* (Spring 1996), 5.

London Street Games (1916), 'Cinderella-Umberella' listed amongst play songs. Kent, 1934, 'Ella bella Cinderella'. Wootton Bassett, 1950. Laverstock, 1951, 'Cinderella ella bella'. Alton, 1952. Swansea, 1952. Maryon Park, London, 1952. Birmingham, 1953, as 1916. Wolstanton, 1960, 'Cinderella dressed in yellow, Turn round once'. Oldbury, Worcestershire, 1970, as text. Cf. 'Cinderella dressed in yellow'.

> Lady in a tight skirt can't do *this*!
> Lady in a tight skirt can't do *that*!
> Lady in a tight skirt can't do *this*, *that*, *this*, *that* . . .

'This' and 'that' are forward and sideways astride positions, that is, 'scissors' and 'splits' alternately. Then the skipper chants, 'Lady in a tight skirt *can* do *this*!', and she repeats the actions but with knees held together as if glued, the movements being made entirely with the lower legs. ('Remarkable', said one observer, 'reminiscent of a circus clown'.)

Farnham, Surrey, 1947. Wootton Bassett, 1949. 1950–80, ten places, including York, 1962, 'Lady in a tight skirt, She can do the kicks, She can do the twirly-whirls, She can do the splits.'

> Ducks in the water go 'Quack!'
> Ducks in the water go 'Quack, quack!'
> Ducks in the water go 'Quack, quack, quack!' etc.

A 13-year-old in Cleethorpes, Lincolnshire, 1952, explained, 'While saying "Quack!" the legs open like a duck's beak. This verse is said while skipping fastly.' Less ephemeral than it might seem, the rhyme was also found in Birmingham, 1971, 'Ducks in the water, one, two, three . . .'

> Up and down, up and down,
> All the way to London Town;
> Swish swosh, swish swosh,
> All the way to King's Cross;
> Leg swing, leg swing,
> All the way to Berlin;
> Heel toe, heel toe,
> All the way to Mexico;
> Knees bend, knees bend,
> All the way to Land's End.

'This is very difficult,' says an Ipswich girl. 'You have to do all the actions. For *heel toe*, for instance, you have to touch your heel and toe in the middle of quick skipping.' Other actions include walking-on-the-spot for 'Up and down'; jumping from side to side with feet together for 'Swish swosh'; swinging one leg while hopping on the other for 'Leg swing'. But actions vary, as do the couplets, for example: 'Clap hands, clap hands, All the way to Amsterdam' (Edinburgh); 'Criss cross, criss cross, All the way to Knotty Ash' (Liverpool). 'As this game is very tiring we soon give it up,' remarked a Spennymoor girl. Nevertheless it seems to have been widespread (records from thirty places), and to have been played since the 1920s. 'Swish swosh' is normally a term for swaying the rope (see above).

> I had a little dolly dressed in green,
> I didn't like the colour so I sent it to the queen;
> The queen didn't like it so she sent it to the king,
> The king said, 'Close your eyes and count sixteen.
>   One, two, three, four . . .'

The popularity of this game has been splendidly renewed since the king began commanding 'Close your eyes and count sixteen.' The skipper has to attempt sixteen consecutive jumps with her eyes shut, a feat that is satisfying to perform and an entertainment to watch. When girls skipped in Edwardian days in Soho, the words were more amusing:

> Mary had a baby dressed in green,
> Because she didn't like it she gave it to the queen.
> The queen didn't like it because it wasn't fat,
> She cut it up in slices and gave it to the cat.

But even when the cat was given the slices with 'Salt, mustard, vinegar, pepper', the game had little to distinguish it from other games. The words had almost certainly been borrowed from some other entertainment. Alice Gillington, in 1909, knew it as an extension of the singing game 'Monday Night, Band of Hope'; and in 1853 Halliwell had printed a similar verse in the fifth edition of his *Nursery Rhymes of England*:

> Hector Protector was dressed all in green,
> Hector Protector was sent to the Queen.
> The Queen did not like him,
> No more did the King:
> So Hector Protector was sent back again.

When the number of variations that have gained currency in the present century are considered, in which the baby- or doll-owner may be a Dutch girl, a Scotch girl, or a Girl Guide, and the doll becomes a teddy bear, a golliwog, a fairy, or a letter, it will be apparent that Hector Protector, in his turn, was not necessarily the original character 'dressed all in green'.

St Anne's Soho Monthly, June, 1907. *Old Surrey Singing Games* (1909), 'Clara had a baby dressed in green'. *London Street Games* (1916), 'I had a dolly dressed in green'; also 'Jumbo had a baby' listed amongst skipping games. Surrey, *c*.1916, 'Mother had a baby'. Versions from thirty-six places since 1950. Variants include Aberdeen, 1952, for ball-bouncing, 'Mary had a dolly . . . The queen didna like it so she gave it to the Pope, The Pope didna like it so he tore it all up' (or, 'The pup didna like it so it ate it all up'). Forfar, 1954, 'My green dolly, dressed in green, Put her in a matchbox and sent her to the queen. The queen took her out and spanked her little doup, My green dolly, dressed in green'. Bristol, 1960, and elsewhere, 'I had a teddy bear it was pink, I didn't like it so I gave it to the sink; The sink didn't like it so he gave it to the drain; The drain said, "Shut your eyes and spell your name."' Frodsham, Cheshire, 1960, 'I'm a little Girl Guide dressed in green, Put me in a matchbox and send me to the queen. I jumped out, she did shout: "Does your mummy know you're out?"'

> Once you're in you can't get out
> Unless you touch the ground;
> Once you're in you can't get out
> Unless you twirl around;
> Once you're in you can't get out
> Unless you do the kicks;
> Once you're in you can't get out
> Unless you do the splits.

The words of this chant are contradictory. The skipper may not stay in the rope unless she touches the ground, twirls around, does the kicks, and then leaps in the air to do the splits, landing with a foot either side of the rope. She will then attempt to touch the ground twice, to twirl around twice, and so on. In the 1970s logicality was sometimes attempted, with children chanting, 'You can't come in, you can't come out, unless you touch the ground', and 'You cannot go into the rope unless you touch the ground'. But by this time the skipper already *is* in the rope.

Britain: Midlands and north country only. Apparently well known in the 1940s. Versions from eight places since 1950.

Australia: Melbourne, 1956, 1967, 'Once you get in you can't get out, Until you say your boy's name.' Sung to tease a girl who is already in the rope (Turner MSS).

> Oliver Cromwell lost his shoe
> At the battle of Waterloo,
> Left, right, left, right,
> Attention! Halt! One, two.

The Protector's discomfiture at Waterloo, though unmentioned in the history books, has been kept in memory for more than eighty years. At 'Left, right' the skipper pretends to march, and at 'One, two' she does bumps.

Britain: versions from nineteen places including Clifton, Manchester, *c*.1910, 'Oliver Cromwell lost his shoe, In the battle of Waterloo, Sent his wife to Botany Bay, Where she sang Ta-ra-ra-boom-de-ay'. Leeds, *c*.1915. Somerset, 1922, 'Lord Nelson lost his shoe'. Enfield, 1951, 'The Prince of Wales he lost his shoe'. Langholm, 1960, 'Cinderella lost her shoe, On the banks of Waterloo'. Oldbury, 1970, 'Oliver Cromwell lost his shoe, In a bottle of water loo'.

Australia: Turner (1969), 35, 'Oliver Crumple lost his shoe', Melbourne, 1967, similar to Clifton, *c*.1910.

New Zealand: couplet used for counting out (Sutton-Smith (1959), 72).

> Cowboy Joe from Mexico,
> Hands up, stick 'em up,
> Drop your guns and pick 'em up,
> Cowboy Joe from Mexico.

Skipping game in which the skipper mimics the cowboy as well as she can while skipping, holding her hands in the air, pointing them in front of her and, at 'pick 'em up', touching the ground. The chant, with its staccato rhythm and he-man character, is much enjoyed by the girls. By 1960 it appeared to be known throughout Britain. Occasionally, as at Ilkley, other verses are added:

> Buffalo Bill from Faraway Hill,
> Hands up, stick 'em up,
> Drop your guns and pick 'em up,
> Buffalo Bill from Faraway Hill.

In the early 1950s the chant was more in the form of 'Jelly on a plate' (see above), from which it may be an offshoot:

> Cowboy *Joe*, Cowboy *Joe*,
> Stick 'em up, *bang! bang!*
> Cowboy *Joe*.
> Indian *chief*, Indian *chief*,
> Walla walla, Walla walla,
> Indian *chief*.

The skipper either did bumps on the stress words, or pointed two fingers in gun-like fashion and clapped hands for the gun's reports. Cowboy Joe is, perhaps, the 'high-fa-luting-scooting, shooting son-of-a-gun from Arizona, Ragtime Cowboy Joe' of the song by Grant Clarke, copyrighted in 1912.

Britain: seventy-five recordings.

Australia: Turner (1969), 22–3, six versions including 'Old Black Joe from Mexico'.

Canada: Edmonton, 1961.

> Poor old Nelson lost one arm,
> Poor old Nelson lost the other arm;
> Poor old Nelson lost one eye,
> Poor old Nelson lost the other eye;
> Poor old Nelson lost one leg,
> Poor old Nelson fell down dead.

One of the more difficult of the action skipping games, which usually becomes a favourite when it first arrives in a street or playground. The girl in the rope has first to put one arm behind her back, then the other arm; then close one eye, then the other; then she has to skip on one leg, and when Nelson dies she has to run out of the rope. In the 1950s Nelson used sometimes to be resuscitated. He lost only one arm, one eye, one leg, and when he 'fell down dead', the skipper crouched. Then, at Spennymoor and Radcliffe, for instance, the game continued:

> Nelson in the last war found one eye [eye opened]
> Nelson in the last war found one arm [arm brought forward]
> Nelson in the last war found one leg [skip on both legs]
> Nelson in the last war rose from the dead [stand up].

Few skippers were skilful enough to complete this sequence.

The same loss of limb occurred in extensions to the singing game 'Romans and English', and from these the skipping game is presumably derived. For instance, in Cecil Sharp's *Children's Singing Games*, set IV (1912), after the game proper the children march round in a ring singing 'Now we've only got one arm . . . leg . . . eye . . . ear.'

Britain: versions from twenty-eight places including Bootle, 1950, 'Lord Nelson lost one eye'. Swansea, 1957, Staines, 1969, and Leicester, 1974, 'Captain Cook he lost one arm'. Wolstanton, 1961, and Edinburgh, 1975, as text. Stornoway, Isle of Lewis, 1975, 'Old, old Nelson lost one arm'. Australia: Turner (1969), 32, Canberra, 1959.

> I can do the tango, the tango, the tango,
> I can do the tango, just like this.
> I can ride a scooter, a scooter, a scooter,
> I can ride a scooter, just like this.
> I can ride a bicycle, a bicycle, a bicycle,
> I can ride a bicycle, just like this.
> I can ride a dobbie horse, a dobbie horse, a dobbie horse,
> I can ride a dobbie horse, just like this.
> I can do the whirlabouts, the whirlabouts, the whirlabouts,
> I can do the whirlabouts, just like this.

The actions are suggested by the words.

London Street Games (1916), 95, among girls' songs, 'I can do the tango'. Swansea, 1936. Farnham, Surrey, 1947. N. Staffordshire, 1953. Wolstanton, 1960, as text.

> Oliver jump, Oliver jump, Oliver jump, jump, jump,
> Oliver kick, Oliver kick, Oliver kick, kick, kick,
> Oliver stride, Oliver stride, Oliver stride, stride, stride,
> Oliver Twist, Oliver Twist, Oliver Twist, Twist, Twist.

The skipper is required to do bumps, followed by kicks, 'splits' with wide apart legs, and turn-arounds.

Britain: Enfield and Taunton, 1951. Swansea, 1953, 'Pussy round the corner makes me jump, jump, jump . . . kick . . . twist'. Birmingham, 1953 and 1966, 'O Johnny, jump, jump, jump . . . kick . . . stride . . . twirl'. Edinburgh (Ritchie, *Golden City* (1965), 142), 'Very, very popular'.
 USA: Maryland, 1949, 'Little Orphan Annie does the split, split, split'.

> Charlie Chaplin went to France,
> To teach the ladies how to dance;
> First he did the rhumba,
> Then he did the kicks,
> Then he did the turnabouts,
> Then he did the splits.

Now popular for skipping, although the second part of the song comes from the yet more dominant skipping game, 'I'm a little Scots Girl' (see p. 221), and the actions performed in the rope are much the same. The song has, in fact, been repaired so often over the years that, like Sydney Smith's carriage, 'the Immortal', no part of it is original. In the early 1920s it seems to have been known chiefly as a ball-bouncing game, the player chanting:

> Charlie Chaplin went to France,
> To teach the ladies how to dance,
> This is what he taught them:
> Heel, toe, alaira,
> Heel, toe, heel, toe, heel, toe, alaira.

At 'alaira' (alternatively, 'over you go') a leg was passed over the ball, as in 'One, two, three, alaira' (p. 136). Versions of this ball game have coexisted with the skipping game to the present day, particularly in Scotland, the chief variations being in the last line. Thus, at St Andrews:

> Charlie Chaplin went to France,
> To show the ladies how to dance:
> First your heel and then your toe,
> Buckle up your skirts and down you go.

'The ball is thrown against the wall and at *first your heel* place one foot on the heel, at *then your foot* point your toe. Then lift up your skirt and throw the ball under the leg onto the wall' (girl, 13).

  Charlie Chaplin entered juvenile mythology during the First World War, usurping the places of several weaker characters in rhyme and song; and there seems little doubt that the verse the comedian has taken over here was one, much loved in the second half of the nineteenth century, that celebrated the polka, the dance which became the rage in Britain in 1844:

> My sister Jane she went to France,
> To teach to me the polka dance:
> First the heel, and then the toe,
> That's the way the polka goes.

Whether or not this verse had antecedents, in its turn, is unknown, but it seems possible. In the preface to the 1692 edition of *The Scotch Hudibras* the author, Samuel Colvil, argues that it is better to write poor verse that makes people laugh, than good verse that is solemn. 'Where one [person] laughs at the Poems of Virgil, Homer, Ariosto, Du Bartas, &c. twenty will laugh at those of John Cockburn, or Mr Zachary Boyd.' The example he gives of Cockburn's verse (the reading of which, he suggests, would cure any hypochondriac) is the following:

> Samuel was sent to France,
> To learn to Sing and Dance,
> And play upon a Fiddle:
> Now he's a Man of great esteem,
> His Mother got him in a Dream,
> At Culross on a Girdle.

These lines have more to them than meets the eye. They are, it appears, about Samuel Colvil himself, whose mother, Lady Colvil or Colville of Culros, was the author of a notorious poem in *ottava rima*, published in 1603, describing a dream in which Christ rescued her from hell. Looking at Cockburn's squib it seems probable that the first three lines, which have little significance, were an adaptation of lines already current in the seventeenth century; and it may be noticed that when Kirkpatrick Sharpe repeated them in 1817, in his edition of Kirkton's *History of the Church of Scotland*, p. 181, he gave them a rhythm that more accords with the verse children repeat today:

> Samuel Colvill's gone to France
> Where he hath learnt to sing and dance
> And play upon a fiddle.

Britain, for ball-bouncing: Axminster, Kilmington, and Seavington, 1922, as quote (Macmillan Collection). Leeds, c.1925 (Kellett MSS). *Street Games of North Shields*, 2nd ser. (1930), 24. Cardiff, 1935. Further recordings in Opie, *Lore* (1959), 110.

   Variants include: Dunoon, 1962, 'Heel, toe, stamp, Gibraltar, That's what Charlie taught them'. Greenock, 1975, 'Elvis Presley went to France'.

   Britain: for skipping: London, 1950, transitional version, 'Charlie Chaplin went to France, To teach the ladies how to dance, This is how he taught them: First you do the heel-toe, Then you do the kicks, Then you do the roundabout, Then you do the splits'. Versions from eighteen places thereafter, including Swansea, 1957, 'He taught them the Rock 'n Roll, He taught them the splits'. Oldbury, 1970, 'Norman Wisdom went to France'.

   USA: Poughkeepsie, New York, 1926, ball-bouncing (*JAFL* 39 (1926), 82). Anna, Illinois, 1954, jump-rope, 'Charlie Chaplin went to France, To teach the girls a hula-hula dance; First on their heels, Then on their toes, Do the splits, And around you go. Bow to the Captain, Bow to the Queen, Turn your back on the dirty submarine' (A relic of the First World War). Abrahams, no. 65, many additional references.

Australia: Adelaide, 1957, 'Oliver Cromwell went to France'. Canberra, 1959, 'Donald Duck went to town, To teach the shielas how to dance' (Turner MSS).

> Policeman, policeman, do your duty,
> Here comes [Linda] the American beauty.
> She can wiggle, she can waggle, she can do the splits.
> I bet you half a dollar she can't do this.

The skipper wiggles, waggles, and attempts to do the splits. The chant is apparently post-war, and as American as it sounds. A version current in 1970 in Coventry possibly reveals the rhyme's roots:

> Postman, postman, do your duty,
> Deliver this note to the sleeping beauty.
> She wiggles, she waggles, she does the kicks,
> She wears her dresses right up to her knicks.

The custom of addressing messages to the postman to speed him in his delivery, particularly on St Valentine's Day, goes back at least 180 years (see Opie, *Lore*, p. 236). One verse children inscribe on the envelopes of their Valentine cards, in Britain and America alike, is the obvious basis for the skipping game:

> Postman, postman, do your duty,
> Take this to my loving beauty.

USA: Abrahams, no. 466, nine references, 1947 onwards. Correspondent, 1951, 'Sailor, sailor, do your duty'.

England: Penrith, 1957, 'Policeman, Policeman, do your duty, Here comes —— the bathing beauty. She can do the rhumba, She can do the splits, And she can lift her skirt Higher than her hips.' Later recordings from Coventry, Croydon, and Workington.

> Aunt Sally sells fish,
> Three ha'pence a dish,
> Cock yer leg up,
> Pull yer skirt up,
> Aunt Sally sells fish.

> [Betty Morgan] sells fish
> Three ha'pence a dish,
> Don't buy it, don't buy it,
> It stinks when you fry it,
> [Betty Morgan] sells fish.

This was heard being sung, while the children skipped, to the tune 'Happy Birthday' (in London) and 'Ash Grove' (in Swansea). The two verses are not always sung together, and both seem to be relics of a street jeer aimed at itinerant fishmongers.

Sheffield, *c*.1900, 'Sam Laycock sells fish, Three ha'pence a dish, Will you have a bit? Will you have a bit? Sam Laycock sells fish' (Kellett MSS). Belgrave, Leicestershire, *c*.1920, similar to 'Betty Morgan' version (EFDSS MS QR5). M. and R. King, *Street Games of North Shields Children*, 2nd ser. (1930), 35, as Hackney, 1965. Hethe, Oxfordshire, *c*.1945, 'Cock yer leg up, Pull yer drawers up, Aunt Sally sells fish'.

    Versions from nine places since 1950, amongst them: London, 1952, 'Old Ikey sells fish, Three ha'pence a dish, Cockalorums, cockalorums, Old Ikey sells fish'. Swansea, 1956, as texts, also, 'Mrs Whirly sells pop, Three ha'pence a drop, Cocklecup, cocklecup, Mrs Whirly sells pop'. Jarrow-on-Tyne, 1957, '[Judy Terrett] sells fish, At three ha'pence a dish. If you want any more, Go to [Judy]'s back door'. Hackney, 1965, 'Mrs Murphy sold fish, Three ha'pence a dish, Cut their heads off, Cut their tails off, Mrs Murphy sold fish'.

❖

> Charlie, Charlie, stole some barley,
> Out of the baker's shop;
> The baker came out and gave him a clout,
> Which made poor Charlie hop, hop, hop . . .

The fun of this game is the hopping. At the end of the chant the skipper either has to do bumps, or jumps on one foot until she is out, or she hops out and the next player takes her place. Alice Gillington described a version of the game in 1909: 'The girl who is skipping has to hop out of the rope and round one of the girls who holds the rope, without leaving off [or] changing her foot. If she makes any blunder, she is out.' The chant, also known as a nursery rhyme, goes back to the nineteenth century, its first appearance in print being *Mother Goose's Nursery Rhymes* (1877), 258.

A. E. Gillington, *Old Surrey Singing Games and Skipping-Rope Rhymes* (1909). *London Street Games* (1916), 57. Farnham, Surrey, 1947. Versions from six places thereafter, including Spennymoor, 1952, 'Peter Marley stole some barley'.

> My Maw's a millionaire,
> Blue eyes an' curly hair,
> Sittin' among the Eskimoes,
> Playin' a game of dominoes,
> Wouldn't get up to blow her nose—
> My Maw's a millionaire.

Popular in Scotland, sometimes for balls but mostly for skipping. It is sung to the tune 'Let's all go down the Strand'. 'When the person jumping comes to the line "Sittin' among the Eskimoes", they jump in a crouch position as if they were sittin'.' In England—for instance in Liss—only an inferior four-line version is known:

> My mother's a millionaire,
> Dark blue eyes and golden hair;
> She can play a violin,
> Sitting on a dustbin tin.

Yet these lines seem to have had some currency, perhaps through cross-fertilization with a rhyme that was not new, apparently, when children were found skipping to it in London in 1953:

> I had a little teddy bear,
> Blue eyes and fair hair;
> I put him in a treacle tin
> And pressed his little belly in.

The rhyme-words *fair*, *hair*, *chin*, and *in*, occur in a ring game known in Maryland in 1944 (Howard MSS):

> Here stands a young lady so neat and fair,
> Sky blue eyes and golden hair,
> Rosy cheeks and dimpled chin,
> Please, kind sir, won't you step in?

Whether or not this conjunction of rhyme-sounds and the line 'Sky blue eyes and golden hair' is a coincidence, it is worth remarking that these lines had, in their turn, more than eighty years' of play-use behind them. E. S. Backus (*JAFL* 14 (1865), 296) recorded a game-song, 'Sailing in the boat when the tide runs high', known at Ashford, Connecticut, in 1865:

> Here she comes so fresh and fair,
> Sky-blue eyes and curly hair,
> Rosy in cheek, dimple in chin,
> Say, young man, but you can't come in.

Scotland: versions from fifteen places since 1950.
  England: south-east London, 1953. Bishop Auckland, 1961. Graveley, Stevenage, 1974. Liss, 1975.

> Fudge, fudge, call the judge,
> Yer muvver's having a baby;
> It isn't a boy, it isn't a girl,
> It's just an ordinary ba-by.
> Wrap it up in tissue paper,
> Send it down the escalator:
> First floor, miss!
> Second floor, miss!
> Third floor, miss!
> Fourth floor, kick it out the door,
> Yer muvver hasn't got a little baby any more.

'At "miss!" you put your feet each side of the rope.' Usually a skipping game, although one 8-year-old told us, 'the girls in Pimlico learnt it me for two-balls'. The verse undoubtedly comes from America, and only became a real favourite in Britain in the 1970s.

USA: Poughkeepsie, New York, 1926, 'Oh fudge, tell the judge, Mama's got a baby', four lines only (*JAFL* 39 (1926), 83). East Orange, New Jersey, 1934, 'Double Dutch, Double Dutch, My mother had a baby. Not a girl, not a boy, Just a little lady'; Rome, New York, and East Orange, 1934, for ball-bouncing, 'Judge, judge, tell the judge, Mamma's got a baby. It's a boy, full of joy, Papa's going crazy. Wrap it up in tissue paper, Send it down the elevator, How many pounds did it weigh? One, two, three . . .' (Howard MSS). Common thereafter.
   Britain: Hampstead, 1949. Golders Green, 1957. Cumnock, 1961, 'Wrap it up in tishy paper, Send it to the alligator, See how many tons it weighs, One, two, three . . .'. Birmingham, 1966. Alton and Selborne, 1970. Common thereafter.
   South Africa: Johannesburg, 1973. Pietermaritzburg, 1974.

> Soldier, soldier, stand to attention,
> Soldier, soldier, stand at ease,
> Soldier, soldier, salute to the officer,
> Soldier, soldier, bend your knees.
>
> Little Miss Muffet sat on a tuffet,
> Eating her curds and whey;
> Along came a spider and sat down beside her,
> And frightened Miss Muffet away.

The player in the rope has to stand stiffly as if at attention, then with legs apart and hands behind back, then has to salute. Finally she must bend her knees—'coupy down' as they say in Bristol, or 'get doon on her hunkers' as they say in Glasgow—and must continue to skip in this position throughout the recitation of 'Little Miss Muffet' (no easy feat, as anyone who has tried it will affirm). The words of the game vary little from place to place (though in Bristol it is 'Salute to the captain', and in Glasgow 'Salute to the king'); and during the quarter-century for which we monitored the game the 'soldier' lines were almost always followed by 'Little Miss Muffet'. The only reason for this seems to be that the soldier ends by squatting and Miss Muffet begins by doing so.

Kirkcaldy, 1951. Sale, Manchester, and Radcliffe, Lancashire, 1951, 'Little tin soldier, stand at attention'. Glasgow, 1952, 1961, 1975. Birmingham, 1952, 1953. Swansea, 1952, 1956. North Staffordshire, 1953. Bristol, 1960. Norwich, 1961. Dunoon, 1962.

> Two little sausages frying in a pan,
> One went pop and the other went *bang*!

The rope is swayed until '*bang*!', when it is pulled taut so that the skipper can come down with her feet on either side. The chant, which has been current since the 1920s, was seemingly more popular in the 1950s than in the 1970s. In a few places, for instance in Swansea and in Brora, Sutherland, it starts with as many sausages as there are skippers in the rope, their number being reduced, in the manner of 'Ten green bottles', until there is only one:

> Five little sausages frying in a pan,
> One got burnt and the other said 'Scram'.

> Four little sausages frying in a pan,
> One got burnt and the other said 'Scram'.

> Andy pandy, sugar and candy,
>   French almond rock;
> Bread and butter for your supper,
>   That is all your mother's got.

A skipping game that is widespread, has been played since the nineteenth century, and is everywhere played in much the same way. The player in the rope skips ordinarily up to the word *rock*, crouches during the second couplet while the rope is turned over her head, and on the word *got* ('this is rather tricky') jumps up and skips ordinarily again. Correspondents confirm the game was being played like this, and with these words, in London before the First World War. However, in some places, notably in the north country and East Anglia, the children chant only the first two lines and, as a Spennymoor girl explained, they skip and crouch alternately each time through:

The rhyme is said while the nobby-ender skips ordinary. Then the rhyme is repeated, but this time the nobby-ender must be 'little man', that is, bend right down while the rope is being twined above the skipper's head. When the enders come to *rock*, the nobby-ender must get up again and be right on time with the rope.

This manner of playing the game is also described by Alice E. Gillington in *Old Hampshire Singing Games and Trilling the Rope Rhymes* (1909). In Soho in 1907, however, the Revd T. Allen Moxon observed that on the word *rock* 'the skipper jumps right up into the air while the rope goes twice or more times under her feet' (*St Anne's Soho Monthly*, p. 154).

That the jingle 'Andy pandy, sugardy candy, French almond rock' was already popular for skipping at the end of the nineteenth century (Alice Gomme printed a Deptford version in 1898) seems evident from a song Marie Lloyd used to sing, 'Buy Me Some Almond Rock', written and composed by Joseph Tabrar in 1894. If the rhyme did not arise from the song, which seems unlikely, it must surely have been well enough known for the refrain to have had point when Marie Lloyd sung of herself as a then-notorious 'giddy girl' invited to a ball:

> Only fancy if Gladstone's there,
> And falls in love with me,
> If I run across Labouchere,
> I'll ask him home to tea.
> I shall say to a young man gay,
> If he treads upon my frock,
> Randy pandy, sugardy candy,
> Buy me some almond rock.

In fact, the words may possibly stem from the rhyme,

> Handy spandy, Jack-a-Dandy,
> Loves plum cake and sugar candy,

which was familiar to children early in the eighteenth century (see *ODNR*, no. 259) and is itself sometimes used for skipping in the present day.

Britain: few variations other than 'Hanky panky, sugar come candy' (Spennymoor), 'Yanky, Panky, sugar and candy' (York); but in Quarry Bank they have a completely updated version:

> Salt, mustard, vinegar, pepper,
> French almond rock.
> Bread and butter for your supper,
> That's all mother's got.
> Fish and chips and Coca-Cola,
> Put them in a pan.
> Irish stew and ice-cream soda,
> We'll eat all we can.
> Salt, mustard, vinegar, pepper,
> Pig's head and kraut,
> Bread and butter for your supper,
> O-U-T spells *out*!

Ireland: Dublin, 1970s, 'Handy Andy sugary candy, Fresh enarmel [carmel] rock, I spy a lark, Shining in the dark, Echo, echo, G-O, Go!' (Eilís Brady, *All In! All In!* (1975), 87).

USA: 'Amos and Andy, sugar and candy, I pop in, Amos and Andy, sugar and candy, I pop down', etc.; after the radio programme 'Amos 'n' Andy' (Abrahams, no. 10).

CHANTS WITH IMPROVISED ACTIONS

> German boys they are so funny,
> This is the way they earn their money:
> Whoop-a-la-la, Whoop-a-la-la,
> Whoop, whoop, whoop.

The actions during the whooping are performed according to personal preference or local custom, the skipper kicking, flouncing around, or landing on the third whoop with feet astride the rope. In the only recording we have of this game in England (Blackburn, 1952) the chant ended 'Oroom-pah-pah, oroom-pah-pah', and the likehood is that the chant originally began 'German bands are so funny' and was sung to make fun of the German bands which were a familiar feature of street life in the years before the First World War.

Britain: versions from eight places in Scotland since 1950, and one in England.
  Australia: Melbourne, 1958 (Turner MSS).

By 1990 the chant had spawned a little trick featuring a Chinaman:

> Chinaman is very funny,
> This is the way he counts his money,
> Uncha, uncha,
> Turn around, punch ya!

(e.g. London, 1990; Stockport, 1996)

The acting ends with the actor turning round and punching the nearest spectator.

> There came a girl from Italy,
> There came a girl from France,
> There came a girl from Germany,
> And this is how she danced:
>   Oosha, la la la,
>   Oosha, la la la,
>   Oosha, la,
>   Oosha, la,
>   Oosha, la la la.

This game is in the ascendant, is played in different ways in different places (for instance the girls may cock a leg in the air during the 'ooshas'), and may be prolonged with verses from other games, such as 'Lady, lady, touch the ground' and an adaptation of lines from 'I'm a little Girl Guide dressed in blue'; while in Scotland, since at least 1975, 'Knees up Mother Brown' has been firmly attached to it, sometimes followed by,

> Birly in a ring,
> Birly in a ring,
> Birly, birly, never twirly,
> Birly in a ring,

or other formulas that give the skipper opportunities for fancy kicks and spins.

Britain: Kent, 1934, 'I'm a girl from yonder, I'm a girl from France, I'm a girl from Italy, This is how we dance, Ushi la la la . . .'. London, c.1940, 'I met a girl from Italy, I met a girl from France, I met a girl from the USA, And this is how she came'. Versions from twenty-nine places since 1950, including variants: Swansea, 1957, 'The king came home from Italy, The queen came home from France, The Prince came home from Germany, And this is how they danced: Osh a la la la . . .'. Liss, 1963, 'I wish I went to Italy, I wish I went to France, I wish I went to Switzerland, And this is how they danced: The queen does the curtsy, The kings do the bow, The girls show their knickers [back of skirt flipped up] And the boys go Wow!'. Edinburgh, 1975, 'There came a girl from France, She didn't know how to dance, And all the dance that she could dae Was knees up Mother Brown.'

Cf. *St Anne's Soho Monthly*, June 1907, 'I've got a bloke down hopping, I've got a bloke down Kent, I've got a bloke down Pimlico, And this is how he went: Kuta, lula, te, heigho'. *London Street Games* (1916), very similar to 1907. Given as skipping game by both sources.

USA: cf. Santa Rosa, California, 1938, 'A negro went to Germany, A negro went to France, A negro went to Italy, To buy a pair of pants' (Howard MSS).

Australia: cf. Melbourne, 1967, 'There was a girl from France, She wore some fancy pants, Danced and danced and danced and danced, And this is how she danced. Hopping on one foot . . .', continues as in Scotland (Turner MSS).

> Mother bought a chicken, she thought it was a duck,
> She chased it round the garden with its legs tied up.
> There's a boy over there and he winks his eye,
> He thinks I love him, but he's telling a lie,
> For his hair won't curl, and his shoes won't shine,
> He's got no money, so he won't be mine!

Sung to a tune like 'John Peel', the rope is usually turned fast, and the skippers enact the words, or cross and uncross their legs in imitation of the chicken, or do bumps at the end.

Versions from fourteen places since 1950. Particularly popular in London. Sometimes begins 'There's a boy over there'.

> See those girls in Russian boots, Russian boots,
> With their garters hanging loose, hanging loose,
> And their skirts above their knees, above their knees,
> Marching as to war.

The skipper taps the ground with toe and heel, touches her leg or imaginary garter, then lifts her skirt and does a bump. 'Russian boots' were the craze in 1926, and the tune was the popular song of the time, 'The Toy Drum Major'. Rowland Kellett remembers a parody current in Leeds not long afterwards:

> See them marching down Boer Lane,
> With their garters hanging loose,
> And their skirts above their knees,
> Marching along to Woolworth's
> To buy some new spattees.

'Spattees', another fashion craze, were knitted leggings for women and girls, produced mostly in tartan designs.

Hackney, 1952 and 1965. Maryon Park, London, 1953. Laverstock, 1953. Bromley-by-Bow, London, 1956, 'See those girls in mushroom bootees . . .' (Alan Smith notebook, Folklore Society Archive). Newcastle under Lyme, 1965. Leeds, 1966, 'Going down to London town to buy a penny whistle.' Adam McNaughtan recorded a version in Glasgow, 1984, 'See those girls in their Russian boots, Russian boots, See those girls in their swimmin' suits, swimmin' suits, See those girls in their skirts up on their knees, And they're marching on to vic-tor-y'.

My wee dolly dressed in blue,
Died last night at half-past two.
I put her in a coffin,
And she fell through the bottom,
My wee dolly dressed in blue.

The skipper enacts the scene: touching dress, laying doll in coffin, and falling through. Even more scope for comedy is given when, as is often the case in Scotland, the chant begins, 'My wee shoe dressed in blue'. The words are sometimes used for ball-bouncing.

Britain: versions from fifteen places since 1950, the majority Scottish.
   USA: Washington, Pennsylvania, 1938, 'I got a little shoe dressed in blue, Got it last night at half past two'; Maryland, 1948, 'Old Miss Pink dressed in blue, Died last night at a quarter to two. Before she died she told me this: Darn that rope that made me miss' (both Howard MSS). Cf. the catch collected by W. H. Babcock in Washington, DC: 'Tell story? Who? My old shoe, Dressed in blue, That came walking down the avenue' (*American Anthropologist*, 1 (1888), 271).

Joe, Joe, broke his toe,
On the way to Mexico;
Coming back he broke his back,
On a great big railway track.

The skipper holds a foot with one hand and, still hopping, has to place her other hand on her back.

Scotland: Kirkcaldy, 1952.
   USA: St Louis, Missouri, c.1900, as a jeer, 'Joe, Joe, bumped his toe, In the battle of Mexico' (*JAFL* 60 (1947), 36). East Orange, New Jersey, c.1934, and elsewhere (Howard MSS). *Gafia Poetry Leaflet*, no. 2 (1951), 'Skip Rope Rimes'.

Policeman, policeman, don't take me,
Take that boy behind the tree;
He stole sugar, I stole tea,
Policeman, policeman, don't take me.

Particularly popular in Scotland. The player in the rope mimes the story, alternately pointing at herself and the boy.

Britain: Edinburgh, c.1900, '. . . I stole brass, he stole gold—Policeman, policeman, don't take hold' (*Rymour Club* 1 (1911), 106). Versions from fifteen places since 1950, three of them being for Two-balls, including, Aberdeen, 1952, '. . . He pulled my hair and broke my comb, But I'll tell my mother when I get home' (these lines were being sung in Aberdeen *c*.1895, see Opie, *Singing Game*, no. 95). Annesley, Nottinghamshire, 1954, '. . . He stole an apple, I stole a pear, Little fatty policeman, I don't care'. Dudley, 1969, '. . . He stole money, I stole none, Little fatty policeman it's half-past one.' Cf. Liverpool, 1911, repeated during Orange Day confrontations, 'Bobby, Bobby, don't take me, Take that fellow behind the tree; He belongs to King Billee, I belong to Popery.' Last two lines reversible.

USA: H. C. Bolton *Counting-Out Rhymes*, (1888), from New York, '. . . He stole gold and I stole brass, P'liceman, p'liceman, go to grass'; and from Greenville, South Carolina, 'Watchman, Watchman, don't watch me, Watch that nigger behind the tree. He stole whisky, and I stole none; Put him in a calaboose for fun'. St Louis, Missouri, *c*.1895, first couplet, 'a chant of no particular purpose' (*JAFL* 60 (1947), 35). Kansas?, *c*.1915, 'Teacher, teacher, Don't whip me, Whip that nigger behind that tree. He stole money and I stole grass, Teacher, teacher, kiss my ——'; Indiana, *c*.1938, for jumping rope, 'Teacher, teacher . . . He stole peaches I stole none; Put him in the calaboose just for fun' (*Southern Folklore Quarterly*, 3, p. 178).

❖

> Horsie, horsie, don't you stop,
> Just let your feet go clipperty clop;
> Let your tail go swish and your wheels go round,
> Giddy up, you're homeward bound!

The popular song 'Horsey! Horsey!', written and composed by Box, Cox, Butler, and Roberts, seems to have jumped into the skipping-rope almost as soon as it was released. It was copyrighted by the Sun Music Publishing Co. Ltd. in 1937, and was reported as being used for skipping at Mayhill Junior Girls' School, Swansea, before the outbreak of the Second World War in 1939. Subsequent enquiries showed that it was not a momentary craze. In 1952 it was still a skipping song in four districts of Swansea, and was also known elsewhere. A 12-year-old girl in Cleethorpes, Lincolnshire, gave us her rendition:

When it says 'clipperty clop' I step like a horse on the spot. When 'your tail goes swish' comes I give a swish of my legs. When 'your wheels go round' comes, I turn around and around. On the word 'bound' I run out.

More recently the song has been picked up from a 9-year-old in Glasgow, from 10-year-olds in Manchester, and from a 9-year-old in Oxford.

> I'm the monster of Loch Ness,
> My name you'll never guess.
> I can twirl in a ring,
> I can do the Highland Fling,
> I'm the monster of Loch Ness.

The skipper usually twirls round and does a simplified Highland fling; but the middle lines, and hence their accompanying actions, can vary considerably (e.g. 'For you whirl around, And touch the ground', Kirkcaldy, 1952).

Scotland: versions from five places since 1950. Edinburgh, Ritchie, *Golden City* (1956), two versions, pp. 115, 125.

> My sister Mary had a wooden leg,
> My sister Mary kicked me out of bed.

This is one of the rhymes the Revd Allen Moxon obtained from his small parishioners in Soho (*St Anne's Soho Monthly*, June 1907, p. 154). It was, he said, 'a rhyme the full mysteries of which I have not yet fathomed'. His young contributor wrote, 'The first line is sung while the skipper skips on one foot; the second while she tries to kick. If she fails she is out; she generally does fail, so that I cannot tell what would happen if she chanced to kick without stopping the rope.'

> Wallflowers, wallflowers, growing up so high,
> We are all children and we must die;
> Except *Mary Thomas*, she's the youngest here;
> She can hop, and she can skip,
> And she can turn a candlestick.

Recorded in Gloucester, 1953. Any verse that describes actions re-commends itself as a skipping game. The hopping, skipping, and turning a candlestick (no one is sure what *that* means) have been an integral part of 'Wallflowers' at least since the time it was first recorded, in 1874 (see Opie, *Singing Game*, pp. 244–7). However, it was not in widespread use for skipping, and was found in only five places. In Suffolk, 1953, they played it like this: 'all skip together, singing "Woolly, woolly, wallflower, Growing up so high, We're all one family, And we shall die"; everyone runs out except a named player, and they all sing, while she does the actions, "Excepting so-and-so, She's the youngest girl; she can skip, she can hop, And she can turn a candlestick [turns round and round]."' The version from Swansea, 1953, was like the text but ended, 'Oh, for shame! Fie, for shame! Turn your back upon the game [skipper runs out]'; and in Cumnock, 1961, 'Water, water, wallflower' was similar to the text but the named girl was 'the fairest of them all' and 'she can dance and she can sing, And she can do the Highland fling, Fly, fly, fly away, turn your back to the wall again' (*Bluebells My Cockle Shells*).

DIVINATION

In bygone days girls could forecast their future by the number of times they could hit a shuttlecock ('Shuttlecock, shuttlecock, tell me true, How

many years have I to go through? One, two, three . . .', *N & Q*, 3rd ser.,
3 (1863), 87), or by the number of times they could catch a cowslip ball
('Tisty-tosty, tell me true, Who shall I be married to? A, B, C . . .'
(*Journal of the Dorset Field Club*, (1889), 43). Both these means of divination were replaced by skipping, the basic formula—and also one of the
earliest skipping rhymes—being:

> Rosy apple, lemon tart,
> Tell me the name of your sweetheart.
> A, B, C, D . . .

or,

> Raspberry, strawberry, gooseberry jam,
> Tell me the name of your young man.
> A, B, C, D . . .

This game is known variously as 'Sweethearts', 'Fortunes', 'Consequences', 'Jam Tarts', 'The Wedding', or 'The Great Day'. The rope is
turned for normal skipping while the tarts or jams are invoked, and fast
when the alphabet begins. The letter the skipper trips on is the initial of
the first name, and of the surname only if the procedure is repeated. Hot
debate on the identity of the sweetheart follows.

Thereafter enlightenment may be sought on the young man's social
standing ('Tinker, tailor . . .'),[24] on the date of the wedding and, in particular, on the style of the wedding, such questions being asked as 'What
shall I be married in?', 'What colour will the dress be?', 'How many
bridesmaids?', 'How many pageboys?', 'How many carriages?', 'What
weight will the cake be?', 'Where will we go for the honeymoon?' and,
almost as an afterthought, 'Do I love him?'

This form of gyromancy may be conducted as a personal oracle in the
short rope, in which case the skipper can take her time and even enquire
into such practical matters as 'Will I have to work after I am married?'
and 'At what age will I retire?' In the long rope, communal values count
for more. The concern is with the wedding and the raising of a family,
and the questions (some fundamental, some trivial) are likely to include:

When will you marry him?
   This year, next year, sometime, never.
   (Player drops out if the answer is 'never'.)
What month?
   January, February, March, April . . .
What day of the week?

---

[24] This enquiry can be the subject of a separate game, see 'Mother, father, tell me true' (p. 249).

Monday, Tuesday, Wednesday, Thursday . . .
What will you wear?
    Silk, satin, muslin, rags.
    (A well-established formula. In Wrecclesham, Surrey, in 1892, it was
      'Silk, satin, cotton, rags.')
What shoes will you wear?
    High heels, low heels, button-ups, clogs.
    (In 1904 it was recorded as 'Boots, shoes, slippers, clogs.' In 1974 an
      informant heard 'Stilts, wedges, platforms, clogs.')
What kind of ring?
    Gold, silver, copper, brass.
    (Or, 'Gold, silver, platinum, wire', Zinc, copper, silver, gold', etc.)
What will you be taken in?
    Coach, carriage, wheelbarrow, dung-cart.
    (This is the most common, as it was in the 1890s. Alternatives are
      'Coach, carriage, car, bike', or, 'Taxi, coach, dustbin, car.')
Where will you be married?
    Church, chapel, cathedral, abbey.
    (Or, 'Church, chapel, cathedral, workshop', or 'Church, chapel, Sal-
      vation Army.' Recorded 1900 as 'Church, chapel, cathedral, hall.')
Where will you live?
    House, palace, pigsty, barn.
    (Or, 'Village, mansion, cottage, workhouse', or, 'Palace, castle,
      farmhouse, barn', or, in Ruthin, 'Palace, ty [house], cwtmochyn
      [pigsty], petty', or, since the 1890s, 'Little house, big house, pigsty,
      barn.')

Each of these formulas is repeated until the skipper tangles in the rope.
As an 11-year-old remarked, 'The game takes a long time. If there are too
many players you get tired waiting for your turn.' The game is indeed the
longest of the skipping games. Generally not more than four girls will
play together, and then only when there is unlimited time. Sometimes
the consequences of marriage are also explored:

How many babies will you have?
    One, two, three, four . . .
    (In Newcastle upon Tyne an old spell began the count at three with
      the assertion: 'One will be lonely, two's not enough, give her three,
      four, five . . .')
What colour will they be?
    Black, yellow, blue, purple, white.
    [In Soho, 1907, 'Black, white, black, white . . .'. In Leeds, c.1952, the

rhymed couplet was, 'What colour will the babies be? Black, white,
yellow, khaki.')
How many nappies will you wash?
    Five, ten, fifteen, twenty . . .

The fantasies children indulge in today are, apparently, no more
detailed than were those of their grandparents and even great-
grandparents in the nineteenth century. In 1888 W. H. Babcock observed that little girls in Washington, DC were 'constantly practising
augury with the skipping-rope', the chief topic of their enquiries being
'the incidents or consequences of wedlock':

'Silk, satin, velvet, calico, rags!'—they cry, keeping time to the words, and the
one who marks a failure in leaping foretells the nuptial apparel of the girl who
fails. The same test is applied to equipage, social position, and even the tint of
the children.

Britain: *Traditional Games*, ii (1898), 195, 'Black currant, red currant, raspberry tart, Tell me the
name of my sweetheart', for shuttlecock. Maclagan, *Argyleshire* (1901), 227, 'Gooseberry, raspberry,
strawberry jam, Tell me the name of your young man'. *St Anne's Soho Monthly*, June 1907, p. 153,
as 1898. A. E. Gillington, *Old Hampshire Singing Games* (1909). *London Street Games*, as 1898. M.
and R. King, *Street Games of North Shields Children* (2nd ser., 1930), 30, 'Raspberry, strawberry,
marmalade jam'; also from Castle Eden, 1943; Helensburgh, 1952; Golspie, 1952; Spennymoor,
1960. No significant regional differences evident in versions collected since 1950 from sixty-four
places, e.g. 'Apple crumble, apple tart' (Chelsea), 'Jam tarts, raspberry tarts' (Plympton St Mary),
'Jam, jam, strawberry jam' (Norwich), 'Black currant, red currant, strawberry jam' (Swansea). Ayr,
1975, 'Ice cream soda, lemonade tart', cf. USA.
    Australia: Toowoomba, Queensland, *c*.1905, 'Bread, butter, marmalade, jam'. Brisbane, *c*.1935,
'Bread and butter and marmalade jam' (Howard MSS); Turner (1969), 20, similar, Melbourne,
1956, Sydney, 1958, with two other versions.
    USA: Cincinnati, 1908, 'Strawberry blonde, Cream of tartar, Tell me the name Of your sweet
daughter' (*JAFL* 40 (1927), 42). Carlisle, Pennsylvania, 1929, 'Ice cream soda, cream of tarts'
(*JAFL* 47 (1934), 385). Kalamazoo, Michigan, 1938 (Howard MSS). *Christian Science Monitor*, 9
April 1949, p. 8, from Negro children in the South, 'Strawberry Jane, cream of tartum, Tell me
initial of your sweetheartum'. Abrahams, no. 214, many references. *New Yorker*, 11 March 1974, p.
29, 'Ice-cream soda with a cherry on top, tell me the name of your sweetheart.'
    South Africa: Johannesburg, 1969, 'Rosey apple, lemon tart'. Pretoria, 1974, 'Rosy, rosy, apple
tart' (Kellett MSS).

                    Raspberry, strawberry, gooseberry pie,
                    Tell me when I am going to die.
                        Five, ten, fifteen, twenty . . .

A Dumfries girl, describing this as her favourite skipping game, confirmed that when the counting begins, the rope-turners 'do pepper',
turning the rope so fast 'the person in the middle has to jump very hard
to keep up'. They also had the rule that only if the skipper gets past
'twenty' is she allowed to continue. The further questions which, with
justification, exercised the minds of the young in previous centuries, are

here supplied by a girl in Pontypool. When the approximate year of death has been determined, they ask:

Which month? January, February, March . . .
What time? One o'clock, two o'clock . . .
What will you be buried in? Night clothes, day clothes, best clothes, old clothes.
What will your coffin be? Gold, silver, copper, wood.
Where will you go to? Heaven, hell, above, below.

Britain: Pontypool, 1954. Dumfries, 1975.
  USA: Cincinnati, 1908, 'Apple, peach, pumpkin pie, How many days before I die?' (*JAFL* 60 (1908), 41). Maryland, as 1908, also 'Strawberry shortcake, huckleberry pie, How many kisses until I die?' (Howard MSS).
  Cf. *Traditional Games*, ii (1898), 195, Deptford, for shuttlecock, 'Grandmother, grandmother, Tell me no lie, How many children before I die?'

> Mother, father, tell me true,
> Who shall I be married to?
>   Tinker, tailor, soldier, sailor,
>   Rich man, poor man, beggar man, thicf,
>   Lord Mayor, millionaire,
>   Cowboy, ploughboy, Indian chief.
>   Tinker, tailor . . .

A 12-year-old girl in Dunoon, explaining that the 'coying' (rope-turning) becomes faster and faster as the chant proceeds, emphasized that she herself was sceptical about the result: 'The one you stop at is *supposed* to be the one you will marry.' The 'Tinker, tailor' formula is, nevertheless, highly popular, as is the equally popular alternative:

> Architect, lawyer, medicine, church,
> Army, navy, left-in-the-lurch.

And in Scotland, particularly Aberdeen,

> A laird, a lord, a lily, a leaf,
> A piper, a drummer, a hummer [or hangman], a thief.

Only in the 1960s, when the Beatles represented the height of romantic ambition, were girls to be heard chanting a more specific refrain:

> Mummy, Daddy, tell me true,
> Who should I get married to?
> Paul, John, Ringo, George,
> Paul, John, Ringo, George . . .

Britain: Herstmonceux, Sussex, 1898, 'Ipsey, Pipsey, tell me true, Who shall I be married to? A, B, C, etc.' (*Traditional Games*, ii. 202). Versions from fifteen places since 1950, including: Swansea,

1952, 'Yellow, yellow, choose your fellow, A, B, C . . .'. Cumnock area, 1965, Beatles version (*Those Dusty Bluebells*, p. 6). North London, 1972, 'Gipsy, gipsy, please tell me, What my husband's going to be'. Hillside, near Montrose, 1975, 'Incy, bincy by blue, Who is it? Not you. What'll you have? Queen, King, Princess, Prince.'

USA: Westchester County, New York, 1938; Maryland, 1944, 'Red and yellow, Who's your fellow? A, B, C . . .' (Howard MSS). Abrahams, no. 165, seven references.

For a note on 'Tinker, Tailor' see *ODNR*, pp. 404–5.

> Mrs Moore she lives on the shore,
> She has children three or four,
> The eldest one is twenty-four,
> And she got married to a—
>   Tinker, tailor, soldier, sailor,
>   Rich man, poor man, beggar man, thief.

At the start the rope is turned ordinarily, but when the prognostication begins, the rope is turned fast, and at Radcliffe in Lancashire, Spennymoor in County Durham, and doubtless elsewhere, the turners begin walking in a circle. Mrs Moore sometimes lives 'next door', has children 'by the score', or occasionally becomes Mrs Brown, Mrs Jones, Old Mother Lollipop, or Mrs Sippy (a joke favoured in the USA); but in general children repeat the chant exactly as it was seventy years ago.

Britain: *St Anne's Soho Monthly* (1907), 153, '*Mabel Turner* lives on the shore; She has children three or four; The eldest one is twenty-one, And she got married to a tinker, tailor, soldier, sailor, Richman, poorman, beggarman, thief'. *London Street Games* (1916), 67, 'Lady, lady, on the sea-shore'. Swindon, *c*.1925, 'Old Mrs Moore'. Soham, Cambridgeshire, *c*.1930, game continued with further prognostications. Versions from twenty-five places 1950 onwards, including Swansea, 1952, 'Thief, thief, stole a leg of beef, He got married to a—Tinker, tailor, soldier, sailor . . .'. Welshpool, 1963, 'Mrs Evans . . . She has children big and small. How many?'

USA: Westchester County, New York, 1934, 'Mississippi lives on shore, She has children sixty-four, The oldest one is twenty-four, Which one shall I marry? Rich man, poor man . . .' (Howard MSS). Patricia Evans, *Rimbles* (1961), 24–5, details of wedding foretold.

Australia: Turner (1969), 34–5, '"Old Mrs Moore" . . . one of the most widely distributed Australian children's rhymes'.

> Gypsy Lee,
> She told me
> What my boyfriend's name would be:
> A, B, C, D . . .
> Tinker, tailor, soldier, sailor,
> How many children will I have?

This was heard in Edinburgh, in 1990, and is yet another example of the closeness of Scottish and American lore. In 1971 F. K. Plous Jun. interviewed children in Chicago for a newspaper article on jump-rope rhymes, and elicited:

Gypsy, gypsy, please tell me
What my husband is going to be:
A rich man, a poor man, a beggar man, a thief,
A doctor, a lawyer, an Indian chief.

(*Chicago Sun-Times*, syndicated to *Los Angeles Times*,
10 November)

It should also perhaps be mentioned that the formula 'Lady, baby, gypsy, queen, Elephant, monkey, tangerine', used for counting out, or for fortune-telling 'with the buttons on your frock, to see what you will become', has also been employed for skipping since *c*.1950.

All the girls in our town lead a happy life,
Except for *Anna Humphreys*, who wants to be a wife.
A wife she shall be, and a–courting she shall go,
Along with *Stephen Smith*, because she loves him so.
She kisses him, she hugs him, she sits upon his knee,
She says, 'Oh my darling, won't you marry me?'
Yes, no, yes, no . . .

This game became popular for skipping when it shed its link with the old singing game 'All the Boys in Our Town' (see Opie, *Singing Game*, pp. 130–3) and turned into a jokey sport for personal prognostication. 'The person who's chosen to be their husband must be someone they really *hate*,' explained two young players in north London, 'and when you do "Yes, no, yes, no" it goes very fast, and what they're out on is the answer.' The only certain record we have of 'All the Boys' as a skipping game before the Second World War is the splendid version the Revd Moxon found in the streets of Soho in Edwardian days:

All the boys in our town lead a happy life,
Excepting *Freddy* ——; he wants a wife.
A wife he shall have, and a courting he shall go
Along with *Florrie* ——, because he loves her so.
He huddles her and cuddles her
And sets her on his knee
Says, Darling do you love me?
I love you and you love me;
Next Sunday morning the wedding shall be.
Up comes the doctor,
Up comes the nurse;
Up comes the devil in a dirty white shirt.
Down goes the doctor, up comes a cat,
Up comes the devil in a dirty straw hat.
Salt, mustard, vinegar, pepper.

(*St Anne's Soho Monthly*, June 1907, p. 156)

Norman Douglas, in *London Street Games* (1916), 59, does not say whether the chant he gives is for skipping or not.

In the 1950s and early 1960s the chant still retained, in some places, the atmosphere of a courtship game, despite being sung in the rope. For instance in Swansea in 1962:

> All the boys in Morriston lead a happy life,
> Except *Peter Lewis*, he needs a wife.
> A wife he will get and a-courting he will go,
> Kissing *Catherine Jones* on the front door.
> He loves her, he kisses her, he sits her on his knee,
> He says, Darling *Cathy*, will you marry me?
> Yes, *Peter*, yes, *Peter*, I will marry you,
> And they got married on the 25th of June.

By this time, however, the modern form of the game, in which a girl may involuntarily commit herself to some ill-favoured youth, was already coming to the fore, and is now a source of laughter throughout the land (or at any rate from Staines to Stornoway) as is also an updated version:

> All the boys in our school aren't very nice,
> Except for *Philip Harding*, he's all right.
> He took me to the pictures,
> He sat me on his knee,
> He said, My darling *Diane*,
> Will you marry me?
> Yes, no, yes, no . . .

> The wind, the wind, the wind blows high,
> The raindrops scatter from the sky,
> She is handsome, she is pretty,
> She is the girl from the royal city.
> She goes courting one, two, three,
> Please will you tell me who it is to be.
> *David Franklin* says he loves you,
> All the boys are fighting for you,
> Chase you round the corner, sit you on his knee,
> Says, 'My darling, will you marry me?'
> Yes, no, yes, no . . .
> How many kisses did he give you?
> Ten, twenty, thirty . . .

Aldershot, 1963. This romantic old singing game has been used for skipping since early in the century. The more recent recordings usually incorporate the 'How many kisses?' episode from 'All the girls in our town' (above), and the game is played in the same manner (compare Opie, *Singing Game*, pp. 133–7).

Letter to *Radio Times*, 4 October 1973, 'Before the First World War this skipping game was played in every playground of my native Lancashire . . . "The wind, the wind, the wind blew high; It blew [skipper's name] across the sky; She is handsome, she is pretty, She is a flower of the Golden City; She goes courting, one, two, three, I pray you tell me who it be." The girl skips out, and a boy, sometimes urged against his will, runs in . . . "[Boy's name] says he'll have her, All the boys are fighting for her; Let the boys do what they will, But [skipper's name] loves her still." The girl runs in and . . . they continue skipping together till they are "out" '. Swansea, 1936, 'The wind blows high, the wind blows low, The wind blows scattering down below . . . *Terry Martin* loves his wife, How many kisses did he give?' Versions from sixteen places since 1950; variations include Aberdeen, 1952, '. . . *John Mackenzie* says he loves her, In his bosom he shall hug her, A hush and a hush and away we go, Off to the land of E-I-O'; Cumnock, 1961, 'The wind . . . blew high, Out popped *Ann* from the sky, She said she would die, If she never saw the boy with the big blue eyes. Johnny, Johnny, is her lover, All the boys are asking for her, Johnny took her to the fair, And how many kisses did she get there? Five, ten . . .'; Coventry, 1970, 'The wind, the wind, the mighty wind, The wind goes howling down the street. Take her down the garden, Sit her on your knee, Tell her you love her, One, two, three.'

Green gravel, green gravel,
Your age is sixteen,
Your true love sent a letter
Complaining of the weather.
Turn your back you saucy cat
And say no more to me!
All right *Kathleen*,
I'll tell your mother
That you kissed *Duncan*
In the parlour last night.
How many kisses did you give him?
Five, ten, fifteen . . .

Lambourn, Wiltshire, 1972—yet another tattered remnant of a singing game that has been fitted out, for the purpose of skipping, with an ending that introduces the desired question, 'How many kisses did you give him?' 'Green gravel' is undoubtedly one of the older courtship games, although it was not written down before about 1835 (see Opie, *Singing Game*, pp. 239–42). The new ending ('All right so-and-so') is the same as the second half of 'I know a nikker boy, he's double-jointed', enlivened by the pleasingly acerbic, 'Turn your back you saucy cat, and say no more to me!';[25] this had already been appearing in the ring game 'Green gravel'—appropriately, since 'Green gravel' is a game of turning one's back to the centre of the ring.

[25] The epithet 'saucy cat' was in use *c*.1898. It occurs in Charles Deane's song 'There's 'Air', when the newly-wed hears his wife talking in her sleep: 'Then I heard her say—the saucy cat! whatever made me marry him?'

London Street Games (1916), 92, amongst a stream of game fragments, 'Turn your back, you saucy cat, And say no more to me—.' Alton, 1947, 'Greengages, greengages . . .'. Ten versions since 1950, including Hindon, Wiltshire, 1965, 'Green gravel, green gravel, Your heart is so sweet . . .'. Four Marks, Hampshire, 1971, 'You jump over the rope when it's going from side to side and sing "Green gravels, green gravels, Your sweetheart is dead, He sent you a letter, Containing the weather", then you start skipping and sing, "Turn your back . . ."' etc. as text.

Most of the 'divination' games are not at all serious, and aim at laughable and unlikely results. They chant 'My age is one, two, three, four . . .' or 'My age is one, my age is two, my age is three, my age is four . . .' and count until the skipper trips in the rope (in the USA, 'I was born in a frying pan, Just to see how old I am'). They may skip to 'How many messages can you carry? One, two, three . . .' (Edinburgh) and 'What is the time you go to bed? One o'clock, two o'clock, three o'clock . . .' ('You can go on till you are tired', one player remarked aptly). More daringly they enquire, 'How many kisses did you get last night? One kiss, two kisses, three kisses . . .' or 'How many nappies did you wet last night?' (Dunoon). Having skipped to give themselves a false age, they determine the age of King Neptune when he lost his crown (Alton) or the age of Sinbad when he lived on an island (north Staffordshire) or the age of their teacher when she joined the Armed Forces (Cumnock):

> My teacher's barmy, she joined the Army,
> My teacher's crazy, she joined the Navy
> When she was one, two, three, four . . .

Some longer counting chants, designed to give ridiculous or exciting answers, are:

> Cinderella dressed in yellow,
> Went upstairs to see her fellow.
> How many kisses did he give her?
> Five, ten, fifteen, twenty . . .

'Cinderella dressed in yellow' (or 'yella') is often described as 'one of our best games'. During the counting the rope is turned fast, and if the skipper trips on sixty-five 'that means', they announce breathlessly, 'sixty-five kisses'. The popularity of the game seems to have increased since the launching, in 1967, of the otherwise unrelated pop-song 'Cinderella Rockafella'; and skippy-boppers, aged 8 and 9, now possess a repertoire of variants, not all of them romantic:

> Cinderella dressed in yellow,
> Went upstairs to kiss a fellow,
> Made a mistake, kissed a snake,
> How many doctors did it take?

And:

> Cinderella dressed in yellow,
> Went down town to buy some mustard,
> On the way her girdle busted,
> How many people were disgusted?

These verses seem to have come from America where, already in the 1930s, Cinderella's suitor was transformed according to the colour of her gown—and the requirement of rhyme. If Cinderella was 'dressed in brown' she went to town to buy a gown and on the way she met a clown. If 'dressed in pink' she went down town to buy a drink and on the way she met a Chink. If 'dressed in blue' she went down town to buy some glue and on the way she met a Jew. In each case the question was 'How many kisses did she receive?' Rhyme-words, or supposed rhyme-words, for Cinderella have long been familiar in Britain. In Yorkshire and elsewhere memory lingers of a verse, having reference presumably to Guy Fawkes, that was chanted by children in the nineteenth century when 'cob-coaling':

> Umberella, down the cellar,
> There I spy a little feller;
> Save his body, save his soul,
> Please will you give us a cob o' coil?

Britain: versions from thirty-seven places since 1950, including, appropriately, 'Cinderella, Cinderella, All I hear is Cinderella, Washed the dishes, did the ironing, Scrubbed the floor' (London, 1953), and 'Cinderella at the ball, Who's the fairest of them all? How many fellows did she kiss? One, two, three . . .' (Ilford, 1969).

USA: *JAFL* 39 (1926), 'Went down town to meet her fellow, On the way she bought an umbrella'. *Evening Sentinel*, Carlisle, Pennsylvania, 1929, 'Went up stairs'. Abrahams, no. 80.

Canada: *Victoria Daily Times Weekend Magazine*, 30 May 1970, p. 8, 'Cinderella dressed in yella, This is the way she treats her fella, She hugs him, she kisses him, She kicks him in the pants, And that is the end of their romance.' This is an updating of the singing game 'All the Boys in Our Town' (see Opie, *Singing Game*).

Australia: Canberra, 1959, and ten subsequent recordings (Turner (1969)).

Rhodesia: Fort Victoria, 1972.

> Grace, Grace, dressed in lace,
> Went upstairs to powder her face.
> How many boxes did she use?
> One, two, three, four . . .

Well known in the United States since about 1910; but despite the attraction of 'climbing the stairs' (raising one foot after the other) and the squandering of the face-powder, it is uncommon in Britain, being reported only from Pendeen, Cornwall, in 1952.

USA: Arkansas, *c.*1910 (*Midwest Folklore*, 3, p. 78). New York, 1926 (*JAFL* 39 (1926), 38). Pennsylvania, 1929 (*JAFL* 42 (1929), 385). Abrahams, no. 158.

> Mrs Mason broke her basin,
> How much did it cost?
> Penny, tuppence, threepence, fourpence . . .

'Mrs Mason' (or, 'Old Mother Mason') occasionally breaks her basin 'On the way to the railway station' (Sale, Manchester, 1951) or 'On the way to London Station' (Swansea, 1952, and Flotta, Orkney, 1961) or on the way to some more local station; and since about 1950 she has often 'bought a basin' instead of breaking it. The money is, naturally, totted up in 'as many quick skips as the person can do', and the aim is to make the basin astonishingly expensive.

At the end of the nineteenth century 'Mrs Mason' was a counting-out rhyme (for instance in *Little Folks*, 41, January 1895, p. 20, 'Mrs Mason Broke a basin: How much will it be? Half a crown For Mr Brown, Oh, dearie me! One, two, three, O-U-T'). It first appears as a skipping rhyme in *St Anne's Soho Monthly*, June 1907, p. 153, and *London Street Games* (1916), 93.

> Early in the morning at eight o'clock,
> You will hear the postman knock.
> Up jumps [Katie] to open the door,
> How many letters on the floor?
> One, two, three, four . . .

A moderately popular game, perhaps more so in the 1950s than in the 1970s, in which the player named joins the postman in the rope, and the pair skip together while the rope is turned for 'peppers' to ascertain the number of letters delivered. A variant, known by a correspondent in Bath before the First World War, is also well established:

> Every morning at eight o'clock,
> I always hear the postman knock,
> He drops a letter,
> I pick it up,
> Every morning at eight o'clock.

Here the girl in the rope has a small object, usually a stone, which she puts down while skipping and has to try to pick up. The game appears to come from the song 'The Postman's Knock', written by L. M.

Thornton, with music by W. T. Wrighton, published in 1855. The song began:

> What a wonderful man the Postman is,
> As he hastens from door to door,
> What a medley of news his hands contain
> For high, low, rich, and poor:
> In many a face he joy doth trace,
> In as many, he grief can see,
> As the door is op'd to his loud Ran-tan,
> And his quick delivery.

The chorus became almost proverbial:

> Ev'ry morn, as true as the clock,
> Somebody hears the Postman's knock,
> Ev'ry morn, as true as the clock,
> Somebody hears the Postman's knock.

These lines had been adapted to skipping before the end of the nineteenth century. Alice Gomme found them in Marylebone (*Traditional Games*, ii (1898), 202), 'the girl named running out, and another girl running in directly':

> Every morning at eight o'clock,
> You all may hear the postman's knock.
> One, two, three, four. There goes *Polly*.

Alice Gillington found them in Surrey, in 1909 (*Old Surrey Singing Games*):

> Early in the morning at eight o'clock,
> You can hear the postman's knock!
> Up jumps *Ella* to open the door,
> One letter, two letters, three letters, four!

An 11-year-old Glaswegian girl gave words almost identical to these in 1975—which is perhaps remarkable to no one but ourselves. In recent times versions have been collected throughout Britain, and also in Australia and New Zealand. Sometimes the words are used for ball-bouncing.

Britain: Parkstone, Dorset, *c.*1914, 'Early in the morning at eight o'clock, I sometimes hear the postman knock. Out goes *Mary* answering the door, In comes *Jane* one, two, three, four'. *London Street Games* (1916), 77. Macmillan Collection, Somerset, five versions. Thirty-six recordings since 1950.

    Australia: Turner MSS, five versions.

    New Zealand: Sutton-Smith, p. 80.

> I know a nikker boy, he's double-jointed,
> He gave me a kiss and made me disappointed.
> He gave me another to match the other.
> All right, *Karen*, I'll tell your mother,
> For kissing *Terry*, down by the river.
> How many kisses did he give you altogether?
> One, two, three, four, five, six, seven . . .

A provocative skipping game that can arouse passionate excitement among the players ('You know who my boyfriend is,' said a small girl fiercely, as she entered the rope, 'so don't you *dare* choose anyone else'). The words Norman Douglas recorded in *London Street Games*, in 1916, could have been collected up to the late 1970s:

> I had a black man, he was double-jointed,
> I kissed him, and made him disappointed.
> All right, *Hilda*, I'll tell your mother,
> Kissing the black man round the corner.
> How many kisses did he give you?
> One, two, three . . .

Despite the chant's Victorian music-hall flavour (in 1883 Fred Coyne was singing 'Oh, very well Mary Ann, I'll tell your Ma'), the outrageous nature of the subject-matter and of the rhyming continues to delight the young. The few metrical embellishments found in thirty-four recordings of the game are mostly borrowings from other games:

> I had a boy and he was double-jointed,
> I kissed him and made him disappointed.
> That one died, and I had another one,
> Bless his little heart, he's better than the other one.
> Now, now, *Gilly*, I'll tell your mother,
> Kissing *Martin* around the corner.
> How many kisses did he give you?
> Five, ten, fifteen, twenty . . .
>
> (Barnes, Oundle, Swansea.)

> I know a nikkie boy, he's double-jointed,
> He gave me a kiss and he made me disappointed.
> He gave me another to match the other.
> Oh, *Amelia*, I'll tell your mother
> For kissing *Michael* down by the river.
> He took me down the garden,
> He put me on his knee,
> And said 'Pretty darling,

Will you marry me?'
Yes—no—maybe so,
Yes—no—maybe . . .

(Bedford)

Increasingly in the 1970s it became the custom for further questions to be asked about the skipper's future, presumably in the knowledge that kisses are only the beginning of an interesting story. Subsequent lines of enquiry may include 'Do you love him?', 'When did you marry?', 'Where will you go for your honeymoon?', 'What colour wedding dress?', 'How many babies?', 'What colour babies?', 'How many nappies?' All this needs an audience, who shout in amazement as the answers come up. If the skipper has been married in green, and has given birth to 108 purple babies who have only five nappies between them, so much the better. It might seem to feminists that the girls are being conditioned to expect a career of domesticity and child-rearing, but in fact they are making fun of such a life, whether they expect to live it or not.

Old Mother Ink she fell down the sink,
How many inches did she fall?
Two, four, six, eight . . .

The skipper does peppers or bumps when the counting begins. The words are also often used for counting out.

Britain: Cardiff, 1914, for counting out. Taunton, 1922, for ball bouncing. 1950 onwards, versions from fifteen places including Enfield, 1952, 'Old Mother Riley fell down the sink, How many gallons did she drink?' and Aberdeen, 1959, 'Little Mrs Inky fell doon the sinky', both for counting out. Oldbury, Worcestershire, 1970, 'Mrs Ink fell down the sink'.
    Australia: Melbourne, 1958, and elsewhere, 'Little Miss Pink fell down the sink, How many miles did she fall?' 'One of the most widely used counting-out rhymes' (Turner (1969), 15 n.).

Charlie Chaplin sat on a pin,
How many inches did it go in?
Two, four, six, eight . . .

Ordinary skipping until the counting begins, then bumps or peppers.

Britain: versions from seven places since 1951, including Cleethorpes, Lincolnshire, 1952, 'Humpty Dumpty sat on a pin'.
    USA: New York, 1926 (*JAFL* 39 (1926), 83). Carlisle, Pennsylvania, 1929 (*JAFL* 49 (1929), 386). Variants include 'Little Orphan Annie swallowed a pin', 1948; 'Charley McCarthy sat on a pin', 1950; 'Rin Tin Tin sat on a pin', 1956; 'Fatty Arbuckle sat on a pin', 1968.
    Australia: Sydney, 1958, 'Jarlie, Jarlie, sat on a pin' (Turner (1969), 31).

Up the pole, down the pole,
The monkey chews tobacco.
How many ounces does he chew
In one whole day?
   One, two, three, four . . .

Ordinary skipping followed by fast skipping for the counting. Similar formulas are much used for counting out, and have been since the 1920s.

Stratton-on-the-Fosse, Somerset, 1922, and Pontypool, 1924, both for counting-out. Versions for skipping from eight places since 1950, some beginning 'Up the ladder, down the ladder'. Nottingham, 1952, 'Up the pole, down the pole, Twirly whirly round the pole, How many people in London town? One, two, three . . .'.

Salt, mustard, vinegar, cider,
How many legs has a great big spider?
   Two, four, six, eight . . .

Ordinary skipping, with peppers for the counting.

Britain: Weybourne, Surrey, 1947. Market Drayton, c.1950. Swansea and Welwyn, 1952. Coulsdon, Surrey, 1960. Covent Garden, London, 1974.

   USA: Funk and Wagnalls *Dictionary of Folklore* (1950), ii. 1016, 'Sugar, salt, pepper, cider, How many legs has a bow-legged spider?'. *Pennsylvania Jumpster* (1974), 'As grandma used to chant: Pepper, salt, mustard, cider, How many bullets killed the Kaiser?'

I had a sausage, a bonny, bonny sausage,
I put it in the oven for my tea.
I went out to play, I heard the sausage say:
'*Judy, Judy*, come in for your tea.
How many bites will you take out of me?
   One, two, three, four . . .'

Skipper runs out of rope 'to play', runs back in again when called, and then has to skip fast for the counting. Sung to 'I Love a Lassie'. Versions from ten places in the Midlands.

CALLING IN

If a girl in a single rope (a rope which she herself is turning) wants someone to join her, she starts 'calling in'. She may simply say 'I call in my very best friend' (naming her); or in Scotland, sometimes, 'I call in my sister' (naming whom she wishes); or chant,

> I call in my very best friend
> And that is *Tracy Parker*.
> Please come in by I count three,
> One—two—three.

Or she may make it clear that the appellation 'best friend' is only a
formality:

> I call in my very best friend,
> E-legged, I-legged, bandy-legged *Kate*.

But if, as is usual, several players want turns, the chant is adapted to get
rid of the guest fairly quickly, and—either in a short or long rope—the
formula is:

> I call in my very best friend,
> Her name is *Mandy Hickman*.
> I send out my very best friend (*or* very worst enemy)
> Her name is *Mandy Hickman*.
> Out—she—*goes*.

Or the chant may end with a question, such as 'How many years have we
been friends?', or 'How old is she when she gets married?', or 'How many
babies does she want?', and the pair then skip peppers until one or other
trips in the rope. In Hackney they chant:

> Callings in and callings out,
> I'll call *Katie* in.
> *Katie*'s in and won't go out,
> I'll have to push her.
> How many pushes shall I give her?
>   One, two, three, four . . .

This chant can be compared with one Norman Douglas collected for
London Street Games (1916), 59:

> Callings in and callings out—
> I call *Rosie* in.
> Rosie's in and won't go out—
> I call *Maudie* in.

Very possibly both Rosie and Maudie skipped together with the caller, as
girls still do, Rosie skipping in front of the caller and Maudie joining in
behind her.

   The player who is being called in, whether into a short rope or a long
one, cannot expect her sensibilities to be respected:

I call in my very best friend,
And that is *Lisa Truman*,
Not because you're dirty,
Not because you're clean,
Because you've got the whooping
cough,
Get out you dirty thing.

> (St Martin's, Guernsey. In Penrith:
> 'Because you've got the whooping cough
> and measles in between.')

I call in my very best friend,
E-legged, I-legged, *Rosemary*,
My friend likes sitting
On her boyfriend's knee.
Does she like it?
Wait and see.

(Girls, 9–10, Liss)

I call in my very best friend,
And that is *Janet Saunders*,
Please *Janet Saunders*
Turn your back to me;
Not because you're dirty,
Not because you're clean,
But just because you kiss the boys,
Upon the village green.

(Bishop Auckland)

I call in my pal *Jeanie*,
Is she in? Yes, she is.
Does she like coffee?
Does she like tea?
Does she like sitting
On her boyfriend's knee?

(Girl, 12, Paisley)

Compare the calling in at the end of the singing game 'On the mountain stands a lady' (see Opie, *Singing Game*, p. 177):

> So call in my *Susan* dear, *Susan* dear, *Susan* dear,
> So call in my *Susan* dear, while I go out to play.

> I like coffee, I like tea,
> I like *Mary* in with me.
> I don't like coffee, I don't like tea,
> I don't like *Mary* in with me.

A simple calling-in and bawling-out game in the manner of 'I call in my very best friend' (p. 261), which has become increasingly popular since the 1940s and is now widespread. The chant has been known in several earlier forms and can be traced back to Georgian days. About 1830, or possibly earlier, James Catnach of Seven Dials, near Leicester Square, produced a small chap-book for the young, *Nurse Love-Child's New Year's Gift*, which contained the rhyme:

> One, two, three,
> I love coffee,
> And Billy loves tea.
> How good you be,
> One, two, three,
> I love coffee,
> And Billy loves tea.

Some ditty such as this was clearly once popular, and even proverbial. In *Punch*, 25 August 1860, a bloater is said to have struck up 'the plaintive stave':

> Some like coffee, some like tea,
> Some like Herrings just like Me!
> I once was white; I now am red:
> Just think of this when you go to bed.

W. H. Babcock saw a circling game in Washington, DC, in 1886, which had the words, 'I like coffee, and I like tea, I like the boys and the boys like me. I'll tell my mother when I get home, The boys won't let the girls alone' (*Lippincott's Magazine*, 38, p. 328). And Cecil Sharp picked up a snippet of a song, beginning, 'Some loves coffee, some loves tea, Some loves money, but they don't love me', in North Carolina in 1918. In Britain by this time the lines had become part of a skipping game. At Wrecclesham, Surrey, about 1892, girls were singing:

> I love coffee, I love tea,
> I love sitting on my sweetheart's knee.

At Spinkhill, in Derbyshire, in the early years of the twentieth century, they sang:

> I like cocoa, I like tea,
> I like sitting on the blackman's knee.
> Down in the cellar, see my Auntie Bella,
> Sitting on a blackman's knee.

And in Scotland, for some long while, girls have expressed a cheerful independence in the skipping-rope:

> I love coffee, I love tea,
> I love the lads and the lads love me.
> I wish my mither wud haud her tongue,
> For she had a lad when she was young,
> [or, 'I wish my mither wud shut her face,
> For she did the same when she was my age.']

Or, as a 12-year-old in Forfar sang it in 1954:

> I love coffee, I love tea,
> I love the nigger wi' the big blue e'e.
> I wish my mother would haud her tongue,
> For she had a boyfriend when she was young.

In the 1950s in the Midlands and north country, 'I love coffee' was often used for action skipping:

> I like coffee, I love tea,
> I like sitting on a blackman's knee,
> And this is what he said to me:
> Lady, lady, touch your knee,
> Lady, lady, touch the ground,
> Lady, lady, twirl around,
> Lady, lady, show your foot,
> Lady, lady, take your hook.

'I like coffee' is now a calling-in game nearly everywhere, for short rope or long, and with the words fairly standardized (reported from twenty places in England and Scotland). But the words found in Covent Garden, London, are interesting for their relationship to the chant 'Callings in and callings out':

> I like coffee, I like tea,
> I like *Julie* in with me.
> Now she's in she won't go out,
> So I'll 'ave to push 'er out.
> How many pushes shall I give 'er?
>   Five, ten, fifteen, twenty . . .

'I like coffee' may have been known in Scandinavia as a song, and certainly it was a skipping rhyme in Norway by about 1915: 'Jeg liker ikke kaffe, jeg liker ikke te, Jeg liker ikke Kari, men kom inni for det. God dag, god dag, adjo, adjo' (Svein Magne Olsen, *Lek og Sang* (1993), 28).

Most calling-in games are for the long rope only. They have the useful function of limiting each player's turn, often offering her, in compensation, the pleasure of choosing the next player. The game must not be held up, so the person in the rope must decide early in her turn whom she is going to choose to take her place; she must hiss the name to the enders in good time, enabling them to sing it out loud and clear when the right moment comes. But the skipper does not always choose her successor. Often the girls line up and the enders call them in by name when it is their turn, which makes for a smoother game.

When skipping to the accompaniment of songs was growing in popularity, early in the twentieth century, a variety of songs were pressed into service for the various kinds of skipping routines. The old courtship games had the most obviously suitable mechanism for calling in a player and getting rid of her when her turn was over, and since their true function as love-agencies for youths and maidens had long ago passed, some of them were now given new life as skipping songs. The next four skipping games were originally ring games or parts of ring games.

> On the mountain stands a lady,
> Who she is I do not know.
> All she wants is gold and silver,
> All she wants is a fine young man.
> So call in my dearest friend,
> My dearest friend, my dearest friend,
> So call in my dearest friend,
> And I'll be on my way.

One of the most popular and poetic of skipping games, being a straight adaptation of a ring game that has been current for certainly a hundred years, and has roots in English folk-song (see Opie, *Singing Game*, pp. 174–8). One player jumps in the rope while the others sing. The next player runs in at 'dearest friend'; or may be called in by name:

> So come to me *Lynn* dear,
> *Lynn* dear, *Lynn* dear,
> So come to me *Lynn*, dear,
> While I go out to play.

The first player then runs out and the newcomer has her turn. The game is, in general, played like this throughout Britain; but in some places the aim is to learn the initial letter of the young man, or to find out the number of kisses exchanged; or the song may end abruptly 'All she wants is—I don't know!' and the skipper, or two skippers, land with their feet astride the rope. The game has, in fact, so many narrative and traditional associations that variations are inevitable, and, as may be seen, combinations are made with fragments of other games; one wit has, with some success, 'improved' the lines (see 1974–5 version, below).

There's a lady on the hill,
Who she is I do not know,
All she wants is gold and silver,
All she wants is a nice young man.
So call in *Barbara*.
Barbara likes whisky,
Barbara likes rum,
Barbara likes kissing the boys,
So I'm going home.

(North Kent, 1953. The ending is as in 'I wish tonight was Saturday night')

Stood a lady on the mountain,
Who she is I do not know.
All she wants is gold and silver,
All she wants is a nice young man.
Right, I'll tell your mother,
For kissing *Stephen Grove*.
How many times did you kiss him?
One, sir, two, sir, three, sir . . .

(Salford, 1970 and 1975)

Upon a hill there stands a lady,
Who she is I do not know.
I'll go and court her for her beauty,
She will answer yes or no.

> (Cleethorpes, 1952. Folk-song version)

On the mountain stands a lady,
Who she is I do not know,
All she wants is gold and silver,
All she wants is a very fine man.
Lady, lady, touch the ground,
Lady, lady, turn around.
Lady, lady, do the kicks,
Lady, lady, do the splits.

> (Coventry, 1970)

On a hill stands a lady,
Who she is I do not know,
All she wants is gold and silver,
All she wants is a nice young man.
Madam, will you walk,
Madam, will you talk,
Madam, will you marry me?

> (Forfar, 1975. Cf. 'Keys of Heaven',
> Opie, *Singing Game*)

On the mountain stands a lady,
Who she is I do not know.
All she wants is gold and silver,
All she wants is a nice young man.
So call in *Richard*.
She kisses him, she cuddles him,
She sits him on her knee,
She says, My darling, will you
    marry me?
Yes, no, yes, no . . .

> (St John's Wood, London NW8, 1971
> and 1973)

On the mountain stands a castle,
And the owner's Frankenstein;
And his daughter Pansy Potter
Hopes to get a Valentine.
So call in *Fiona* dear,
Fiona dear, *Fiona* dear,
So call in *Fiona* dear,
And I'll go out till next New Year.

> (Versions from twelve places in Scotland,
> 1974 and 1975)

There's a lady on the mountain,
Who she is I do not know,
All she wants is gold and silver,
All she wants is a nice young man.
So call in *Jamie Seymour*.
How many kisses did she give him?

> (Wereham, Norfolk, 1975)

The distinctive tune became familiar to viewers of light comedy when it was adopted as the theme music for the BBC television series about Merseyside, *The Liver Birds*.

Britain: London, 1910. Leeds, *c*.1925. Northumberland, *c*.1935. Versions from sixty-five places since 1950.
   USA: Abrahams, no. 446, twelve refs from 1934 onwards.
   Australia: Turner (1969), 36, Melbourne, *c*.1957, 'All she wants is gold and silver, And a jewel comb in her hair', and two other versions.

Down in the valley where the green grass grows,
There's a pretty maiden she grows like a rose,
She grows, she grows, she grows so sweet,
She calls for her lover at the end of the street.

> *Christie, Christie*, will you marry me?
> Yes, love, yes, love, half-past three.
> Ice cakes, spice cakes, all for tea,
> And we'll have a wedding at half-past three.

The rope-turning may be ordinary, or French, or Waves; and a second skipper, nominally but rarely a boy, may join the first skipper when the name is made public. This game is another with words borrowed from a singing game (see Opie, *Singing Game*, pp. 127–30). However, the chant benefits from the discipline of the rope. Whereas the singing game is susceptible to accretions from other games, which slow it down like barnacles on a ship, the skipping chant seldom consists of more than eight lines (versions similar to the above exist in Aberdeen, Edinburgh, St Andrews, Kirkcaldy, and Plymouth); and almost all versions maintain their pace and please the ear. For instance in Cheltenham:

> Down by the river where the green grass grows,
> Where little *Susan* bleaches her clothes.
> She sang, she sang, she sang so sweet,
> That she sang *Johnny* across the street.
> *Susan* made a dumpling, she made it so sweet,
> She cut it up in slices and gave us all a piece,
> Saying, 'Take this, take this, and don't say no,
> For tomorrow is my wedding day and I must go'.

And in Dunoon, Argyllshire:

> Down in yonder meadow where the green grass grows,
> Where *Linda Robson* bleaches all her clothes.
> She sang, she sang, she sang so sweet
> That she sang *Robbie Findlay* across the street.
> She huddled him, she kissed him, and sat him on her knee,
> And said, 'Dear Robbie, will you marry me?
> Agree, agree, I hope you will agree,
> Tomorrow is your wedding day at half-past three.'

Elsewhere the maiden is found 'hanging out her clothes', or 'sits washing her clothes', 'sits washing her toes' (Pendeen in Cornwall), 'sits and sews' (Dundee), 'preaches and crows' (Jedburgh). She may grow not so much like a rose as 'so tall' (quite commonly) or 'so stout'. In north Staffordshire she 'grows so sweet' but leaves her 'dear little' partner 'with sweaty feet'.

In some places in Britain the game now ends, as it has long ended in the United States, with a tally of the kisses. In Bristol:

> Down in the valley where the the green grass grows,
> Dear little *Julie* she grows like a rose.
> She grows, she grows, she grows so sweet,
> That she calls for her lover down the street.
> *Paul, Paul*, are you coming out tonight?
> *Paul, Paul*, the moon is shining bright.
> Put your hat and coat on,
> Tell your mother you won't be long.
> How many kisses did you get last night?
> One, two, three, four . . .

And in Coventry:

> Down in the valley where the green grass grows,
> There stands *Gillian* growing like a rose.
> She grows, she grows, she grows so stout,
> She calls for her boyfriend down the street.
> Meet him on the corner at half-past eight,
> Put your hat and coat on, say you won't be late.
> How many kisses did you give him at the gate?
> One, two, three, four . . .

Possibly the emotive nature of the chant, rather than its poetry, accounts for its popularity. (One observer has noted how regularly the girls are both embarrassed and excited when their names are linked with a particular boy, and that this is so whether or not the friendship is a reality.)

'Down in the Valley' was already a skipping game at the beginning of the twentieth century, being reported as such both by A. E. Gillington in *Old Surrey Singing Games and Skipping-Rope Rhymes* (1909), and by the Revd T. Allen Moxon in his parish magazine, *St Anne's Soho Monthly*, June 1907.

> I come from Chinky China, my home's across the sea,
> I wash my clothes in China, for two-and-six a day.
> Oh *Mary, Mary, Mary*, you ought to be ashamed,
> To marry, arry, arry, a boy without a name.

Popular in Scotland, particularly as a lead-in to 'Tinker, tailor, soldier, sailor, rich man, poor man, beggar man, thief', the song seems to have come from the USA. In Cincinnati in 1908, children playing the ring game 'Down in the Valley' used to sing:

> Down in the valley where the green grass grows,
>   There sat *Felici* as sweet as a rose,
> And she sang and she sang and she sang so sweet,
>   Down came *John* and kissed her on the cheek.

O *Felici*, O *Felici*, you ought to be ashamed
  To marry a boy instead of a man.
I'm a boy, I'm a boy, I'll soon be a man,
  I'll work for my living as hard as I can.

(JAFL 40 (1927), 22)

The second verse is not ordinarily attached to 'Down in the Valley' and bears no obvious relationship to it; it can be presumed to be of separate origin. These scolding lines were found independently, as a jump–rope song, in Westchester County, New York, in 1938 (Howard MSS):

I come from Chinky China,
My home is by the sea,
I wash the dirty dishes
For forty cents a week;
And over, over, over,
You ought to be ashame
To marry, marry, marry,
A fellow without a name.

In Glasgow, 1952:

I come from Chinky China, my home's across the sea,
I get my clothes from China at two-and-six a piece.
Oh, *Georgina, Georgina, Georgina*, you ought to be ashamed
To marry, marry, marry, a boy without a name.

In St Andrews, 1974:

I come from Chinky China, I come from across the sea,
I do the lady's washing for all you one-and-three.
Oh *Cathy, Cathy, Cathy*, you ought to be ashamed
For kissing *Tony Bell* and giving us the blame.
How many kisses did you give him?
  Five, ten, fifteen, twenty . . .

But in Wigan in 1960, according to an 8-year-old, the song formed the basis of a ring game:

We all stand in a ring and a girl or boy skips around and we sing,

Oh my name is Chinky Chinaman, I live in Chinatown,
I do the washing up for six and half a crown.

When you have sung that you will be near somebody in the ring. Get their arm and swing them round and, suppose their name is Ann, you will say,

Ann, Ann, Ann, you ought to be ashamed
To marry, marry, marry, a man without a name.

Then you stand where the other girl was standing, and she does the same till they have all had a go.

Britain: versions from twelve places since 1950.

    USA: Abrahams gives references to Connecticut, 1947; Pennsylvania, 1945; and California, 1955.

    Cf. Bishop Auckland, 1961, for skipping: 'I wish I live in Switzerland, A dear Switzen town, To skate a senorita, And whirl around the town. Oh *Judith, Judith, Judith*, you ought to be ashamed, To marry, marry, marry, a man without a name.'

> Here comes Solomon in all his glory,
> Riding on a lily-white pony,
> Up the hill in all his glory,
> Because it's *Mary's* wedding day.
> Om pom om pom Susianna,
> Om pom om pom Susianna,
> Om pom om pom Susianna,
> Because it's *Mary's* wedding day.

When the skipper has called the second girl into the rope, they hold each other by the right hand, so that they are facing opposite directions, and skip round 'like a propeller' to the end of the chant, as in the old singing-game version (see Opie, *Singing Game*, pp. 337–9)—no mean feat when the rope is going over and under them at the same time.

Versions from twelve places since 1950, all but one from the north of England and Scotland, where the words are often, 'Here comes Mrs Macaroni, Riding on a snow-white pony, Through the streets of Aberdoni, This is *Mary's* wedding day'. Newcastle upon Tyne, 1953, 'Old King Solomon [or, Good King Solomon] in all his glory'.

> Vote, vote, vote for *Karen Parker*,
> Call in *Donna* at the door;
> For *Donna* is the one who gives you all the fun,
> And we don't need *Karen* any more—
> Shut the door!

The second player enters the rope early in the chant and the first leaves at the end, so for most of the time both players are in the rope together. Generally the girls are called by turn, rather than being chosen, and in Wythenshawe, Manchester, the chant ends not 'Shut the door!' but 'Go back in the queue, Thank you'.

    The tune is 'Tramp, tramp, tramp, the boys are marching', the American Civil War song, copyrighted 5 January 1865 by George Frederick Root, the composer of 'The Battle Cry of Freedom'. The words of 'Vote, vote, vote' are themselves almost a footnote to history. Over the course of the twentieth century boys at election time have adjusted the words to the advantage of the candidate they favoured and bawled them in the streets (see Opie, *Lore*, pp. 348–9). The words a correspondent remembers being shouted in the Vauxhall Bridge Road *c.*1905 were of typical hopefulness:

> Vote, vote, vote for Burdett-Coo-oots,
> Vote, vote, vote for Burdett-Coots.
> For if you get him in
> You'll have pockets full of tin,
> Vote, vote, vote for Mr Coutts.

The American-born William L. Burdett-Coutts (1851–1921) was the Member for Westminster from 1885 until his death.

Amongst the earliest political recollections of Michael Foot was the song sung in 1919 when he was 6 and his father was standing (unsuccessfully) as the Liberal candidate in Plymouth against Nancy Astor:

> Who's that knocking at the door?
> Who's that knocking at the door?
> If it's Astor and his wife
> We will stab them with a knife,
> And we won't have the Tories any more.

At West Leicester in 1923, on the other hand, the tide of opinion was for the Labour candidate of suffragette fame, and against Winston Churchill:

> Vote, vote for Pethick-Lawrence,
> Chuck Winston Churchill in the sea.
> If you vote for him
> You'll be happy as a king,
> And we'll chuck old Winston in the sea.

And the lines remembered by Clifford Hanley from his Gallowgate days in Glasgow, perhaps during the same election, were—despite their inappropriateness for a pacifist—also in support of Labour (*Dancing in the Streets*, p. 84):

> Vote, vote, vote for Campbell Stephen,
> Vote, vote, vote for a' his men;
> And we'll buy a penny gun
> And we'll make the Germans run
> And we'll never see the Germans any more.

Since 'Vote, vote, vote' was already being used as a skipping song in the First World War (see below), its decline as a political song was predictable, quite apart from the fact that expressions of partisanship tend to be more subtle today. Nevertheless, at the 1951 election children in Swansea did not hesitate to express their views in traditional form:

Vote, vote, vote for Winston
  Churchill,
Here comes Attlee at the door,
Churchill is the one that we all like
  best
And we don't want Attlee any more

Vote, vote, vote for Mr Attlee,
Don't mind Churchill and his lot,
For we'll put 'em in the pot
When the water's getting hot,
Then we won't see them any more.

And in 1961, in Swansea, as well as voting for Elvis Presley, they observed 'Here comes Macmillan at the door'.

The following chants, some of them closely similar to the political past, have been heard in the skipping rope:

Vote, vote, vote for *Billy Martin*,
Chuck old *Ernie* at the door,
If it wasn't for the law,
I would punch him on the jaw,
And we won't see *Billy Martin* any
  more.

(*London Street Games* (1916). Very similar
words current in Swansea in the 1950s)

Vote, vote, vote for *Mr Teacher*,
In comes *Sally* at the door,
We will buy a penny gun
And we'll shoot him up the bum,
And we won't see *Mr Teacher* any
  more.

(Several places. Cf. Clifford Hanley's
chant of the 1920s)

Vote, vote, vote for *Jeanie Walker*,
Who's that knocking at the door?
If it's Hitler and his wife,
We will stab him with a knife,
And we won't see Hitler any more.

(Kirkcaldy, 1952. Cf. Michael Foot's
song, 1919, above)

Vote, vote, vote for *Debbie Johnson*,
Punch old *Pauline* in the eye.
If it wasn't for the King
We would do the blighter in,
And she won't come voting any
  more.

(London, 1952)

Vote, vote, vote for dear old *Diane*,
Here comes *Amanda* at the door,
  Amanda is the woman
To make the batter pudding,
We don't want *Diane* any more.

(Ipswich, 1953)

Vote, vote, vote for little *Sheila*,
Who's that knocking at the door?
  Knock, knock.
If it's *Janice* let her in,
With a dimple on her chin,
And we don't want *Sheila* any
  more.

(Thurso, 1975)

Vote, vote, vote for *Bridget Wrigley*,
Call in *Patricia* at the door.
For *Patricia* is a lady
And she's going to have a baby,
So we won't vote for *Bridget* any
  more.

(Now widespread, but not found before
the 1970s)

> There's somebody under the bed,
> Whoever can it be?
> I feel so shocking nervous
> I call *Anne* in to me.
> First she lights the candle,
> Then she lights the gas;
> Get out, you fool, get out, you fool,
> There's no one under the bed.

The rope is turned slowly, the player in the rope calling in whom she wishes. The second player, sometimes designated a 'sister', shoos out the first, then herself becomes nervous and calls in someone else, and so on. The game seems to have the right mixture of scariness and humour to be taken up enthusiastically when it arrives at a school. It then disappears and in general is only moderately well known (thirty-six recordings). The variations are no more than might be expected. Sometimes there *is* somebody or something under the bed: a bogeyman, a burglar, a little rat, a pussy cat, 'a monkey [or a mouse] up your leg' or, when the rescuer lights the candle, 'underneath she goes, Then she sees a skeleton that makes her blow her nose'. The chant may end:

> *Anne* lit the candle,
> *Anne* blew it out,
> *Anne* gave a hell of a shout
> And knocked the burglar out.

Or the companion may 'light the oil lamp', or suchlike, and 'set the house on fire'. Norman Douglas listed 'Someone's under the bed' in *London Street Games* (1916).

> I know a little girl, sly and deceitful,
> Every little tittle-tat she goes and tells the teacher,
> Long nose, ugly face, put her in a glass case,
> If you want to know her name, her name is *Sandra Wilson*.
> Oh, *Sandra Wilson*, keep away from me,
> I don't want to speak to you, nor you to speak to me.
> Once we were friends, now we disagree,
> Oh, *Sandra Wilson*, keep away from me.

This well-established teasing verse (see Opie, *Lore*, p. 175) has made an exhilarating calling-in game. The girl in the rope names anyone she likes (or dislikes), and when the newcomer comes in she turns her back on her, or keeps ostentatiously away from her, until the end of the song, when she runs out. The simulated hostility is fun to act out and, say the girls, 'It makes us giggle'.

Newcastle, *c.*1900. *London Street Games* (1916), 84. North Somerset, 1916. Whitchurch, nr. Cardiff, 1928. Bradford, *c.*1935. Widespread since 1950.

Cf. Durham, *c.*1925, 'Oh *So-and-so* thinks herself a treat, Long skinny banana legs and umbrella feet [player named enters rope] Oh, *So-and-so*, keep away from me, I neither want to speak to you nor you to speak to me'. Edinburgh, 1975, 'Oh *Carolyn Donaldson*, I'm ashamed of you, For leaving *Jamie Lindsay*, across the road from you. His heart is nearly broken, he's dying for a kiss, Oh *Carolyn Donaldson*, I'll tell your mother this.'

> I wish tonight was Saturday night,
> Tomorrow will be Sunday,
> I'll be dressed in all my best
> To go along with *Wendy*.
> *Wendy* likes whisky,
> I like brandy,
> *Wendy* likes kissing the boys,
> Whoops-ee-diddle-ee-dandy.

Another of the call-in-a-friend-and-nip-out-yourself games that has not changed since Edwardian days.

Portobello, Edinburgh, *c.*1912, 'I wish tonight was Saturday night, Tomorrow would be Sunday, I'd be dressed in all my best to go along with ——. —— likes whisky hot, And I like brandy, I like kissing the boys, Hooch! sugary candy'. *London Street Games* (1916), 89, fragment only, 'O tonight is Saturday night, Tomorrow will be Sunday'. Camberwell, *c.*1935. Versions from twelve places since 1950; as text in the south, but in the north ending '*Kristie* likes whisky, *Kristie* likes rum, *Kristie* likes kissing the boys, Ee bai gum.'

> My schoolmaster is a very nice man,
> He tries to teach me all he can,
> A-reading, a-writing, a-rithmetic,
> And he never forgets to use his stick.
> When he does he makes me dance
> Out of England into France,
> Out of France into Spain,
> Over the hills and down the lane,
> Down the lane and into school.
> In that school there is a stool,
> On that stool there sits a fool,
> Her name is *Tracy Robinson*.

A composition that has had many roles in its time, including nursery rhyme (first printed 1795, see *ODNR*), derisive school rhyme, counting-out rhyme, and accompaniment to 'Kiss in the Ring'. In recent years it has been increasingly drawn into the orbit of the skipping-rope. By using an ending borrowed from 'In Liverpool there is a school', the version above embodies calling in. The player named at the end takes the place of she who named her a fool, and the song is repeated. Alternatively, or in addition, the skipper may have to jump out of the rope and in again at the words, '*Out* of England *into* France, *Out* of France *into* Spain'; and, if the song continues 'Round the world and back again', she has to run out of the rope, around one of the enders, and into the rope from the other side. Yet again, the song can, as an 11-year-old explained, 'tell you when you're going to have the cane'. The rope is turned ordinarily up to 'He never forgets to use his stick', and then fast as the enders chant, 'January, February, March . . .'. The fashion in Scotland is to lead into the song with the words, 'Bluebells, cockle shells, eevy, ivy, over'.

Britain: Midgley, Halifax, *c.*1900. Endings vary, e.g. Dudley, 1970, 'Out of Spain into Germany, All round the world and back again. How many times did I go round? Five, ten, fifteen, twenty . . .'.
   USA: *Children Magazine*, 12 (1927), 21, 'Mister Brown is a very nice man, He teaches the best he can, What day does my birthday come?' (Abrahams, no. 108). Westchester County, New York, 1938 (Howard MSS).

> House to let, apply within,
> Lady put out for drinking gin;
> Drinking gin is a very bad thing,
> So *Jean* goes out and *Margaret* comes in.

This calling-in game, with its whiff of sin and sanctimoniousness, is one of the favourites of its genre. It is better known in Scotland than England, and its already brisk turnover of players is usually made even brisker by being shortened to, 'House to let, apply within, When I go out Mrs——comes in'. In Nelson, Lancashire, and West Bromwich they chant, 'House to let, enquire within, If you can't pay your rent, Mrs—— will come in'.
   In Shrewsbury and in Radcliffe, Lancashire, the words are unusual:

> I know a woman with a big red nose,
> She lives in a house where nobody knows;
> A house to let, apply within,
> When I go out Mrs —— comes in.

In London, and in the south and west generally, the children often play what is in fact a different game, in which they skip bumps on the last word in each line:

> A house to let,
> No rent to pay,
> Knock at the door
> And run away.

Variations include:

House to let,	A house to let,
House to sell,	With a TV set,
Knock at the door,	A house to sell,
And ring the bell.	With a knocker and bell.

In the United States 'House for rent, inquire within' is well known, and so too, for the past fifty years, has been a chant that is similar to the Scottish chant, but more charitable:

> Room for rent, inquire within,
> A lady put out for drinking gin.
> If she promises to drink no more
> Here's the key to her [or to named player's] back door.

Versions of this chant have, in fact, been used in the USA for jumping rope, and also for counting out, since before the First World War. In Britain correspondents report, from the beginning of the century, both 'A house to let, no rent to pay, Knock at the door and run away' and 'House to let, enquire within, When I go out *Nellie Smith* comes in', both apparently for skipping, though today they are sometimes used for ball-bouncing or fivestones.

Obviously the words stem from the old practice of putting placards in windows, 'House to Let—Inquire Within' (referred to by Mrs Craik in *Our Year*, 1860); and G. F. Northall in *English Folk-Rhymes* (1892), reported that in Warwickshire the following derisive lines were repeated in front of houses displaying such notices:

> House to let, enquire within,
> Men turned out for drinking gin,
> Smoking tobacco and pinching snuff,
> Don't you think that's quite enough.

> Little fatty doctor, how's your wife?
> Very well, thank you, she's all right.
> She won't eat a bit of fish,
> Nor a bit of liquorice,
> O-U-T spells out.

A counting-out rhyme that has been adopted for skipping. The skipper runs out of the rope at *out*, and the enders continue:

> O-V-E-R spells over.

The skipper runs back over the turning rope and out the other side.

U–N–D–E–R spells under.

The skipper runs *under* the turning rope.

I–N spells in.

The skipper re-enters the rope, and skips while the others chant:

> Once you're in you can't get out
> Until you show your knees—Whee!

The skipper 'shows her knees' (whether or not they have been visible all the while) and runs out. Then the next player enters the rope.

'Little fatty doctor' has long been popular (versions from twenty-four places since 1950) and can be traced back by stages to Regency days. The text version above (from seven places) is how the game is commonly played today. Earlier recordings include:

Little white doctor, how's your
  wife?
Very well, thank you, quite all
  right.
Can't eat a bit of fish,
Nor a bit of liquorice.
O–U–T spells out
With a bottle of stout.
I–N spells in
With a bottle of gin.

  (Worthing, 1960)

Doctor, doctor, how's your wife?
Very well, thank you, she's all
  right.
Can she eat a tuppenny pie?
Yes, sir, yes, sir, so can I.
How many pies can she eat?

  (Peppers till skipper trips. York City,
  1952)

Maister Munday, hoo's yer wife?
Very sick and like to die.
Can she eat ony meat?
Yes, as muckle as I can buy.
A platefu' o' parritch very thin,
A pund o' butter melted in;
Black fish, white troot,
Eerie, orrie, you're oot.

  (Edinburgh, *c*.1850. *Rymour Club*, I
  (1911), 164)

Old Mr Fat Belly, how is your
  wife?
Very well, thank you, can't save her
  life.
Can't eat a bit of flesh,
Not a bit of licoresh.
Out goes she.

  (Counting out. Swansea, 1939)

Fat-faced doctor, how is your wife?
Very bad, very bad, can't save her
  life;
Can she eat a bit of fish?
Yes, she can, yes, she can.
Can she eat a penny pie?
Yes, she can, and so can I.

  (Dialogue between player in the rope and
  the rest. Chaffcombe, Somerset, 1922)

Doctor! doctor! how's your wife?
Very bad, upon my life.
Can she eat a bit of pie?
Yes, she can, as well as I.

  (Poole, Dorset, *c*.1820. Set dialogue when
  boys meet. *Longman's Magazine*, 13,
  p. 518. Cf. Chaffcombe, Somerset, 1922)

❖

In the following games the newcomer pushes, or chases, the original skipper out of the rope:

> Mother's in the kitchen,
> Doing a bit of stitching,
> In comes a bogeyman
> And pushes her out.

One of the most popular of skipping games (collected in eighty-two places in Britain, and current since the 1920s). The reason for its popularity is obvious: 'We all get a quick turn and we like pretending to do the shoving.' The skippers line up and the words are chanted with one player in the rope. As soon as the chant comes to an end—sometimes sooner— she is chased out of the rope by the next player, 'the bogeyman', who quickly becomes apprehensive because she herself is about to be chased out. It is thus a game which not only makes jumping into the rope extra fun but then makes the player *want* to leave the rope quickly. The game is much the same everywhere, but in some places 'Fanny' is in the kitchen, in others 'Granny', 'Jenny', 'Mary', 'Nancy', 'Polly', or 'Sally'. The chant also commonly starts, 'As I was in the kitchen'. The heroine may then be frightened out, scared out, popped out, or pushed out ('She really jumps out by herself, because if she was pushed she might fall and get grit into her knees'); and the scary one may not be a bogeyman but a burglar, or a black man (in Cwmbran, 'Black Joe'), or a golliwog ('who hit me on the googey-gog'—Blaenavon).

Extensions to the chant are uncommon, but in some places the drama is sustained in a second verse, for example:

Polly in the kitchen,
Doing a bit of stitchin',
In came a burglar
And knock Polly out.

Burglar in the kitchen,
Doing a bit of pinchin',
In came a policeman
And pushed the burglar out.

(Spennymoor)

As I was in the kitchen,
Doing a little stitching,
In came bogeyman
And I ran out.
The bogeyman kissed her,
And made a big blister,
In came *Shirley*
And out ran she.

(New Cumnock)

I was in the kitchen
Doing a bit of stitching,
In came a black man
And I ran out.
He chased me round the parlour,
He chased me up the stairs,
He chased me in the bedroom,
And made me say my prayers.

(Birmingham)

Granny's in the kitchen,
Doing some stitching,
In comes the bogeyman
And out goes she.
Granny comes in later,
The bogeyman ate her—
One, two, three.

(Bressay, Shetland Isles)

Variants are quite common, and one Scottish observer has noticed that some girls, when they think themselves on their own and want to be naughty, take delight in improving on the words, for example:

> Polly in the kitchen, pickin' up sticks,
> Along comes a bogeyman and pulls down her knicks.

And the following two variants have been noted which have been influenced by other verses:

<table>
<tr><td>

As I was in the cellar,
Mending an umberella,
In came a bogeyman
And pushed me out.

(Radcliffe, Lancashire)

</td><td>

Mickey, Mickey Mouse,
In a haunted house,
Along came a bogeyman
And pushed Mickey out.

(Dundee)

</td></tr>
</table>

The standard chant has been current certainly since the early 1920s (Macmillan Collection, 1922), and probably earlier. One of the games Norman Douglas printed in *London Street Games* (1916), 68, seems to have been played in the same manner:

> I-N spells in—
> I was in my kitchen
> Doing a bit of stitching,
> Old Father Nimble
> Came and took my thimble,
> I got up a great big stone,
> Hit him on the belly bone—
> O-U-T spells out.

The game is popular in Australia, but not as well known in the United States as might be expected (only two American references in Abrahams, no. 22). In Canada, according to the *Victoria Daily Times*, 20, no. 22, 30 May 1970, a more sophisticated version exists: 'Someone in the kitchen, chewing bubble gum. In comes Susie and asks for some. No you dirty rascal, no you dirty bum. You can't have any 'cause you asked for some.'

> Old Mrs Knockabout knocked her cat about,
> Outside *Margaret*'s door;
> Out comes *Margaret* with a big stick
> And knocks poor *Susie* out.

The Revd T. Allen Moxon pointed out, as early as 1907, that the fun of the game is that 'the skipper constantly changes according as the name of one or other of the girls is mentioned'. The words have changed little.

St Anne's Soho Monthly, June 1907, p. 155, 'Old mother Knockabout knocks her cat about outside *Milly*'s door. *Sally* comes out and gives her a clout, and lets her know what for'. *London Street Games* (1916), 79, 'Old mother roundabout Knocking all the kids about . . .'. Swansea, 1952 and 1962, as text.

> *Rita* had a toffee shop,
> *Joan* came a-buying;
> *Rita* gave her a skinny bit
> So *Joan* knocked her flying.

Second player enters the rope and knocks or pushes the first player out, and so on.

Current *c.*1945. Blackburn, 1952. Radcliffe, 1952. Swansea, 1952.

> *Denise* is on the telephone, calling on *Jacqueline*,
> *Jacqueline*'s on the telephone, calling on *Mary*,
> *Mary*'s on the telephone, calling on *Pauline* . . .

A neat system for achieving a fast changeover in the rope. Only from Dundee, 1975.

> Goodbye, *Ann*, while you're away,
> Send me a letter to tell me you're better;
> Goodbye, *Ann*, while you're away,
> Don't forget your old pal *Valerie*.

The words of this calling-in game come from a farewell song which, in Liverpool, Manchester, and the north Midlands generally, was (and perhaps still is) sung to a child leaving school, or going into hospital for a while. Her friends sing:

> Goodbye, *Caroline*, while you're away,
> Send me a letter every day,
> And don't forget your old pals at ——.

In 1916 Norman Douglas found girls skipping to the valediction (*London Street Games*, p. 84), which had acquired a flavour, in the second and third lines, of the old singing game 'Isabella' (see Opie, *Singing Game*, pp. 171–3):

> Goodbye (May), while you're away,
> Send a letter, love,
> Say you're better, love,
> Don't forget your dear old (Nell).

Versions, similar to text, from Accrington, Liverpool, Shrewsbury, and Workington.

A less usual type of calling-in game is in the form of a dialogue. While the restrictions that continued after the war were still not forgotten, a skipper might call out: 'I say, *Jenny*, stick it.' Jenny ran into the rope, while the other girl ran out and demanded 'What with?' The first girl ran back into the rope to say, 'With jam', and Jenny ran out. Then they changed places again and Jenny said, 'Jam's on the ration.' Jenny ran back into the rope, and the dialogue continued, with the players continuing to run in and out.

> So is Spam [tinned spiced ham]
> How shall we manage?
> The best way we can.

In 1960–1 the game was still played in the same way, but the dialogue had been modernized:

> I say, *Linda*, stick it.
> What with?
> With shampoo.
> Shampoo is soapy.
> You are dopey.
> How is that?
> Because you're fat.
> Who said that?
> The big fat cat.

And further development offered an alternative dialogue:

> *Linda*, I want you.
> What d'you want me for?
> I want to tell you something you've not heard before.
> What is it?
> Try and guess.
> Is it stick-it?
> No, it isn't [though it could have been, in
>   which case they would have continued as above]
> Is it ditto?
> Yes, it is.

Then Linda would say, 'Ditto one', and the first player would say 'Ditto two'. They would change places and Linda would say, 'Ditto three', and the pair would continue like this until one or the other missed the rope.

Bishop Auckland, 1952 and 1961. York City, 1952. Spennymoor, 1960. Leeds, 1966.

> All in together, girls,
> Never mind the weather, girls,
> When I count twenty
> The rope must be empty.
> Five, ten, fifteen, twenty!

This, the major calling-in game, has an uncharacteristically pedagogical flavour. Since very few of the skipping songs were originally intended for skipping, it is fascinating to speculate how and where this one was launched; and even so the words bear traces of the convergence and divergence to which oral lore is subject. It is certainly an effective skipping game. All the players enter the rope at once, and all are expelled at once. This is easier said than done, and its accomplishment, or otherwise, evokes much merriment. For this reason, probably, it is one of the games most often recalled by the elderly, who confirm it was played in the same way in Victorian times as it is today, those players who are caught in the rope, or are last out, having to 'take ends'. Early versions of the chant usually referred, for some reason, to 'Caroline' or 'Peter':

All in together
All sorts of weather
I saw a pig's tail
Hanging out the window.
Shoot, bang, fire!

       (Wrecclesham, Surrey, c.1895)

All in together
Frosty weather,
I see Peter
Sitting on the window sill.
Caroline, Caroline, shoot!

       (Stockport, Cheshire, c.1905)

All in together,
This fine weather,
I saw Peter
Knocking at the window,
Caroline, Caroline, shoot.

    (F. Kirk, *Rhythmic Games* (1914).
    Probably Bradford)

All in together, girls,
Best fine weather, girls,
I saw Peter
Looking out the window.
Shoot, bang, fire!

       (Kingston, near Taunton, 1909. Swansea, c.1914)

All in together,
All sorts of weather,
When the wind begins to blow
Shoot, bang, fire!

       (North-east London, c.1910)

All in together,
This fine weather,
One, two, three,
Busy, busy, bee.
I saw Peter,
Hanging out the window.

       (Cardiff, c.1910)

These chants may be compared with a song for playing 'Ring o' Roses', from Sheffield, which S. O. Addy gave Alice Gomme (*Traditional Games*, ii (1898), 110):

> Windy, windy weather,
> Cold and frosty weather,
> When the wind blows
> We all blow together.
> I saw Peter!
> When did you meet him?
> Merrily, cherrily,
> All fall down.

Lines similar to the first four, above, turn up in Norman Douglas's *London Street Games* (1916), 88 ('All in together—all sorts of (*or* frosty) weather—When the wind blows we all go together'), amongst rhymes of miscellaneous use; also, in Dublin, in the 1950s, three children could walk along the street singing 'Windy weather, Frosty weather, When the wind blows We all go together', and the end ones would crash together on the last words; also in Dublin, similar lines are sung in a fast-whirling, crashing circle (Eilís Brady, *All In! All In!* (1975), 13). Perhaps this form of words has come from a sea-chanty tradition, for 'The Fishes' has a chorus of 'Windy weather, stormy weather, When the wind blows we're all together', and the chorus to 'Windy Old Weather' is very much the same.

The further development of the chant may be seen in versions from the 1920s and 1930s:

All in together,
Frosty weather,
When the wind blows
We all run out.

> (M. and R. King, *Street Games of North Shields Children* (2nd ser. 1930), 29[26])

All in together,
This cold weather,
I see a pistol
Hanging on the wall;
Fire! Fire! Fire!

> (Cairnbulg, Aberdeenshire, *c.*1920)

All in together, girls,
This frosty weather, girls,
Put your hats and coats on,
Tell your mother you won't be
   long.
Shoot, bang, fire!

> (Liss, *c.*1935. Similar in Swansea, 1962; Leeds, 1966)

All in together, boys,
This fine weather, boys,
I saw Peter
Hanging out the window.
He had a pistol,
I had a gun.
Shoot, bang, fire!

> (*Soham, Cambridgeshire, c.*1930)

---

[26] Cf. the words in Montrose in 1974: 'All in together, Like a bunch of heather, When the wind blows We all jump out' (girl, 10).

In the majority of cases the chant is on the lines of the first version above; but a number of variants exist, some of them leading off to other rhymes:

The day is clear,
The sky is empty,
By the time I count twenty
This rope must be empty.
Five, ten, fifteen, twenty.

                   (Pontefract)

Never mind the weather,
Never mind the rain,
As long as we're together—
Whoops! we'll go again.

     (Swansea. Influenced by a popular song)

All in together, boys,
The cows are in the meadow, boys,
By I count twenty
The bough must be empty.

    (Spennymoor; Worcester; Penrith, 'We always say boys even when there are only girls skipping')

I saw the teacher
Looking out the window,
And she said—
All in together, girls,
Never mind the weather, girls,
Blish, blash, fire!

        (Newcastle upon Tyne; Bishop Auckland)

All in together,
This fine weather,
Along came a nanny goat
Hanging out its petticoat.
Shoot, bang, fire!

        (Uckfield. Similar in Sydney, Australia)

All in together,
To see Cinderella,
Before we count twenty,
The rope must be empty.
Five, ten, fifteen, twenty.

                 (Ruthin)

All in together, girls,
Never mind the weather,
See Mrs Jones
Hanging out the clothes.
By the time I count twenty
The rope must be empty.
Five, ten, fifteen, twenty.

            (Haverfordwest)

All in together,
This fine weather,
I see a picture
Hanging on the wall.
When I count twenty
The rope must be empty.

     (Cumnock area. *Those Dusty Bluebells* (1965), 13)

In places where skipping is of a high standard, they not only manage to skip all together in the long rope but to 'do actions according to the instructions in the rhyme'. In Cleethorpes, Caistor, Manchester, Birmingham, Leeds, London (Hackney), Langholm (in Dumfries), and Market Rasen, Lincolnshire, and probably in many other places in the 1950s and 1960s, 11- and 12-year-olds were operating a sequence like this:

> All in together, girls,
> This fine weather, girls,
> By I count twenty, the rope must be empty.
> Five, ten, fifteen, twenty.

When we are counting from five to twenty the people in the rope have to run out. Then when they come in we say—

> By I count six you've got to do the splits,
> Two, four, six,

and they have to get the rope between their legs. Then we get started again—

> By I count five you've got to do a dive.
> Two, four, five,

and they have to run out. Then they come back and stay in till the end. It goes,

> By I count four you've got to touch the floor,
> Two, four.
>
> By I count three you've got to touch your knee,
> Two, three.
>
> By I count two you've got to touch your shoe,
> One, two.
>
> By I count one you've got to be gone,
> One.

They have to go out of the rope.

This and the next form of 'All in together' are well known in the USA (see Abrahams, no. 7).

> All in together, girls,
> This fine weather, girls,
> When it's your birthday
> Please jump out.
> January, February, March, April . . .

A close relation of 'All in together', but nevertheless a game with a form of its own, played since the 1920s and found from Pendeen in Cornwall to Benbecula in the Outer Hebrides with little variation. All the skippers rush into the rope at the beginning, but leave it individually when their birthday month is called. On a second round their birth dates may be called: 'One, two, three, four . . .'; and in Forfar, the day of the week, and the number of their house as well. Alternatively, the game may begin with the players running in one at a time ('When I call your birthday please run in') and then the verse is repeated so that they leave on their birthdays, as before. The game is normally called 'Birthdays', or,

simply, 'All in for January, February . . .'; in Welshpool, 'Birthday
Bumps'.

> All in, a bottle of gin;
> All out, a bottle of stout.

This chant is, or was in the 1950s, as popular as 'All in together, girls',
ignorance of what gin or stout might be proving no hindrance. In some
places the chant is extended:

> All in, a bottle of gin,
> All under, a bottle of thunder,
> All over, a bottle of clover,
> All out, a bottle of stout.

And in some, only one player is called in at a time:

> In, in, a bottle of gin,
> I call my friend *Susan* in;
> Out, out, a bottle of stout,
> I call my friend *Susan* out.

This calling in and sending out of a single player is known to have
been the form of the game as played in Somerset in 1922 (Macmillan
Collection). The 'All in' version has not been found earlier than in
M. and R. King, *Street Games of North Shields Children* (2nd ser., 1930),
29.

SONGS AND RHYMES FOR PLAIN SKIPPING

The majority of skipping rhymes are comic or sentimental songs which
were once popular, or simply 'funny rhymes', forgotten by the adult
world and perpetuated only by children. When they proved pleasurable
to skip to, they became 'skipping rhymes'. Their repetition to the accompaniment of the turning rope fixes them in the children's memories. The
following is the main canon.

> I am a little orphan girl,
> My mother she is dead.
> My father is a drunkard
> And wouldn't buy me my bread.
> I sit upon the window sill
> And hear the organ play,
> And think of my dear mother
> Who is dead and far away.
> Ding dong the castle bells,
> Farewell to my mother,

Bury me in the old churchyard
Beside my eldest brother.
My coffin shall be white,
Six little angels by my side,
Two to kneel and two to pray,
And two to carry my soul away.

(Edinburgh, 1961)

The source of this doleful song is not known; it sounds like a temper-
ance song of the mid-nineteenth century.[27] J. P. Emslie was the first
to record its use by children—or, indeed, at all. On 9 August 1895, he
heard two little girls and a little boy singing it, at Sunderland, County
Durham, while they performed a rudimentary dance (*Folklore*, 26
(1915), 155):

I am a little workhouse girl,
    My mother she is dead,
My father is a drunkard,
    He will not buy me bread.

I travelled through the country,
    I had a mansion fair.
God bless my poor old mother,
She's dead and in her grave.

The last two lines are borrowed, or have been added, from the White
Paternoster (see *ODNR*, no. 346). They were already part of the text
when a Cornish boy repeated to the Revd Baring-Gould a song he had
learned from his aunt (*Songs of the West* (1905), 247):

Ding dong, the parson's [passing] bell,
Very well my mother,
I shall be buried in the old churchyard
By the side of my dear brother.
My coffin shall be black,
Two little angels at my back,
Two to watch, and two to pray,
And two to carry my soul away.
When I am dead and in my grave,
And all my bones are rotten,
Jesus Christ will come again
When I am quite forgotten.[28]

---

[27] 'The Drunkard's Lone Child', in S. Spaeth's *Weep Some More, My Lady* (1927), 191, is a
different song although it has the same story-line.

[28] The last four lines occur in book inscriptions, e.g. in Maclagan, *Games of Argyleshire*, 247,
'Donald Macdonald is my name . . . When I'm dead and in my grave, and all my bones are rotten,
This little book will tell my name, when I am quite forgotten.'

Children continued to use the song as a ring game for many years, see for instance *Rymour Club* 1 (1911), 150, and EFDSS MSS, Gilchrist MS G231 B iv.

It was Norman Douglas who first noted it as a skipping game, in *London Street Games* (1916), 71 ('Not a very cheerful rope-song, you'll say'). In Belfast, in 1933, a correspondent told us, they sang 'See saw Margery Daw, Farewell to my brother, We left him in the old church-yard, Beside my dear old mother'; 'it was a special sort of skipping. We swayed the rope to and fro and jumped over it from side to side.' The song has often seemed to require some 'special sort of skipping', and often it was the swaying action connected (via the 'castle bells' mentioned in the song) with the rhyme 'Bluebells, cockle shells'. In fact, in Cumnock, 1961, the song began, 'Bluebells my cockle shells, Farewell my mother'. In Edinburgh *c.*1949–61 it was 'German Ropes' or 'French Ropes' (see p. 197). A plain-skipping version was known in Kirkcaldy in 1952, and also in Aberdeen, where it ended, 'Ding, dong, Catholic bells, Mary is my mother, Carry me over the ocean bridge, Towards my elder brother.'

> Ten o'clock is ringing,
> Mother may I go out?
> My young man is waiting,
> Waiting to take me out.
>
> First he buys me apples,
> Then he buys me pears,
> Then he gives me sixpence
> To kiss him under the stairs.
>
> I wouldn't take his apples,
> I wouldn't take his pears,
> I wouldn't take his sixpence
> To kiss him under the stairs.
>
> At last I took his apples,
> At last I took his pears,
> At last I took his sixpence
> And kissed him under the stairs.

A song that has been part of street-play (mostly while skipping) in both Britain and America since the nineteenth century. E. W. B. Nicholson was given a version, much like the above, by a 13-year-old girl in Golspie in 1892; and Leah Yoffie recalled the following from her Missouri child-hood, 1895–1900:

> Nine o'clock is striking,
> Mother, may I go out?
> My beau is in the garden;
> He wants to take me out.

In Soho, in 1907, the Revd Allen Moxon found girls skipping to the lines, although sometimes instead of rebuffing their young man by singing, 'I don't want your apples, I don't want your pears; I don't want your sixpence for kissing you on the stairs', they would, by implication, accept his offerings, and would themselves contribute to the feast:

> Half-a-pound of bacon, fry it in the pan,
> No one else should have it but me and my young man.

This couplet was still being sung by girls at Dalton-in-Furness in 1957. In Everett, Massachusetts, in the 1920s, children sang while rope-jumping:

> I don't want your apple, I don't want your pear,
> I don't want your fifty cents to kiss me on the stair.
> I'd rather wash the dishes, I'd rather sweep the floor,
> I'd rather kiss a Chinaman behind the kitchen door.

This variant, too, has survived. On the Isle of Lewis in 1975 a 10-year-old was skipping to a version that ended:

> I do not want the apple, I do not want the pear,
> I do not want the sixpenny bit to take me to the fair.
> I'd rather wash the dishes, I'd rather scrub the floor,
> I'd rather kiss the bogeyman behind the kitchen door.

Britain: Golspie, 1892, singing game, 'Mother, the nine o'clock bells are ringing' (E. W. B. Nicholson, *Golspie*, p. 155. A version from Cheshire is also given, pp. 344–5). *St Anne's Soho Monthly*, June 1907, p. 153. A. E. Gillington, *Old Surrey Singing Games* (1909), 24, 'Skipping Rope Song', 'Eight o'clock bells are ringing'. Newport, *c*.1910, 'Ten o'clock is striking, Mam, may I go out? My young man is waiting To take me all about. He took me to the seaside, He took me to the sea, He took me to his father's house To have a cup of tea'. *Rymour Club*, 1 (1911), 153, Edinburgh ring-game, 'Jingle bells are ringing, Mother let me out'. Tune given as an adaptation of 'Hey Rickety Ba-loo, Cock-a-doodle doo'. *London Street Games* (1916), 69. Leeds, *c*.1925, similar to *c*.1910. Enfield, 1952. Aberdeen, 1952 and 1960, 'Jump or ball'. Featherstone, near Pontefract, 1954. Pontefract, 1954, similar to *c*.1910. Matching Green, Essex, 1954, 'Eight o'clock bells are ringing, Mother may I go out? My young man's awaiting To kiss me on the snout'. Dalton-in-Furness, 1957, with tune (*English Dance and Song*, 22, p. 116). Stornoway, 1961 and 1975. Swansea, 1962.
Cf. *Traditional Games*, ii (1898), 179–80, 'Salmon Fishers' at Rosehearty.
USA: St Louis, Missouri, 1895–1900 (*JAFL* 60 (1947), 33). Poughkeepsie, New York, 1926 (*JAFL* 39, pp. 84–5). Everett, Massachusetts, 1929 (*JAFL* 42, p. 306). Washington, Pennsylvania, 1938; Maryland, 1947 (Howard MSS). San Francisco, *c*.1950, starts 'Johnny gave me apples, Johnny gave me pears' (P. Evans, *Jump Rope Rhymes* (1954), 18).

❖

> Green coffee grows on the white oak tops
> And the river flows with brandy;
> Go choose someone to roam with you
> As sweet as lasses can be.

Reported from Forfar for skipping and two-balls. It was probably introduced to the playground from a session of square-dancing, and is of interest only as a curiosity, since the verse keeps recurring in the English-speaking world. The editor of *The Gilwell Camp Fire Book* (1957), for instance, gives a version 'introduced at the 1955 Gilwell Reunion and immediately popular', beginning:

> Love grows under the wild oak tree,
> Sugar flows like candy.
> The top of the mountain seems like gold
> When you kiss your little honey sort of handy.

Ian Turner, in his collection of Australian children's play-rhymes, *Cinderella Dressed in Yella*, gives the following, collected in Melbourne in 1967, as 'a rhyme, apparently of childish origin, of some poetic merit':

> Love grows under a wild oak tree,
>   Sugar grows like candy;
> The top of the morning shines like gold
>   When you have a little fellow kind of handy.

However, should the verse be of juvenile origin, the author was unlikely to have been alive in the 1960s. A Tennessee version was recorded in *JAFL* 2 (1889), 104, as part of a game-song, 'My Pretty Little Pink', the verse in question being, it is true, not so poetic as the more recent renderings:

> Where coffee grows on a white-oak-tree,
>   And the rivers flow with brandy,
> Where the boys are like a lump of gold,
>   And the girls as sweet as candy.

A better version (*JAFL* 28 (1915), 187) portrays this transatlantic land of Cockayne as a place where 'Rocks all shine with a glittering gold, And the girls as sweet as candy'.

The song, as recorded in 1889, began:

> My pretty little pink, I once did think
>   That you and I would marry,
> But now I've lost all hopes of that,
>   I can no longer tarry.

The age of the song is indicated, perhaps, by the rejected lover saying he has now placed his knapsack on his back, his musket on his shoulder, and is about to go to Quebec Town 'to be a gallant soldier'. Newell (1883), in his article on 'My Pretty Pink', said, 'In another version, not gained in full, "Mexico" was substituted for "Quebec"'; and this was exactly the version that Dr Dorothy Howard collected *c*.1950 in Allegany County, New York:

> I'll put my knapsack on my back,
> > My rifle on my shoulder,
> And away I'll go to Mexico
> > To be a Union soldier.

The earliest recording of the 'Little Pink' verse appears to be that in *The Only True Mother Goose Melodies*, published in Boston *c*.1845:

> My little Pink
> I suppose you think
> I cannot do without you,
> > I'll let you know
> > Before I go
> How little I care about you.

> Away down east, away down west,
> > Away down Alabama,
> The only girl that I love best
> > Her name is Susianna.
> I took her to the ball one night
> > And sat her down to supper,
> The table fell and she fell too
> > And stuck her nose in the butter.
> The butter, the butter,
> > The holy margerine;
> Two black eyes and a jelly nose
> > And the rest all painted green.
> My father died the night before,
> > He left me all his riches,
> A box of stones, a feather bed,
> > And a pair of rubber breeches.
> Breeches, breeches,
> > The holy margerine,
> Two black eyes and a jelly nose,
> > And the rest all painted green.

Skipping song in Edinburgh. The words are sung to the tune 'Polly Wolly Doodle', or one very like it, although formerly, or perhaps elsewhere, the tune seems to have been 'The girl I left behind me', otherwise known as 'Brighton Camp'. Despite being a composite song, with verses from at least three sources, the lines have remained together with remarkable steadiness. About 1910 the Duchess of Sutherland observed children in the village schools of Helmsdale and Port Gower singing the following words as they went round in a ring:

> My uncle died a week ago,
>   He left me all his riches,
> A wooden leg, a feather bed,
>   A pair of leather breeches,
> A tobacco box without a lid,
>   A jug without a handle,
> A coffee pot without a spout,
>   And half a farthing candle.
> I travelled east, I travelled west,
>   I came to Alabama,
> I fell in love with a nice young girl,
>   Her name was Susy Anna.

They then all knelt, and continued:

> I took her to the ball one night,
>   And also to the supper,
> The table fell, and she fell too,
>   And stuck her nose in the butter.

The lines that itemize the sobering legacy were already old, perhaps, when they headed a song, 'The Bumkin's Wife; or, Fashion in its Glory', printed in London in the 1820s:

> My father died the other day,
>   And left me all his riches,
> His gun and volunteering cap,
>   Long sword and leather breeches.

The source of the lines telling of Susianna's misadventure, or misbehaviour, when taken to the ball, is unknown, although the incident recurs in versions of the American play-party song 'Turkey in the Straw':

> Took my gal to Old Joe's house,
>   We stayed there for the supper;
> She stubbed her toe on the table leg
>   And stuck her nose in the butter.

But the pedigree of Susianna herself is traceable. 'Black-eyed Susiannah' was one of the songs of the Christy Minstrels. It appeared in a collection of their songs printed in London in 1857:

> I've been to de east, I've been to de west,
>   I've been to Indiana;
> And ob all de gals I lub de best
>   My black-ey'd Susiannah.
> She am black, dat am fac',
>   My black-ey'd Susiannah.

This song had earlier been sung by the Nightingale Serenaders, possibly in Philadelphia, 'Brack Eyed Susianna' being entered 'according to act of Congress' in 1846. Yet the lines are not necessarily wholly American. About 1825 a halfpenny songster was printed in Ireland entitled 'A New Song called Cockibendy'. The song commenced:

> As I came east, as I came west,
>   As I came thro' St Johnson,
> The prettiest lad that e'er I saw,
>   Was a ploughboy dancing;
> He had silk stockings on his legs,
>   Silver buckles glancing.
> A nice blue bonnet on his head,
>   And O but he was handsome.

Even that is not the beginning of the story, for in *Wit and Mirth: or, Pills to Purge Melancholy* (1700), appeared 'A New Scotch Song', of which the tune, if not the words, was composed by Thomas D'Urfey. The first three verses are:

> Walking down the Highland Town,
>   There I saw Lasses many;
> But upon the Bank in the highest Rank,
>   Was one more gay than any:
> I Look'd about for one kind Face,
>   And I saw *Billy Scrogy*;
> I ask'd of him what was her Name,
>   They call'd her *Catherine Logy*.

> I travelled East, and I travelled West,
>   And I travelled through *Strabogy*;
> But the fairest Lass that e'er I see,
>   Was pretty *Catherine Logy*.

> I Travelled East, and I Travelled West,
>   And Travell'd through *Strabogy*;
> But I'd watch a long Winters Night,
>   To see fair *Catherine Logy*.

Popular song: *Paul Pry's Merry Minstrel* (*c*.1825). *Brack Eyed Susianna* (Philadelphia and New York, 1846). Christy's *One Hundred Negro and American Songs* (1857).
  Game: *Rymour Club*, 1 (1911), 192, and 2 (1919), 7. Edinburgh, 1954, 1961, and 1975, mostly skipping, sometimes two-balls. Bellshill, Lanarkshire, 1975.

The night was dark, the war was over,
The battlefield was shed with blood,
And there I spied a wounded soldier,
A-lying dying as he said:
God bless my home in bonnie Scotland,
God bless my wife and only child,
God bless the men that fought for freedom,
A-holding up the Union Jack.

This must surely be a sentimental song from the nineteenth century, but in spite of much searching and asking around among the experts, no original has been found. It seems to be known only in Scotland, and can also be used for two-balls.

Edinburgh, 1949, ending 'God bless my wife and only child And tell the people I am dying For I've won the Union Jack: That's that!' (N. Mclsaac, J. Ritchie, and R. Townsend, *The Singing Street* (1951), no. 21). Aberdeen, played in lundies (with two ropes), 1952 and 1960; two-balls, 1984. Dunoon, 1962. Glasgow, 1975 and 1984, two-balls (recorded for Adam McNaughtan's BBC1 *Collectors* programme, 1984; the tune is 'Bobby Shaftoe').

I've a laddie in America,
I've a laddie in Dundee, aye-ee, aye-ee,
I've a laddie in Australia,
And he's coming home to me, aye-ee, aye-ee.

First he took me to the pictures,
Then he set me on his knee, aye-ee, aye-ee,
Then he went away and left me
Wi' three bairnies on my knee, aye-ee, aye-ee.

One was sitting by the fireside,
The second was sitting on my knee, aye-ee, aye-ee.
One was sitting at the doorstep
Crying, Daddy, please come home to me.

In Aberdeen they play balls as they sing this song, but in general in Scotland and the north of England they caw the ropes, and the skipper may have to turn round at each 'aye-ee'.

Edinburgh, 1949, BBC Record Library, no. 13869; also 1959, and J. T. R. Ritchie, *The Golden City* (1965), 125. Aberdeen, 1960. Glasgow, Moss Park and Bothwell, 1961, 'First he took me for ma wedding, Then he took me for ma tea'. Cumnock, 1961, and another recording in *Those Dusty Bluebells* (1965), 9. Dunoon, 1962. Newcastle upon Tyne, 1966, F. Rutherford, *All the Way to Pennywell* (1971), 38–9. Bellshill, Lanarkshire, 1975. Edinburgh, 1975.

I'll awa' hame tae ma mither, I wull,
I'll awa' hame tae ma mither, I wull,
There's the bairns tae feed and the pirns tae full,
I'll awa' hame tae ma mither, I wull.

I winna' bide wi' ma grannie nae mair,
I winna' bide wi' ma grannie nae mair,
She skelps ma dowp an' she pu's ma hair,
I winna' bide wi' ma grannie nae mair.

Dundee, 1956. Sung to a version of 'The Laird o' Cockpen'. '*The pirns tae full*', the bobbins to fill.

Oh what a life, a weary, weary life,
It's better to be single than to be a married wife.

When I was single I used a powder puff,
Now I'm married I can't afford the stuff.
Oh what a life, a weary, weary life,
It's better to be single than to be a married wife.

One calls Mummy, gie me a piece on jam;
The other calls Daddy, lift me oot the pram.
Oh what a life, a weary, weary life,
It's better to be single than to be a married wife.

Sung by the girls at Dean Orphanage, Edinburgh, in 1955, when playing 'Two Bally'. It was earlier collected at Norton Park School, Edinburgh, by Ritchie, McIsaac, and Townsend, and printed in their booklet *The Singing Street* (1951), no. 14. The girls there sung it while skipping, and had an extra verse:

One shouts Mammy, put me to my bed,
The other shouts Daddy, scratch my wooden leg.

Katie Barety had a coo [cow],
It was yellow, black, and blue;
Open the gates and let it through—
Dance, Katie Barety!

Paisley, 1975, girl, 12: 'Rule. If just one person playing then you jump right through it and do the dance at the end putting in your own name. If more than one the first person goes to "coo" and jumps out. The next person to "blue" and jumps out. The next person to "through" and jumps out. The last person does it to the end, remembering the dance and to put in your own name.'

In Cumnock and Dunoon the heroine is 'Kitsy Katsy'; in Glasgow, 'Katie Bardie' or 'Katie Birdie'. In all three places 'coo' has become 'canoe'. Sometimes the skipper is required to end each repetition of the verse with a different action: in Cumnock 'Kitsy Katsy' must 'dance', 'hop', and go 'out'; in Dunoon she must 'skip', 'dance', 'climb', 'sit', and 'biff'.

Once again girls in Scotland have kept an old song alive in the skipping-rope. 'Katie Beardie had a coo' certainly goes back to 1824, and probably much further (see *ODNR*, no. 98).

> I'll tell Mum of Mary Ann,
> Walking down the street with a nice young man,
> High-heeled shoes and a feather in her hat,
> I'll tell Mum on that saucy old cat.

The game is played differently in different places, with the skipper doing bumps, or skipping with eyes shut, or touching shoes and head in the third line, or with two players running in and out of the rope as if chasing each other. But considering the Edwardian music-hall flavour of the chant and its wide distribution, the similarity of the texts is remarkable.

London Street Games (1916), 94 and 95, 'I'll tell mother, Mary Anne' given as name of song, and 'Little Mary Anne who lives up stairs, With high legged boots and a feather in her hat—That's the way she meets her chap' as part of skipping song. Versions from seventeen places since 1950 including St Helens, Lancashire, *c*.1955, 'I'm going to tell me ma on our Mary Anne, Walkin' down the street with a nice young man; Laced up boots and feathers in her cap, That's the way she gets a chap'. Aberdeen, 1960, for two-balls, 'I am telling my mother on my sister Mary Anne, She's out walking with a nice young man, High-heeled shoes and a feather in her hat, So what shall we do with a sister like that?'
   The chant sometimes continues with 'Policeman, policeman, don't take me'; see following.

> Policeman, policeman, don't take me,
> I have a wife and fa-mi-lee.
> How many children have you got?
> Twenty-four and that's the lot.

Sometimes these lines follow 'I'll tell Mum of Mary Ann' and the players skip to find out the number of children.

London Street Games (1916), 54, 'Policeman, policeman, don't touch me, I have a wife and family. How many children have you got? Five and twenty is my lot'. Leslie Daiken, *Out Goes She* (1963), 27, *re* Dublin, *c*.1929. Farnham, Surrey, 1947. Enfield, 1951. Chudleigh, Devon, 1955. North London, 1972, as text.
   Canada: *Victoria Daily Times Weekend Magazine*, 30 May 1970, p. 8, known in 1940s.

> When I was young I had no sense,
> I bought a fiddle for eighteen pence;
> The only thing that I could play
> Was 'Over the hills and far away'.

This verse was well known to children in the middle of the nineteenth century, was popular for skipping at the turn of the century, and continues, in a different form, to be popular in Australia. The mystery is why it should no longer be popular in Britain.

Britain: c.1855, not for skipping, 'And th' only tune that I could play Was "Nix my dolly, pals, fake away"' (*N & Q*, 11th ser., 11 (1915), 35; also 10 (1914), 515). Wrecclesham, Surrey, 1892, for skipping, 'Dancing Dolly had no sense, She bought a fiddle for eighteen pence, And all the tunes that she could play Was "Sally get out of the donkey's way"'. G. F. Northall, *English Folk-Rhymes* (1892), 'well known throughout the midlands'. Deptford, 1898, similar to Wrecclesham (*Traditional Games*, ii. 203). Walmersley, Lancashire, 1901, 'Old Mrs Bence' (Crofton MS). *London Street Games* (1916), 53, 'Was "Take my dolly and fire away"'. Crewkerne, 1920, 'Punchy Ireland'. M. and R. King, *Street Games of North Shields Children* (2nd ser., 1930), 36. Glasgow, 1952, as text. Birmingham, 1952, 'Charlie Chaplin had no sense'. See also Opie, *Lore*, pp. 107 and 109.

Australia: Canberra, 1955 and 1959; Adelaide, 1957; Sydney, 1957 and 1958, 'Dancing dolly hasn't any sense, She bought four eggs for eighteen pence, The eggs went bad and dolly went mad', etc. (Turner MSS).

New Zealand: Sutton-Smith, p. 80, 'Dancing Dolly', similar to Australia.

> Gipsy, gipsy, living in a tent,
> She has no money to pay the rent;
> The rent man came and threw her out,
> Now she's living on a roundabout.

Ordinary skipping, with a bump on each rhyme-word. But if the words are as follows, from Worcestershire, the skipper runs out at the end, and the next player has a turn.

> Hipsy Gipsy lived in a tent,
> Couldn't afford to pay the rent.
> When the rent man came next day
> Hipsy Gipsy ran away,
> Over the fields and far away.

The chant, which is well known, and as often as not used for ball-bouncing, seems to derive from a traditional taunt. About 1905 the Lancastrian folk-song collector Anne Gilchrist noted in one of her music notebooks (EFDSS MSS G231 Bv) the now-familiar words:

> Gypsy, Gypsy, live in a tent
> Can't afford to pay your rent!

The musical notation gives the falling cadence of a standard jeer; and it seems clear that children today preserve in their play the words with which the Romany used to be tormented.

In the United States and Canada children have been heard chanting:

> Indian, Indian, lived in a tent,
> Indian, Indian, never paid rent.

> Down by the river, down by the sea,
> Johnny broke a bottle and blamed it on to me.
> I told me ma, me ma told me pa,
> Johnny got a licking, ha ha ha.

A favourite in Britain and America for skipping, this song about the joys
of retribution is also used for ball-bouncing and for tacking onto the end
of clapping sequences. Skipping versions from twenty-eight places from
1950, some beginning 'Johnny on the water, Johnny on the sea', and
some beginning 'Over the ocean, over the sea'. Johnny usually breaks
a bottle, or, occasionally, a window. In consequence he may be given
a 'spanking', 'beating', 'thrashing', 'hiding', 'whipping', 'licking',
'leathering', 'skelping' (Cumnock), or 'hammerin''.

USA: Everett, Massachusetts, 1929, 'Down by the ocean in the reeds, Johnny broke a tea pot and
blamed it on to me', etc. (*JAFL* 42 (1929), 305). Abrahams, no. 291, gives many references.

> Up in the north, a long way off,
> The donkey's got the whooping cough;
> Sixteen doctors came and said
> 'Your little donkey's nearly dead.
> The only thing to make him better
> Is salt, mustard, vinegar, pepper.'

Ordinary skipping until the last line, which ends with peppers. The
words appear to go back to the first half of the nineteenth century.
Norman Douglas recorded the first two lines in 1916 as part of a skipping
song, and evidence that this couplet, or one similar, had already been in
existence for some while comes in the early black minstrel song, 'Jim
along Josey', published 1840:

> Now way down south not very far off,
> A Bullfrog died wid de hooping cough,
> And de oder side of Mississippi as you must know,
> Dare's where I was christen'd Jim along Joe.

The writers of Jim Crow songs regularly scrambled traditional doggerels
to their own taste, and it is an open question whether the Mississippi
couplet or the couplet preserved by London schoolchildren is the older.

London Street Games (1916), 92. Staple Fitzpaine, Somerset, 1922 (Macmillan Collection). Versions
from ten places since 1950, England only, including text version from Laverstock, Wiltshire.

> Jumbo, Jumbo, sitting on a dustbin
> Eating mouldy cheese, ha! ha!
> Along came a fly, and hit him in the eye,
> And made poor Jumbo sneeze, ha! ha!

Played in various ways, according to local custom.

Huyton, Liverpool, *c*.1910. Lincoln, *c*.1920, 'Sambo, Sambo, sitting on a rainbow, Eating bread and
cheese, Up came Kelly and hit him in the belly, And made poor Sambo sneeze'. London, *c*.1930.

Sheffield, 1949, as Lincoln. Versions from thirteen places since 1950, five of them having the couplet
'Along came a copper, Knocked him on the napper'.
  Australia: Turner (1969), 56, Sydney, 1956, 'Jumbo, Jumbo, sitting on a matchbox'.

> Pounds, shillings, and pence,
> The monkey jumped over the fence.
> The fence gave way, he had to pay
> Pounds, shillings, and pence.

Text from Workington in 1960. Almost the identical words for skipping
were recorded by Norman Douglas in 1916. He also reported that an
improper version was current; and such a version is certainly current
today. But the chant is not now much used in the rope. It is sometimes
employed for dipping, sometimes for ball-bouncing, and most of all is
repeated simply for its sound and nonsense. The relationship of the verse
to one of the more popular skipping chants in the USA can be traced
through antipodean variants.

London Street Games (1916), 54. Pontypool, 1924, and later common elsewhere, 'Pounds, shillings,
and pence, The cow jumped over the fence, He caught his nail on a rusty nail, Pounds, shillings, and
pence'. Birmingham, *c.*1930, 'Pounds, shillings, and pence, I met a dirty wench, I pulled her hair
and made her stare, Pounds, shillings, and pence'. Worcester, *c.*1930, 'Pounds, shillings, and pence,
I met a Worcester wench, I gave her a kick and made her feel sick, Pounds, shillings, and pence'.
Lingfield, 1939, and London, 1964, 'Pounds, shillings, and pence, A copper jumped over the fence,
He cut his arse on a piece of glass, Pounds, shillings, and pence'. Enfield, 1952, and elsewhere,
'Pounds, shillings, and pence, A copper fell over a fence, He cut his belly on a lump of jelly, Pounds,
shillings, and pence'. Workington, 1960, as text, for ball-bouncing.
  Australia: Canberra, 1961. Cf. Turner (1969), 102, Western Victoria, *c.*1930, 'Pounds, shillings,
and pence, The elephant jumped the fence, He jumped so high he hit the sky, And didn't come back
till the first of July'.
  New Zealand: Collingwood, Nelson, 'used as early as 1903' for skipping, 'Pounds, shillings, and
pence, The monkey jumped the fence; He went so high, he reached the sky, Pounds, shillings, and
pence' (Sutton-Smith, p. 74).
  USA: Cf. the following with Australia and New Zealand. Auburn, Alabama, 1915–16, 'I give
fifteen cents to see the elephant jump the fence; He jumped so high he touched the sky And never
got back till next July' (Newman I. White, *American Negro Folk-Songs* (1928), 249). T. W. Talley,
Negro Folk Rhymes (1922), 116, 'My mammy gimme fifteen cents'. Hence the jump-rope rhyme
widespread in the USA, 'I asked my mother for fifteen cents, To see the elephant jump the fence,
He jumped so high, He reached the sky, And didn't come back till the Fourth of July' (Abrahams,
no. 211).

> On Saturday night I lost my wife
> And where do you think I found her?
> In a lobby kissing a bobby
> And all the kids around her.

Sung by juniors in York City while skipping peppers. Versions of the
song seem to have been commonly shouted about the streets of Liverpool, Leeds, and London in the 1920s and earlier; whilst a forerunner, or

perhaps an adaptation of a yet older song, was noted down by John
Aubrey in the sixteenth century:

> The Deane of Paule's did search for his wife,
>   and where d'ee thinke he found her?
> Even upon Sir John Selbye's bed,
>   as flatte as any Flounder.

The Dean, according to Aubrey, was John Overall, one of the translators
of the Authorized Version of the Bible, whose wife Anne was said to be
the greatest beauty in England. She was also, says Aubrey, 'not more
beautifull than she was obligeing and kind, and was so tender-hearted
that (truly) she could scarce denie any one'. Since Overall married her
about 1607, and became Bishop of Coventry in 1614, dying in 1619, the
verse appears to belong to the early years of the century.

John Aubrey, *Brief Lives* (Oliver Lawson Dick's edn., p. 226). *Mother Goose's Melodies* (Boston,
1833), 31, 'Once in my life I married a wife And where do you think I found her? On Gretna Green,
in velvet sheen, And I took up a stick to pound her'. Edinburgh, *c.*1900, *Rymour Club*, 1, 94, 'I
lost my wife on Setterday nicht, And cudna tell far to find her; Up in the mune, sellin' shune,
A penny the piece they're a' dune.' Second verse ends 'Ahint the pump I garred her jump; Tally
ho, the grinder!'. London, *c.*1910. Liverpool, *c.*1914, 'Follow your mother to the treacle shop,
And where d'you think I found her? . . .' (Frank Shaw, *You Know Me Anty Nelly?* (1969), 35).
York, *c.*1952, as text. Montrose, 1979, for skipping, 'Left, right, a pennorth of tripe, I lost my
mother one Friday night, And where d'you think I found her? Up a lobby, kissing a bobby, With all
her kids around her', which is almost word for word how Rowland Kellett knew it as a child in
Leeds, 1920s.

> Chew, chew, chewing gum,
> Put me in my grave,
> My mother told me not to chew,
> But still I disobeyed.

The chant is not as modern as it sounds. It was collected thus, in an
Edinburgh playground in 1909:

> Chew, chew, chewing gum,
> Brought me to my grave,
> Mother told me not to chew,
> But still I disobeyed.

And in Liverpool, about 1915, the story the girls believed was that a
gravestone existed in Ford Cemetery bearing the inscription:

> Chewing gum, chewing gum, made of wax,
> Has brought me to my grave at last.

Rymour Club, 1 (1911), 237. Stoke-on-Trent, *c.*1948. Kirkcaldy, 1952, as text. Frank Shaw, *You
Know Me Anty Nelly?* (1969), 34.

> Chip, chop, cherry,
> All the boys in Canterbury
> Can't spell 'Chip, chop, cherry'.
> Chip chip, chop chop, cherry cherry,
> Chip chip, chop chop, cherry cherry.

A strange survival of yesteryear, in which the last line is repeated until the skipper is out. A writer in 1890 recalled that in his childhood in the north of Ireland there was a doggerel rhyme 'sung by our Catholic nurses' which ran:

> Chip, chip, cherry,
> All the men in Londonderry
> Couldn't set up
> Chip, chip, cherry.

But even at that time the meaning was unclear. 'In that hotbed of party feeling', he wrote, they used to have a vague notion that it 'covered some allusion to Orangeism and the siege of Derry'.

N & Q, 7th ser., 9 (1890), 312. Enfield, 1951.

> There was a little nigger
> And he would not grow no bigger,
> So they put him in a wild beast show.
> He jumped through the window
> And broke his little finger,
> And couldn't play the old banjo.

'Double-unders on the last word of each line.' A north-country rhyme, best known in County Durham. No recordings further south than Caistor, and none since 1952. It was also used for dipping, and perhaps began as a Nigger Minstrel song. It has been in the possession of children for many years; in a story about a party, in *Choice Chips*, 2 May 1885, p. 5, one of the boys is said to have 'made up this':

> There once was a little nigger
> Who wouldn't grow no bigger,
>   So they put him in a wild-beast show;
> A boy called Tommy Singer,
> Broke the nigger's little finger,
>   So he couldn't play his old banjo.

USA: Long Island, *c*.1935 (Howard MSS). Abrahams, no. 555, one (debased) recording, from 1941.

> Please Mrs Bunny is your bunny coming out?
> With his hands in his pockets and his shirt hanging out?

A chant that was absurdly popular in the 1950s. In one school sixty
children named it amongst their skipping games. The words seem to
have evolved from a more down-to-earth verse that was formerly a
favourite with the young:

> Oh! Mary Ellen is your sister coming out?
> We've a nice young man to walk her about:
> His hands in his pocket and his shirt hanging out,
> Oh! Mary Ellen, is your sister coming out?
>
> (Stalybridge, Cheshire, c.1910)

South Shields, c.1927 (James Kirkup, *The Only Child* (1957), 150). Versions from ten places in
1950s. Bristol, 1960. Wolstanton, 1961. Swansea, 1962. Birmingham, 1971.

> Mother, mother, I am ill,
> Send for the doctor over the hill.
> Doctor, doctor, will I die?
> I don't know, but do not cry.

Repeated while skipping or ball-bouncing or hand-clapping. A correspondent to *N & Q*, 24 December 1864, asked: 'Where is the following
rhyme to be found beginning—

> Doctor, doctor, I shall die!
> Yes, pretty maid, and so shall I.'

He received no reply. Probably the lines come from a song such as the
following, which seems to have been widely known in the latter part of
the nineteeenth century:

> Oh dear mother what a cold I've got,
> Give me tea and toddy hot,
> Roll me up in a big black shawl,
> And take me to the doctor's hall.
> Doctor, doctor, shall I die?
> Yes, pretty maid, and so will I.
> Take this medicine twice a day
> And that will cure your cold away.
>
> (Lanarkshire, c.1900)

In 1892 children in Golspie were found to be hand-clapping both to a
fragment of the song:

> Oh! what a cold you have got
> Come with me to the doctor's shop;

and to the couplet:

> Oh! dear, doctor, shall I die?
> Yes, my lady, and so must I.

In Somerville, Massachusetts, also in the 1890s, one of the reasons children jumped in the rope was to be assured of the magnificence of their funeral:

> Mother, mother, I am sick,
> Send for the doctor, quick, quick, quick.
> Doctor, doctor, will I die?
> Yes, my darling, by and by.
> How many hacks am I to have?
> One, two, three, four . . .

This game was played in Britain and America at least up to the late 1950s. A further, and yet more wry commentary on the medical profession accompanied an American game of running in and out of the rope, played in the 1920s:

> In goes the doctor, in goes the nurse,
> In goes the lady with the big black purse;
> Out goes the doctor, out goes the nurse,
> Out goes the lady in a big black hearse.

These lines, foreshadowed in versions of the singing and skipping game 'All the Girls in Our Town' (see p. 251), became attached to the more consolatory song; and in the 1950s skippers commemorated a lady with an alligator purse who was no longer, it seems, a patient but a health visitor:

> Mother, mother, I am ill,
> Send for the doctor over the hill,
> In came the doctor, in came the nurse,
> In came the lady with the alligator purse.
> Measles said the doctor, Measles said the nurse,
> Measles said the lady with the alligator purse.

The 'In came the doctor' lines are often part of the clapping game 'The Johnsons Had a Baby' (see Opie, *Singing Game*, pp. 472–3).

A skipping and stotting song commencing with the same couplet, which has long coexisted with the above in Scotland, has probably no connection other than in the summons for the doctor:

> Mother, mother, I am ill,
> Send for the doctor over the hill;
> Over the hill is far too far,
> We'll have to buy a motor car;

A motor car is far too dear,
We'll have to buy a bottle of beer;
A bottle of beer is far too strong,
We'll have to buy a treacle scone;
A treacle scone is far too tough,
We'll have to buy a box of snuff;
A box of snuff will make you sneeze,
We'll have to buy a pound of cheese;
A pound of cheese will make you sick,
Run for the doctor quick, quick, quick.

Compare 'There was a man, he went mad' in *ODNR*, pp. 285–6.

Britain: *N & Q* 3rd ser., 6, 514. *Golspie* (1897), 189–91. Chardstock, *c*.1922, 'Oh dear, Doctor, I feel sick, Yes, my girl, you want the stick', for ball-bouncing. Halifax, 1952, 'Doctor, doctor, give me a pill'. Swansea, 1956, 'How many carriages shall I need?'. Stepney, London, 1956 (*News Chronicle*, 30 April). Penrith, 1957, 'No, my dear, so do not cry'. Opie, *Lore*, p. 34. Windermere, 1960. Workington, 1960. Fife, 1974.

　　Scots rhyme: Aberdeen, *c*.1910. Forfar, *c*.1910. Edinburgh, 1949; also J. T. R. Ritchie, *Golden City* (1965), 88. Golspie and Kirkcaldy, 1952. Langholm, 1957, 'Granny, granny, I am ill'. Cumnock, 1961. Glenrothes, 1974.

　　USA: Somerville, Massachusetts, *c*.1895. Cincinnati, 1908, 'Granny, Granny, I am ill, Send for the doctor to give me a pill', for rope-jumping (*JAFL* 40 (1927), 41). Boston, *c*.1925, as quote. Westchester County, New York, 1934 (Howard MSS). *New York Herald Tribune*, 2 August 1964. Abrahams, no. 353, numerous references.

　　Australia: Turner (1969), 25.

　　New Zealand: Sutton-Smith, p. 81.

In China there lived a very funny man,
His name was Chika-roka-chee-chi-chang;
His legs were long and his feet were small,
Poor Chinaman, he couldn't walk at all.
Chiko-chako-chee-chang, on denora,
On denora E-mite-U-M.
Oko-oko, hit him on the boko,
Enemi-enemi-em-pem-pen.
His servants used to carry him about
On their backs, with a merry old shout;
And when they got to the top of the hill,
They rolled him down like a rolling pin.
Chiko-chako . . .

The outlandishness of the refrain is, of course, the chant's chief attraction. A contributor to *N & Q* in 1904 recollected: 'In West Yorkshire, about 1875, the grooms, stable-boys, butchers' lads, and others . . . had a "Chinese" nominy of which they made a good deal of mystery, and the learning of which they considered quite an accomplishment.' There was a translation, he added, 'which rather gave itself away by using a proper name dissimilar from anything in the original':

> Once in China there lived a man,
> His name was Ramo Tamo Tyrie Tan,
> His legs were long, his feet were small,
> Chinee feller couldn't walk at all.

That this was a verse of a comic song, common to both sides of the Atlantic, can be seen in Louisa May Alcott's *Under the Lilacs*, ch. 21, published in Boston in 1878. The young people are entertained with a pantomime performed by potato puppets, who enact the following story:

> In China there lived a little man,
> His name was Chingery Wangery Chan.
> His legs were short, his feet were small,
> And this little man could not walk at all.
> Chingery changery ri co day,
> Ekel tekel happy man;
> Uron odesko canty, ah, oh,
> Gallopy wallopy China go.

The hero Chan pays court to 'Miss Ki Hi' ('She had money and he had not'). He serenades her with a nonsense song, and is deluged with a wash-bowl of water for his pains.

West Yorkshire, *c*.1875 (*N & Q*, 10th ser., 2 (1904), 507). Brixton, *c*.1914. Sidcup, *c*.1920. Bath, *c*.1945. Helensburgh, 1952, 'His sons were merry and his fate was small, He went to a fountain and he lost them all'. Radcliffe, Lancashire, 1952, refrain ends, 'Giddy boys, giddy boys, see him go', and the player runs out. *Norfolk Boy Scouts Song Book* (13th imp., 1962), 25, 'They took him up a great big hill And rolled him down like a Beecham's Pill'. Swansea, 1966, as text.

> In the school, out the school,
> Pass the 'qualie' and that's the rule;
> You silly wee ass, you didna pass,
> *Mary Smith*'s the dunce of the class.
>
>   (Glasgow, 1974)

The 'qualie' was the qualifying exam, the 11-plus exam which meant a child could go on to grammar school. During the mid-1970s this rhyme was popular in Scotland; in fact, said a Dumfries girl, it was 'one we do nearly every day'. At Bellshill, Lanarkshire, and Greenock, 1975, it began, 'Off to school, off to school'.

> Blue bells, cockle shells,
> Here comes the teacher with the big fat stick,
> Now get ready for your arithmetic.
> Two plus two is four.
> Now get ready for your spelling.

C-A-T spells cat,
D-O-G spells dog.
Now get ready for your hot dogs.

Glenrothes, Fife, 1974. 'Hot dogs' are, of course, 'fast caws'. The rhyme has been used indiscriminately for skipping or ball-bouncing, in America and Scotland, since at least the 1940s. Leah Yoffie knew it for ball-bouncing in St Louis in 1944 (*JAFL* 60 (1947), 46–7); Patricia Evans knew it as a jump-rope rhyme in San Francisco, 1945–53 (*Jump-Rope Rhymes*, p. 19), when it began, as above, with 'Bluebells, cockle shells' and the rope swaying (this is not usual, however). It seems to be related to the finger-squeezing trick called 'Playing Schools'—see Opie, *Lore*, p. 59.

IO

Tops and Tipcat

In the late 1960s or early 1970s, both these venerable and ingenious games died away in Great Britain, after giving centuries of pleasure.

TOPS

The top has fascinated the world since the earliest recorded times. In the British Museum are glazed composition tops from Egypt, *c.*1250 BC, which were spun, not whipped, and a clay model of a spinning top from ancient Greece, *c.*750–700 BC, which must surely have come from a grave. Young women, as well as girls and boys, whipped tops in ancient Greece, and vases often depict the activity: for instance, on a lekythos in the Metropolitan Museum of New York, two women are seen vigorously whipping their tops.[1] It is a pity we do not know more about ancient China; it would be characteristic of Chinese civilization to have practised the skill of spinning tops long before the rest of the world.

When a circular object is spun, it will go on spinning because of centrifugal force; the length of time it spins depends on the ability of the operator, the design of the object, and the effect known as 'wobble', or, more scientifically, 'precession'.[2] A sphere, such as a heavy marble, can be spun; but a steadier spin can be achieved if the object has a point to spin on, and all tops have points, either long or short.

Whipping tops have grooves in their straight sides so that they can be kept spinning almost indefinitely if they are lashed with a whip. The whip consists of a stick, to which a thong is attached, of leather (a leather bootlace, in an emergency), or eel skin (deemed best, because of its rough surface), or string, often with a knot at the end for greater purchase; a girl in Pontnewydd emphasized, November 1954: 'You *must* have a good,

---

[1] Helen McClees, *The Daily Life of Greeks and Romans* (1924), fig. 53.
[2] See the excellent article 'Spinning Tops and Their Relevance to Physics', in *The Times*, 2 Dec. 1966, and the follow-up letter on 7 Dec. The letter recommends John Perry's 'Lecture to the Operative Classes' at the Leeds meeting of the BA in 1890, later published in the 'Romance of Science' series as *Spinning Tops*. Perry had seen, in Japan, an old top-spinner's performance, in which he caught one of his great tops on the point of his sword, or sent it running up the handrail of a staircase into a house by the door and out again by the window.

'Le Sabot' ('whipping top') in Jacques Stella, *Les Jeux et plaisirs de L'enfance*, 1657.

'La Toupie' ('peg top') from the same book as above.

strong knot at the end of your string. If you don't you will hit the top, but it will not spin so good, and will die down very quickly.' Or a piece of rope can be used, frayed out at the ends, called a 'slash' in Argyllshire in the late nineteenth century, a 'lash' in Aberdeenshire, 1930s, and a 'scudger' in Cowie, Kincardineshire, 1930s.[3] In medieval days several thongs were used.[4] A whipping top can be set spinning by twisting it between one's hands from a standing position, or by pressing its iron point lightly into the soil so that it stands upright ready to be whipped, or by holding the top in your hand with the stick-end of the lash beside it, then pulling them apart and dropping the top before it stops turning, or by holding it lightly under your heel or knee and pulling ('but this was the girls' way'). Whipping a top was often said to be for exercise, or for keeping warm on a cold day ('Toppe and scourge to make him hott', Samuel Rowlands, *Letting of Humour's Blood* (1600)). Richard Mulcaster, the first headmaster of Merchant Taylors' School, recommended the top as an exercise in his comprehensive *Positions wherein those primitive circumstances be examined, which are necessarie for the training up of children, either for skill in their booke, or health in their bodie* (1581), 79. Boys at public schools continued to whip tops (and trundle hoops) as a recreation until well into the nineteenth century. The town, or parish, top, 'kept in every village, to be whipped in frosty weather, that the peasants might be kept warm by exercise, and out of mischief, while they could not work', was much referred to, especially by dramatists, in the early seventeenth century.[5] Keeping the top spinning is obviously an end in itself, but *The Boy's Own Book* (1828), 11, points out that one can also play 'Races', or 'Encounters' (knocking down other boys' tops).

Playing with tops is usually thought of as a boys' game and it was at one time considered a hoydenish thing for a girl to do. In *The History of Tom Noddy and his Sister Sue* (*c.*1800), Sue found her brother's top, and 'had a curiosity to make use of this play thing, tho' it was not fit for a girl'. She was discovered whipping it, and 'for this she was herself severely whipped by her aunt'. Yet at about the same period Jane Taylor, aged

---

[3] Maclagan, *Argyleshire*, p. 243; letter to *Aberdeen Press and Journal*, n.d.; E. Christie, *The Empty Shore* (1974), 47.

[4] See e.g. lower border of f.64ʳ, MS Bodl. 264, illuminations finished 1344. The ordinary medieval whip was of this type, whether used for whipping tops or driving animals.

[5] Steevens, note to *Twelfth Night*, I. iii, 'He's a coward and a coystril that will not drink to my niece till his brains turn o' the toe like a parish-top.' Also Beaumont and Fletcher's *Thierry and Theodoret* (1621), ii. 2, 'scourge him hither like a parish-top, and make him dance before you', and *Night Walker* (1640), i. 3, 'dances like a town-top, and reels and hobbles'; Jonson, *New Inn* (1629), ii. 5, 'A merry Greek . . . spins like the parish-top.'

SONG XXI.

On the whipping of Tops.

I.

SEE the Tops on the pavement, they twirl
and they bound, (the ground;
And swift is the course which they take on
The lads all pursuing, each doubles his blow,
And the faster they scourge them, the better
they go. II.
If once the whip ceases to urge their career,
These little gay play-things will heavy appear,
'Tis the lash, when well follow'd, that makes
them to spin,
And the boy that leaves striking, again must
begin. M 2 III.

Tom lov'd playing at top,
And often would stop
To have a long spin in the street,
Though he knew it was wrong
To daudle along,
And fear'd when at school to get beat.

Left: Boys energetically whipping tops in *A Choice Collection of Hymns and Moral Songs ... By several Authors* (chiefly Isaac Watts and Thomas Foxton) published by and for T. Saint, Newcastle, 1781. The lines 'If once the whip ceases to urge their career, These little gay play-things will heavy appear' are followed by the obvious moral.

Right: From *The History of Little Tom Tucker* (London: Printed for the Booksellers), *c.*1835.

about 12, used to whip a top while composing her tales and dramas.[6] Later, the general rule seemed to be 'peg tops for boys, whip tops for girls'.

A great part of the enjoyment of playing with tops, especially for girls, was the custom of colouring patterns on the top, in concentric circles, or in sections:

Sometimes you could get a striking pattern; at others a combination of colours you thought rather good turned out to be quite ordinary. You used coloured paper—sometimes silver paper as well—or you could use crayons. You couldn't do this with one of the old superior polished tops, it had to be a cheap one. The individual patterns also served to identify a top, so that you could pick out your own by the markings.   (Ilminster, Somerset, c.1910)

This notion is far from modern, for in *Nancy Cock's Pretty Song Book* (c.1780), is a verse on a 'Peg-Top' which ends, 'And on its smooth and well turned head, Are painted yellow, blue, and red.' Dr Johnson's friend Mrs Piozzi, writing from Italy in July 1786, used this analogy to illustrate the too-many impressions assailing a traveller:

The more one sees of different Places & People, the less Effect has that diversity upon one's Mind—If you take a Boy's Top, and paint it in stripes of red, Blue, Green, & Yellow only; whip it merrily round—& the general Appearance will be *white*.[7]

Coloured patterns on tops were later used to demonstrate or test scientific principles. In R. Routledge (ed.), *Science in Sport* (1877; 1893 edn., p. 71) the father of the family explains the refraction of light, and shows the children how the colours of the spectrum can be made to blend into a near-white if arranged in the proper proportions and spun on a humming top. Tops made of variable coloured discs were used in experiments on light and human vision; and by anthropologists recording skin-colour in Polynesia.[8]

Whip tops were for sustained spinning, often as an amusement on the way to school. Peg tops (once also called 'casting tops') are far more versatile instruments; pear-shaped, with a long iron 'peg' or foot, they were chiefly used in a game of aggression that has affinities with conkers and ring marbles. The top was set in motion by means of a length of tough cord wound round the grooves of the wooden section. 'Winding a

[6] E. V. Lucas, *Original Poems* (1903), p. vii.
[7] *Thraliana: The Diary of Mrs Hester Lynch Thrale (later Mrs Piozzi)*, ed. K. C. Balderston (1951), ii. 657.
[8] *Scientific Papers of James Clerk Maxwell*, ed. W. P. Niven (1890), pt. 1, 127; anthropologist correspondent.

top was an art in itself,' said Ted Willis, remembering his boyhood in
south Tottenham, *c*. 1925. 'The free end of the cord was held between the
thumb and finger, with a small knot [or trouser button, or flat leather
disc] to aid the grip; then the top was hammered to the ground with a
high wristy action, its peg foremost, while at the same moment the cord
was jerked away.'[9] Each boy launched his peg top into a ring drawn on the
ground, hoping to knock over an opponent's top while it was still spinning, or split an opponent's top lying 'dead' in the ring (when the peg of
the split top was kept as a trophy). Holloway, in his *Dictionary of Provin-
cialisms* (1838), 113, explains that 'if the top, when it has ceased spinning,
does not roll without the circle, it must remain in the ring, to be pegged
at by the other boys', or the player 'redeems it by putting in an inferior
one, which is called a "Mull". When the top does not roll out, it is said
to be "Mulled".' In another game, played for instance in Bermondsey,
south-east London, *c*. 1910, players tried to edge a top, laid on the ground
by the boy who had lost the toss, towards a ring or hole; and the owner
of the last target-top was allowed 'grudges' against the owners of the
previous target-tops—a 'grudge' being the right to jab holes in the
previous target-tops with the iron peg of his own, wedging them into a
crevice in a wall to obtain a better purchase.[10] 'Grudges' was one of the
many games with tops listed by Norman Douglas in *London Street Games*
(1916), 7–8.[11] *The Boy's Own Book* advises the use of a top with a long
peg, 'as it is more calculated to swerve out of the ring after it is spun; a
top that sleeps after it is cast runs the greatest danger, and those that
sleep most are heavy bodied tops with short blunt pegs.' A top is said to
'sleep' when it spins so steadily it is perfectly motionless, hence the
expression 'to sleep like a top'.

   The ability of the peg top to drive other objects before it created many
minor games. 'Chipstone' was the best known. Two lines about 6 feet
apart are marked upon the ground, and some small stones ('not larger
than a bean') are placed midway between the lines. Two boys set their
tops spinning, pick them up on a wooden spoon, and then, with the top

---

[9] *Whatever Happened to Tom Mix?* (1970), 37–8. One of James Beresford's *Miseries of Human
Life* (1806), 47, was the remembered misery of 'Winding up a top badly grooved, so that the string
buckles down over the peg; and, on your attempt to peg it down into the ring—"*volat vi fervidus
axis*"—i.e. it flies into the eye of a play-fellow.'

[10] Is this what young Nehemiah meant (*double entendre* apart) in Richard Brome's *New Academy*
(1659), ii. 2? When he is told he must take a wife he asks, 'Will she play with me at peg-top? . . . And
she ha' not good box and steel, I shall so grull her.'

[11] An example of the continuity that exists in children's games is the expression 'Live O's', found
in Douglas's list. More than a hundred years before, the boys of Sedgley Park School in Staffordshire talked of 'laying a "live O"', meaning to put a spinning peg top onto the ground, probably from
a spoon (F. C. Husenbeth, *History of Sedgley Park School* (1856), 105).

still spinning in the spoon, throw the point of the peg against a stone, so as to chip it out of bounds. While the top continues to spin, a boy may take it up on the spoon as many times as he can. The poet Edward Thomas, who 'played endlessly' with peg tops as a boy in Battersea, c.1885, sometimes ' "chipped" brass buttons or other tops into or out of a ring by taking [his] own top on to the hand while spinning, causing it to sway upon a fixed centre hollowed in the palm, and then casting it at the object with all the rotary force thus gained.' These skills had been known for a long time. In *A Little Pretty Pocket-Book* (1744), a peg top drives a coin out of a ring; in a late eighteenth-century engraving after F. Whatley, one boy spins a peg top on his hand, another spins his on a spoon. (Special wooden spoons were, indeed, sold at toy shops for the specific purpose of taking the top 'up to "sleep"' (*Boy's Own Book* (1828), 12).) The earliest depiction of a peg top is in Brueghel's 'Fight between Carnival and Lent', 1559; it is very probable that those boys in Flanders knew all about chipping coins and stones, and obtaining extra momentum for their tops by spinning them on hand or spoon.

A thesaurus of the names of tops will show how common they were, and how diverse was their shape: big ben (large whip top, Derbyshire, 1905); birken (a top made of birch wood); boxer (a whipping top or peg top made of box wood); carrot top (a whipping top shaped like a dumpy carrot or a large bullet); castle-top (a peg top, ? *ex* 'casting-top'): common in Ireland and in England from Derbyshire northwards. When an edition of Mary Ann Kilner's *Memoirs of a Peg-Top* was published in Dublin c.1785, it was retitled *Memoirs of a Castle-Top*); Colchester top (ancestor of the mushroom top, and rather more squat, pictured in *The Boy's Own Book* (1829), 13, 'a local variety of the whip-top . . . too singular for us to pass unnoticed'); dummy (same as the carrot top; could the name derive from the 'dum-dum' bullet, or from Norman Douglas's 'dumb-bargee'?); farthing-kicker (same shape as the mushroom); gig (a hollow top, usually made of horn; fifteenth century onwards); granny (same shape as carrot); horney-top (made from the tip of a cow's horn); jinny spinner (like carrot, Penrith, 1957); monkey top (like mushroom, London 1908); mushroom (shaped like a mushroom with a long stalk and a hob nail in the bottom, usually made of white wood, with red and blue rings painted on; 'a farthing each—start it by wrapping the string round, fling it out and keep it going by whipping it, it needs a light stick with a light piece of string, otherwise you'd knock 'em off balance, see?'); peerie, pirie, or pear (a peg top, standard Scottish); racing top or racer (same as mushroom); scopperil (made from a button); scourge tops (Hawkins, *Apollo Shroving* (1626); and *Longman's Magazine*, March

The great F Play.

PEG-FARTHING.

SOON as the Ring is once compos'd,
  The Coin is in the Centre clos'd;
And then the wish'd-for Prize to win,
The Top that drives it out must spin.

RULE of LIFE.
Be silent if you doubt your Sense,
And always speak with Diffidence.

KNOCK

PEG-TOP.

"I'll tell you what," said little Hal, one day,
"When I'm a man, I'll be a turner gay:
See now, this top of mine its peg is lost,
To buy a new one will more money cost
Than I can give: because you know it took
My bright half-crown to get poor Tom a
    book;
And then there's Lucy's doll put quite
    away,
Because I broke its leg the other day:

Left: 'The great F Play—PEG-FARTHING' in John Newbery's *A Little Pretty Pocket-Book*, 1744 (1767).

Right: From *Sports of Childhood* (Harvey and Darton), 1839. A peg top could be kept spinning on a wooden spoon sold for the purpose, or on the player's hand.

1889, p. 516, referring to Poole, Dorset, c.1820); Spanish peg top (made of mahogany: 'the peg is very short . . . and rounded, not pointed, at the end . . . they spin best when set up under-handed . . . for playing on flooring or pavement, they are much superior to those made in the English fashion . . . although totally unfit for "Peg in the ring"', *Boy's Own Book* (1828), 14); top (found as 'top' in Old English, c.1060, and c.1325 as 'toup'); totum (a small top for spinning on the table or floor, often homemade from a reel or pirn: E. Christie, *The Empty Shore*, p. 47, referring to Cowie, Kincardineshire, 1930s); turnip top (a real turnip shape, with a knob on top); window-breaker (self-explanatory; the shape is the same as the mushroom).

The game of casting peg tops did not survive the First World War; the game of whip top was nearing its end in the 1960s. It was certainly widespread in the 1950s. ('Most children in March buy whips and tops ready for the whip and top season. You can whip them on the flat surface of High Street. You can have little races down the road': girl, 11, Monmouth, 1954.) Red and blue tops with reinforced tips could be bought in Woolworth's for 3d., in 1955; and whips, also for 3d. By the end of the 1950s they were no longer on sale. A correspondent told us, 1963: 'Whip and top is still played furiously in Yorkshire, around Shrovetide. I purchased whips and tops in Yorkshire chain stores and imported them here [Hyde, in Cheshire] because identical chain stores in Lancashire and Cheshire hadn't them in stock.' Wooden whipping tops were on sale in at least three places in Oxford in 1967. The proprietor of one shop bought them 25 gross at a time. The Worcester wholesale branch of W. H. Smith did not handle tops, but a spokesman said (February 1969), 'They must sell somewhere, we get enough fuss about them in signals—whether there is any lead content in the paint, and whether it comes off and is dangerous and so on.' J. T. R. Ritchie says (*Golden City* (1965), 68), 'the peerie still birls cheerfully in the quieter back streets [of Edinburgh]'. Some children were still decorating their tops and whipping them in Sneyd Green, Stoke-on-Trent, in the spring of 1966; and in Workington, Cumberland, 1968, during the top season it was said that 'Life comes to a standstill. There is nothing else going on. Every street has got them.' An article by a teacher, in *The Guardian*, 27 August 1969, recorded the punctual arrival of the top season in Leeds:

One day a pandemonium of tag, football, and British bulldog, and overnight chaos transformed into an atmosphere of intense concentration, and activity that is considerably more static . . . I watch with fascination the skill and artistry with which boys handle their whips as, with a dextrous flick of the wrist, one spins his own top away from that of another boy, leaving the rival top untouched and still spinning.

Generally, the game seemed still to be well known in the streets of northern cities in the early 1970s, especially in Aberdeen. In some schools, however, not everyone would play whip and top; it might be the province of a handful of enthusiasts. A boy from Tighnabruaich, Argyll, said (July 1966), 'Five or six people at my school do it. You can buy the tops in town.' Numbers of children said their grandmothers had shown them how to whip a top (for instance, a 9-year-old boy in Guernsey, 1962, became quite proficient after 'a couple of tops tucked away in a cupboard helped to pass a rainy afternoon') but even if they took a top to school and showed the other children how to whip it, the game did not catch on.

Tops have not vanished from sight altogether. One can still buy small decorative spinning tops, of wood or plastic, large metal humming tops, magnetic tops, gyroscopic tops, or holograph laser tops, for the pleasure of private spinning. Peg tops have survived only in boxed games, such as the 1972 'Battling Tops', in which the aim was to drive opponents' tops out of an arena. Whips and tops can be bought (along with other bygone toys like diabolos), in craft shops and the gift shops of stately homes. The problem is that the knack of spinning them has been lost; and when the nostalgic gift has been given, the grandparent is mortified to find he is unable to show his grandchild how to spin the thing.

TIPCAT

The essence of the game of 'Tipcat' is the 'cat' itself, a piece of wood about 6, or 4, or 2 inches long, and about as thick as one's thumb; it is sharpened at both ends, so that when struck on one or other point with a cudgel it rises in the air ready to be hit. (It can also be used instead of a ball in games similar to cricket and rounders; hence *any* game in which a 'cat' is used may be referred to as 'Cat', making it difficult to know which game is meant.) This ingenious gadget has been around for some time. Mr Flinders Petrie found wooden 'tip cats' among the ruins of Rahun, Egypt, which flourished about 2500 BC, and when we lived in Alton, Hampshire, in the 1950s, we found a homemade tipcat in the gutter outside our house.[12] The basic tipcat game was best described by Joseph Strutt, in *Sports and Pastimes* (1801), 86:

A large ring [is drawn] upon the ground, in the middle of which the striker takes his station; his business is to beat the cat over the ring. If he fails in so doing so he is out, and another player takes his place; if he is successful he judges with his

[12] See *JAFL* 4 (1891), 233n; one of these Egyptian tipcats is in the Museum of Archaeology, University of Pennsylvania. In Woolworth's, about 1953–5, a WHIZZ TIP-CAT SET could be bought for 6d. The bat was shaped like a cricket bat, about a foot long, with red binding on the handle; the 'cat' was about 4in. long, and painted bright red.

eye the distance the cat is driven from the centre of the ring, and calls for a number at pleasure to be scored towards his game: if the number demanded be found upon measurement to exceed the same number of lengths of the bludgeon, he is out; on the contrary, if it does not he obtains his call.

Was this the game of 'Cat' being played by Bunyan when he heard a voice from heaven? Bunyan 'fell in very eagerly with the Religion of the times' soon after marrying, in 1649. One morning in Elstow church, near Bedford, the parson spoke of the evil of breaking the Sabbath, and Bunyan 'felt what guilt was'; but after dinner he 'shook the Sermon out of my mind' and 'to my old Custom of Sports and Gaming I returned with great Delight':

As I was in the midst of a game at Cat, and having struck it one blow from the Hole, just as I was about to strike it the second time, a Voice did suddenly dart from Heaven into my Soul, which said, *Wilt thou leave thy sins and go to Heaven, or have thy sins and go to Hell?* At this I was put to an exceeding Maze. Wherefore, leaving my Cat upon the ground, I looked up to Heaven, and was as if I had, with the eyes of my understanding, seen the Lord Jesus looking down upon me, as being very hotly displeased with me.[13]

Classic tipcat survived most notably in 'Nipsy', played as a league game in Yorkshire and Lancashire, especially by miners. The 2-inch nipsy, which is tapered only at one end, is teed up on the edge of a building brick, flipped into the air with a blow on the pointed end with the 3-foot nipsy stick, and hit as far as it will go. This is 'not just a case of brute strength, it is as skilful as golf,' said a Mr Joseph Cook of Bentley, near Doncaster, who claimed to have set up a world record by striking the nipsy a distance of 208 yards (*Yorkshire Post*, 3 July 1962). After the strike has been made, the striker challenges an opponent to reach the nipsy in so many strides (called 'jumps'). If he succeeds, he counts the number of strides as points to his credit. If he fails, the striker adds the points to his own score. A revival of the league game in 1950 resulted in a 600-man strong 'South Yorkshire Nipsy League', with teams (roughly seven a side) raised by local pubs (*News Chronicle*, 15 September 1952, and subsequent correspondence); and another revival took place in 1959 under the name of the 'Barnsley and District Nipsy League'; a demonstration game played at Wombwell, near Sheffield, was shown on television in the *World of Sport* series, on 23 October 1959 (*Yorkshire Post*, 12 October 1959). The enthusiasm did not last, however. The landlord of the Pheasant Inn, Monk Bretton, Barnsley, an inn which

---

[13] *Grace Abounding* (1666), paras. 20–2.

The little l Play.

TIP-CAT.

THE *Gamester* here his Art displays,
And drives the Cat a thousand
Ways;
For should he miss, when once 'tis toss'd,
He's out—And all his Sport is lost.

RULE *of* LIFE.

Debates and Quarrels always shun;
No one by Peace was e'er undone.

FIVES.

Left: 'The little l Play—TIP-CAT' in John Newbery's
A Little Pretty Pocket-Book, 1744 (1767).

Below: Tipcat in a London street, showing why the
game was banned. *Phil May's Gutter-Snipes*, 1896.

once had its own team, said (1996) that nipsy had not been played in the locality 'for about thirty years'.

Another method of playing somewhat resembles rounders (and base-ball). Players do not, however, run round and score rounders individually; instead, each player guards a hole, and at a given signal they all change holes. It does not use the tipcat's special properties, and although called 'Cat(s)' is more often played with a ball. In *The History of Cotton College*, Canon W. Buscot gave the rules of 'Cat' as played at Sedgley Park, *c.*1795, the game having being introduced by masters who had played it at Douai. The arm–long cat-sticks were made by the students, and had bottle-shaped ends about 5 inches long. The ball was 'a little larger than a golf ball'; it had 'a core of lignum vitae, around which hemp was tightly twisted, and dipped into tar, and then more hemp and tar till it was of the required size, and was covered with sheepskin tightly drawn and sewn':

There was a ring about 25 yards in diameter with seven holes at intervals, each having a special name, one called the striker's hole; each of the players in the 'In' side stood at a hole, while of the players on the 'Out' side, one was a feeder at the striker's hole, another stood close by the ring, and the rest occupied special places in the field which was of considerable size. The game began by the feeder tossing the ball from the striker's hole to a height indicated by the striker who drove it through the ring into the playing field. The players then ran round the ring clockwise, a caller giving out the number of holes to be run, usually 3, 5, or 7: the player who landed on the striker's hole becoming next striker. When two rounds and five holes had been run they were entitled to try for a *Cross*. The player at the striker's hole was allowed 3 attempts to hit the ball far enough to enable the 'In' side to run to the centre of the ring, to touch their sticks together, and to get back to their holes before the ball was returned and placed in one of the holes. If they succeeded they scored one cross; if they failed the whole side was out, and the players changed places.

This was the game of 'Cats' played at Ushaw College, in County Durham, until the school was closed; now that only the seminary remains the game is played once a year, at the Old Boys' reunion in July. The game was by no means confined to Roman Catholic colleges, though. Strutt describes it as an alternative method of playing 'Tipcat' (pp. 86–7). Edward Moor's *Suffolk Words* (1823), gives it as *Kit-Cat*; and Jamieson (1825), as 'Cat i' the Hole'.

'Cat' was the most common name.[14] F. J. Furnivall points out, in

---

[14] Also in Roman Catholic schools, where it had been imported from France. S. Luce, *La France pendant la guerre de cent ans* (1890), 124, alludes to an act of 1347 in which the game of 'jeu du chat' appears. The name 'cat' spawned many variations. See, in *OED*, 'cat's pellet' (1609; 1648), 'cat's play' (1668); in *EDD*, 'cat and dog', 'cat and kittens', 'cat-frat', 'cat-kidney', 'catty'. 'Cat and dog' is

N & Q, 4th ser., 3 (1869), 169–70, that it appears in a fourteenth-century poem by William of Nassyngton, the *Myrrour of Lyfe*: 'play some tyme ate pe spore, Atte pe beyne, and ate pe cate'. Florio, in his *Dictionarie of the Italian and English Tongues* (1598), has '*Lippo*, a trap or cat, such as children play at'. *The Windsor and Eton Gazette*, 6 March 1626, p. 4, refers to 'playing at Catt in the Parke medow'. Randle Holme listed, among the 'recreations and sports . . . used by our country Boys and Girls' in *The Academie of Armory* (1688), iii, pt. 16, para. 91, 'Kat and Kinney, or striker or cat stick and tip cat': Holme was a Cheshire man. The game was a gift to the bawdy London playwrights of the seventeenth century: the 'cat' and 'stick' are much mentioned by a foolish character in Thomas Middleton's *Women Beware Women* (1622); and in Richard Brome's *New Academy* (1659), the teenage Nehemiah is discovered, ii. 2, 'at his exercise of Armes, With a new Casting top [and] a Cat and Catstick'. 'Cat' continued to be a common name for the game. 'Tipcat', which became the standard English word, was familiar especially in London, the Home Counties, and the Midlands; variations included 'Tibbet', London, 1908; 'Tibby Cat', London, East End, *c*.1910–39; 'Tick-tack', Leicester, *c*.1925.

'Trippet' is another old name, included in an Anglo-Latin lexicon of *c*.1440: 'TRYPET. *tripula*, *trita*.[15] The Revd William Richardson listed 'Trippet' as one of the games he played as a boy in Cumberland *c*.1760.[16] The 1820s quotations, for 'Trippet and Coit' in Brockett's *North Country Words* (1825), and 'Trip' in Hunter's *Hallamshire Glossary* (1829), show the 'trip' (i.e. cat) to have had one point, not two.

'Piggy' or 'Peggy' is predominantly a northern name; its first notice in print in 1861, in C. C. Robinson's *Dialect of Leeds*, does not rule out its being very much older. 'Piggy', traditionally played by Lancashire

the name known particularly in Scotland and in Wales (and its borders); it was sent by many Opie correspondents, including boys from Kirkcaldy and Forfar in 1952 and 1954; however, Jamieson's *Dictionary of the Scottish Language* (1808), ii, describes 'Cat and dog' as a game for three players, in which a bowler tries to get the cat into the holes defended by the batsmen—in fact, an early form of cricket. (This was the same game Lady Gomme observed in Barnes, London, under the name of 'Cudgel'.) In *The Life of the Scotch Rogue* (1722), 7, 'Cat and Doug' is listed among the sports which kept the rogue from his book. 'Cat and bat' (or 'Catty and Batty') was another old-time name given by Scottish correspondents. 'Cat and buck-stick' was played in Wigton, Cumberland, *c*.1880; 'Wop cat' in Reach, Cambridgeshire, *c*.1890; 'Cat and cudgel' in Sundon, *c*.1900 (*Bedfordshire Magazine*, 7, no. 53 (1960)); 'Catty' in Barry, *c*.1900, and Milford Haven, *c*.1920; 'Gât a ganstick' (Cat and cane-stick), in Llangeler, Carmarthen, *c*.1905; 'Kitty cat' or 'Cat and kitten' in Guernsey, *c*.1920; 'Cat and Bat' in Shetland, *c*.1940.

[15] *Promptorium parvulorum sive clericorum, lexicon Anglo-Latinum princeps* (Camden Society, 1843–65), 503–1. It is illustrative of provincial dialects of East Anglia.
[16] Revd Hunter's MSS notes in Brand's *Popular Antiquities*, BL Addit. MSS 24545, p. 29.

miners, 'especially during the Depression when there was nothing else for the men to do all day', is played, like Nipsy, in proper tipcat form, complete with the challenge to an opponent to reach the piggy in so many strides. This, too, had its revival, in Bolton in 1959, when the men who drank in the Rose Hill Hotel challenged the men of the Waggon and Horses, and soon the Farmer's Arms and the Egerton Arms had flung down the gauntlet as well (*Manchester Guardian*, 13 April 1959). When the Sports Advisory Council at Turton, near Bolton, planned a contest to find a 'piggy' champion as part of the urban district's centenary celebrations in August 1972, they were disconcerted to find there were at least half a dozen different sets of rules (*Daily Telegraph*, 4 June 1972).

The game seems to have died out entirely in the mid-twentieth century. Descriptions were received from schoolchildren from only eleven places, although 'Tippet' was listed as a 'favourite' game by very many boys in Swansea, 1960 (the aim was simply to hit the tipcat as far as possible, as it also was in the game of 'Tip cat' in Bristol, the same year). The other places were Kirkcaldy ('Cat and Dog'), 1952; Forfar ('Cattie and Doggie'), 1954; Newbridge, Monmouthshire ('Bat and Catty'), 1954; Spennymoor ('Tip Jack'), 1960; Perth ('Tibby'), 1960; Barrow-in-Furness ('Guinea Pig'), 1960; Barnsley ('Peggy'), 1961, ('Nipit'), 1962; and Oldbury, Worcestershire ('Tip Cat'), 1970. In Kirkcaldy, Spennymoor, and Perth, the game was played on the road, and the 'cat' (or 'tip jack', or 'tibby') had numbers on the sides. An 11-year-old boy in Perth explained:

You play this game on a condie [drain cover]. You hit the wood with the bat and see what number it lands on. If it lands on two, hit it twice. The other boy must try to throw the wood on the condie. If he misses the batter will see how many baby steps to the condie and that's his points.

Although it is sad that this excellent game has vanished from the British sports scene, it has far from disappeared from world sport. The game of tipcat seems to be known in every country on earth, even though in some it is said to be dying out; travellers passing through Asia, Indonesia, India, Russia, South Africa, and Australia, will find their enquiries met with comprehension, and may, in the springtime, see most of the male population of a town engaged in the game—as, for example, in Bukhara in April 1973.

Here should be mentioned two games allied to tipcat, but played with balls; they are very similar, and both employ a sort of machine to lift the ball into the air so that it can be hit.

KNURR AND SPELL, OR SPELL AND KNURR

This was popular with the coal-miners and mill-hands of Yorkshire and Lancashire. The game was still thriving in the 1960s, and has probably never really died out. *The Yorkshire Post*, 8 April 1991, reported the 'first world championship since 1979' at Otley on the 7th, a Sunday.

The 'knurr' is a small ball, originally a natural 'knurr', which is a hardened excrescence on the trunk of a tree, and later made of wood, or of clay or porcelain (when it is called a 'pottie'). The simplest form of 'spell' is a wooden, gallows-like affair, about 2 feet tall; the upright part is stuck in the ground, the arm is adjustable on the upright, and from it hangs a cord at the end of which is a loop just large enough to hold the knurr.

In the game's heyday, *c*.1850–1930, it was felt essential to have a specially made 'spell', foundry-made from iron and steel. This was described in Routledge's *Every Boy's Book* (1897), 566–7:

It has a horizontal base of wood or iron, usually eighteen to twenty-four inches in length, and made fast to the ground by two or more iron spikes. To this is attached a steel spring, having near its free end a little cup to receive the knurr. At a few inches' distance from its fixed end is a thumbscrew, by means of which the extent of its rise, and consequently the force of its throw, can be regulated. At the opposite end is a ratchet with three or four teeth, pivoted at its lower extremity. The spell is set by pressing down the spring, and engaging its free end in one or other of these teeth. The ball is then placed in the cup. A tap on the upper end of the ratchet frees the spring, and throws the ball in the air.

The bat, or 'tripstick', or 'pommel', is a whippy stick about as long as a golf-club, with either a flat boss at the business end, or a mallet-like head smoothed at one side.

The name 'Spell and Knurr' must have been a frequent alternative to 'Knurr and Spell' long before 'Knurr and Spell' actually appeared in print, or it could have been the earlier form; the Revd William Richardson played 'Spell and Orr' (Norr) as a boy in Cumberland *c*.1760.[17] By the time Strutt described the game (1801), 'Knurr [or Nor] and Spell' had already been corrupted to 'Northern Spell', and was thus titled in the nineteenth-century games books, since they all copied him— and each other. *The Boy's Own Book* (4th edn., 1829), 20, used his description and put forward his opinion that 'this pastime possesses but

---

[17] MS note in Brand's *Popular Antiquities*, ii. 29 (BL Add. MS 24545). Grose, in the 2nd edn. of his *Provincial Glossary* (1790), has 'Spel and Knor. The game of trap-ball. North', and J. T. Brockett, *North Country Words* (1846), gives 'Spell and Ore', from Durham. Recordings of regional names are quite fortuitous; 'Knurr and Spell', was not recorded until 1861, as 'Knor and Spell' in C. C. Robinson's *Dialect of Leeds*.

Above: 'Knur and Spell' being
played on the Yorkshire moors,
in George Walker's *Costumes of
Yorkshire*, 1813.

Right: Yorkshire 'Knur and Spell'
players, 1972.

Fifteenth-century youths playing a game like Trap Ball, in the Bodleian Library's MS Douce 62. The pivot of the trap is fastened into a waist-high post.

'TRAP BALL, Played at the Black Prince, Newington Butts', a print published by Carrington Bowles, c.1750.

little variety, and is by no means so amusing to the by-standers as cricket or trap-ball'. In the later books this opinion became stronger: 'It is a very dull, heavy, uninteresting pastime, when compared to trap, bat, and ball'. Men, however, found it highly entertaining. The game is played between individuals—each man for himself; wagers were placed on local champions and money prizes were given (the championship at Stainland, Yorkshire, on 31 March 1970, was offering £200 to the winner: *The Times*, 1 April 1970, p. 10).

TRAP BALL, OR BAT, TRAP, AND BALL

The 'trap' is a wooden object, shaped like a seventeenth-century shoe, with squared-off toe. A pivoted trigger—like a long spoon—operates within the trap. A ball rests on the 'bowl' of the spoon, in the heel end, and the other end of the trigger is thereby slightly raised. The player knocks the raised end sharply downwards with his bat (a bat like a ping-pong bat, though with a longer handle); the ball flies up in the air and is driven to a distance. A fifteenth-century drawing in MS Douce 62, f.122^r, shows two youths playing 'Bat and Trap', where the pivot of the trap is fastened into a waist-high post; a woodcut in John Newbery's *A Little Pretty Pocket-Book* (1744), 'The great L Play', shows three boys playing 'Trap-Ball', with a verse and moral: 'Touch lightly the *Trap*, And strike low the *Ball*; Let none catch you out, And you'll beat them all. MORAL. Learn hence, my dear Boy, To avoid ev'ry Snare, Contriv'd to involve you In Sorrow and Care.' The fifteenth-century rules are unknown. The eighteenth-century rules were probably those Strutt (1801) knew as a boy: 'It is usual . . . to place two boundaries at a given distance from the trap, between which it is necessary for the ball to pass when it is struck by the batsman; he is also out . . . if the ball . . . is caught by one of his adversaries before it grounds . . . and again, if the ball when returned by the opponent party touches the trap, or rests within one bat's length of it . . . if none of these things happen, every stroke tells for one towards the striker's game' (note that in trap ball each player plays for himself). The rules were much the same when bat-and-trap was played on public holidays in Sussex, especially as a Good Friday event on the Level at Brighton (see pp. 6–7). The game was revived as the Canterbury and District Bat and Trap League in 1922 (*Manchester Guardian*, 19 January 1955) and is still played with enthusiasm at Kentish pubs on summer evenings.

'Knurr and Spell' is the northern raise-the-ball-and-hit game; 'Trap Ball' the southern. None of the many references to 'Trap Ball' relate to the north of Britain. The first notice of the game is in a dictionary, as so

The great L Play.

TRAP-BALL.

TOUCH lightly the *Trap*,
    And ſtrike low the *Ball*;
Let none catch you out,
    And you'll beat them all.

MORAL.

Learn hence, my dear Boy,
    To avoid ev'ry Snare,
Contriv'd to involve you
    In Sorrow and Care.

TIP-

Left: 'The great L Play—TRAP-BALL', in John Newbery's *A Little Pretty Pocket-Book*, 1744 (1767).

Below: From Christopher Comical, *A Lecture upon Games and Toys*, pt. 2 (Francis Power), 1789. Young readers were invited to draw 'the following moral simile. The Trap may be compared to the world at large, and we to the players of the game. The Ball may be compared to the various circumstances in human life, which rise from the Trap sometimes in a direct, and sometimes in a crooked line. It is our place to watch its various twistings, and seize the critical moment in which we are to strike it. If we miss the Ball, we may perhaps be incapable of recovering the loss, and thereby lose the game.'

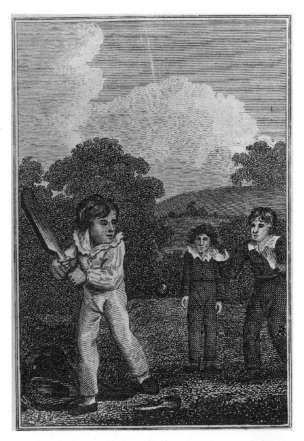

Above: 'Trap Ball', in *The Book of Games* (William Darton Junior), 1818.

Right: Buying 'traps' in a shop, from *The Home First Number and Music Book*, Mrs Charles Butler, 1854.

often with sixteenth-century references; it occurs in Percivall's *Spanish Dictionary* (1591): '*Paleta*, a trapsticke'. In the seventeenth century, theatrical references come thick and fast: 1627, Hawkins, *Apollo Shroving*, p. 5, 'our best game. What? . . . trappe out may hap?'; 1668, in Shadwell, *The Sullen Lovers*, act III, the character Sir Positive At-all is said to be 'so eminent' at trap ball as to offer to play it for a £5,000 wager (Pepys, *Diary* (1976 edn.), ix. 190–1); 1675, Duffett, *Mock-Tempest*, I. 2, 'Angry? no, I'le sooner break my Trapstick'. The eighteenth century saw the game as being for little boys, as in *A Lecture upon Games and Toys*, by Christopher Comical, pt. 2 (1789), p. 42 ('After you have raised the Ball from the Trap, you must be very careful and dexterous in striking it . . .').[18] Trap ball regularly appeared in the boys' games books of the nineteenth century, even in W. H. G. Kingston's *Infant Amusements* (1867), 83, where the game is said to be admirably suited for young children, and it is advised that traps and bats may be purchased at any toy shop for a shilling.[19]

All the batting games resemble each other in some way. The boy who is learning the rules of trap ball, in Tabart's *Book of Games* (1805), 20, enquires: 'We are to try to hit the trap, with the ball, are we not? and you will be out, as soon as we have done so?'; this rule is reminiscent of cricket, and in fact the bat in the accompanying engraving looks very much like a cricket bat. His mentor replies: 'Yes, that is the way *we* usually play; but I believe sometimes the person who is in, guesses how many bat's length off the ball has stopped, and reckons as many as he guesses, if that is less than the real number; but if he guesses more than there really are, he is cast.' This other method, alluded to by Strutt in 1801, as 'played by the rustics in Essex', is in fact nothing but tipcat played with a ball and trap; and thus trap ball can be seen as a later development of the undoubtedly very old game of tipcat.

[18] Printed for Francis Power ('Grandson to the late Mr J. Newbery'); in the Cotsen Library at Princeton University.

[19] They were also obtainable from 'swag' barrows in London streets at that time (H. Mayhew, *London Labour and the London Poor* (1861), i. 448); sets of 'Bat, Trap, and Ball' were still on sale at James Shoolbred & Co., Tottenham Court Road, in 1903, at prices ranging from 1s. to 4s. 6d.

General Index

Index of the Names
of Games

Index of Chants